Inclusive Dance

Inclusive Dance

The Story of Touchdown Dance

Katy Dymoke

intellect
Bristol, UK / Chicago, USA

First published in the UK in 2023 by
Intellect, The Mill, Parnall Road, Fishponds, Bristol, BS16 3JG, UK

First published in the USA in 2023 by
Intellect, The University of Chicago Press, 1427 E. 60th Street,
Chicago, IL 60637, USA

A catalogue record for this book is available from the British Library.

Copy editor: MPS Limited
Cover designer: Tanya Montefusco
Cover image: Tanya Montefusco
Production manager: Laura Christopher
Typesetter: MPS Limited

Print (hbk) ISBN 978-1-78938-869-5
Print (pbk) ISBN 978-1-78938-836-7
ePDF ISBN 978-1-78938-837-4
ePUB ISBN 978-1-78938-838-1

Ray Chung letter;
Originally printed in *Contact Quarterly* Vol 24, no. 1, w/s 1999.
www.contactquarterly.com
Body–Mind Centering® and BMC® are the registered service marks of
Bonnie Bainbridge Cohen, used with permission.

www.intellectbooks.com

This is a peer-reviewed publication.

With thanks to Steve Paxton, Anne Kilcoyne and Lisa Nelson
To all those dancing CI and making it accessible

'Before you learn how to fly you first need to learn how to land'

– Katy Dymoke, March 2022

To Liam and Sam who shared this journey with me

Contents

Figures

Abbreviations

BMC® – Body-Mind Centering®
CI – Contact Improvisation
DCA – Dartington College of Arts
PTW – Performance Training Workshops
RNIB – Royal National Institute for the Blind

Preface

This brief autobiographical account serves to show how the moral sentiment to write this book emerged and how arts practice and research became a vocation and ultimately a huge part of my life. I became dedicated to the practice of delivering opportunities for people usually excluded from the arts, so that, like me, they too had a choice of what to say yes and no to, of taking risks, of finding a way to grow and determine a life of autonomy and self-determination. The decision to write this book was made when I contemplated a Ph.D. by publication and the one thing I had dreamt of doing, without taking it so seriously due to lack of time, was to document the work of Touchdown Dance. This wish has been with me over the past 30 years of being with the Touchdown Dance praxis. The content of the book is coloured by my enthusiasm for the work, and by my wish for dance and arts practice to be inclusively taught, available to all learning needs, and for professional vocational courses to be the same. This moral sentiment required particular circumstances and an epistemic context to form and further lived experiences and encounters to inform and shape it. The overlapping and layered nature of experience underlies the iterative and reiterative nature of the chapters' content.

First, I introduce myself as a dancer. I danced ballet from the age of 3 years old till my late teens when I danced on point shoes most nights a week; it was a 'gift', a natural affinity, I performed in the school productions and then one day I felt the need for something else and never wore ballet shoes again. I desperately wanted to find a surrogate and did so in writing, running and theatre-going, but still yearned to dance, just not ballet. I started to write poetry and plays, and once a student in London, I found the Royal Court Theatre and the Half Moon Theatre were on my doorstep. I did a degree in European Studies (French and Spanish) at Queen Mary College London University, and volunteered at the Half Moon Theatre next door, supporting the youth work, then as a literary manager putting on plays by new play-writes. As a member of the Royal Court Theatre's young people's play-write scheme, I found the wonder of playwriting and getting something on stage. Having languages proved beneficial, as the borders to Europe were opening further in the

Arts. I learned to value words, to deal with not being understood and to value all types of communication.

The radical arts scene at the Half Moon Theatre brought lived experience, my way that took me on a journey of choice and risk-taking, into the world of arts practice and research, creativity, production and performance. The opportunity to participate in a six-week long residency with Dario Fo and Franca Rame at Riverside Studios in 1982 led to an unplanned course of events of theatre training, teaching and production over the next ten years. The chance of a scholarship for an MFA in playwriting at Yale University was shelved, to go, instead, to Europe, to train with Akedemia Ruchu from Poland, with Théâtre du Movement and the School of Philippe Gaulier and Monika Pagneux in France, and to follow Dario Fo and Franca Rame to Umbria and Milan. This path took me into the world of theatre anthropology, of the actor/dancer, the clown and the grotesque buffoon, where sardonic political farce and the ultimate physicality of performance meant that verbal language was less or of equal importance to physical theatre methods. The popular farcical theatre of Fo and Rame was a political tool, a means to create community and to give voice against injustice, oppression, discrimination, corruption and prejudice – Fo showed me video footage of shepherds performing in his early pieces which were crafted for one day only, not to be seen again. He was indeed the anti-Pope, his Mistero Bufo and Franca Rame's Female Parts are iconic pieces of work that speak for human rights, liberation and against societal control. Fo gave me two of his plays, a one-act and a three-act which I translated and produced during a Post Graduate Diploma at Cardiff University Sherman Theatre; I enjoyed teaching actors in his craft and the art of production.

In 1985, I attended a creative hub in Cardiff, called Chapter which is still open today, and started work with Cardiff Laboratory Theatre Company based there. I worked as a research officer for the *Centre of Performance Research* and then with the *Magdalena 1986* project for women in Laboratory Theatre. Here, at Chapter, I finally reconnected with dance. I started to attend dance workshops given by visiting companies, some that I had organised and tasted Contact Improvisation (CI). I had a sense of belonging to a community that shared and co-created uniquely crafted artwork, with discipline and body-focused physical theatre techniques. The laboratory research process for any performance event may start texturally but become physical, embodied rather than scripted, enacted rather than spoken, with a poetic, sensorial intensity that honoured ritual in heightened reconstructions of reality. It was a formidable experience to immerse myself in the wonder of the Magdalena 86 project, where for the first time I experienced exploring creativity with women only, with some amazing women from all over the world. Diverse and powerful, these women affirmed my feminist outlook (particularly Margareta D'Arcy). I then moved to Devon to Dartington College of Arts to support students in their final-year assessment pieces and tutored some on the acting methods of Dario Fo. I continued with CI as it was prevalent there, joined the dance classes of Katie Duck and attended the Dartington Dance Festival to see Steve Paxton and others perform work not so far removed from Laboratory Theatre. In essence, 'live' performance became a craft of the actor as maker and mover, not as speaker of text.

Through the practice of CI, and the teachings of Dario Fo, Philippe Gaulier and Monika Pagneux, I became an artist, teacher and maker. In 1988, I took an opportunity to teach a four-month training in the fool and buffoon work, at the Aarhus Theatre Akademi in Denmark. I then stayed for six years, with my young family, teaching consistently in self-devised performance training workshops and making performances with groups of unemployed, women only or mixed groups, by using funding from the state, social services and artist unions. I also maintained contact with the world of theatre anthropology with Odin Theatre in Holstebro, participating in *Transit*, a 'follow-on' of the Magdalena project which I loved. Then things changed – a group of disabled people saw the buffoon performance work, and associated it, correctly, with the freak shows of the old travelling theatre. With no access to available vocational training, they asked for training and performance experience to grow as performers and to create work that would be valued by audiences and the cultural arena. We co-created Korona, a theatre company of disabled actors. It was awe-inspiring; their belief that we could find a way together, for them to perform publicly and be considered 'professional' performers, was motivating, at a time when no disability arts companies existed in Denmark, and disabled people, though well provided for, were hidden. I had been prepared by my own journey to know that the way forward could be found by saying yes to the unknown; I brought the creative concepts of the previous years into another self-made collectively creative space. This was the start of the next 30 years of inclusive collaborative practice-led research projects.

Touchdown Dance, led by Steve Paxton and Anne Kilcoyne, was one of three companies that we invited to Denmark in 1989 as part of the one-year development project to train and create a touring production with differently disabled communities. After three long weekends, we had enough material to use as the foundation for further practice-led research. A performance group was formed to include two deafblind, a deaf and two disabled actor/dancers, a musician and three non-disabled performers, under the name Performance Training Workshops (PTW). We made the show about the sea culture, with solos and ensemble scenes based on ideas and experiences of the sea. We toured Nordic countries, Faeroe Islands, Iceland, Finland, Sweden and Aaland, as well as Denmark in 1990.[1]

Through the Touchdown Dance experience and the practice of CI, I grew an interest in touch. I had experienced discomfort in some CI partnering experiences, whilst becoming liberated within my moving body, in others. Touch was a challenge for many who were unaccustomed to being in proximity with strangers, let alone to be picked up and carried on a shoulder, or rolled around with. A far cry from ballet, the nature of CI was a magnet; it filled the gap that ballet and other physical training had lacked. CI was also a community for discussion and debate about dancing differently, about the value of this new radical 'form'. CI was revolutionary as women carried men, men carried women and disabled people partnered with non-disabled people; it was about mutuality and reciprocity, and physical contact in the frame of inquiry into how we may touch others in sport or daily encounters – CI was associated with a 'touch revolution' (Nelson 1996).

The moral sentiment that I carry now is at times despondent and indignant, particularly where touch is concerned. Over three decades, I have excavated touch in my personal practice of CI, as a curator of key events and festivals, as a teacher and performer. Touch became more significantly a topic of research when I encountered Body–Mind Centering® (BMC®)[2] in 1992. Like CI, BMC® is a key recurring aspect of my practice which I bring into the Touchdown Dance pedagogy and so to this book. A colleague from the Centre for Performance Research days, Ric Allsopp organised a conference for performing artists, with Scott Delahunta. I attended as a scribe to document a week-long workshop in BMC® as part of the *Connected Body Conference* at the School for New Dance Development in Amsterdam (1992). Another Eureka moment – deep connections and deep relationships evolved in that week, which I cherish today. I discovered the role of movement and touch in human development, well-being, communication and community. It gave credence to my own discoveries with Touchdown Dance, about the language of touch, in particular, the relationship of touch and consciousness, way beyond the institutional binary and societal polarity of erotic and non-erotic touch. My moral sentiment or conscience surges at this point, as I recall the need I felt to challenge the use of the word erotic; it appears on both sides of the binary, so 'erotic' and 'touch' are irrevocably entwined as if no other types of touch exist. In this binary touch and erotic are reduced, and 'vulgarised' (Dymoke 2014), yet as natural to life as drinking water (Dymoke 2000). Taking touch back from the taboo context became a cause worth fighting for and BMC® made this possible through experiential learning methods in which a range of touch qualities are differentiated, relative to the cellular nature of each body system. The value of touch is reiterated and brings awareness to the affective impact that touch has – and to how this requires attentive responsible practice.

Don Hanlon Johnson (2006) has been frequently quoted for describing the complexity of touch in a paper entitled, '*Intricate Tactile Sensitivity; A Key Variable in Western Integrative Bodywork*', implying the intricate and sensitive nature of touch – the tactile sense. Herein lies another moral sentiment, compassion; touch needs careful handling; it is relational and evidently those who use touch with malign intent are aberrant and take away from the benign essential intricacy of human touch and its warranted place in daily life (Dymoke 2014, 2017, 2019). I have continued to research and publish chapters and articles on the value and impact of touch in dance, Dance Movement Psychotherapy and BMC®, in both treatment and teaching contexts (Dymoke 2000, 2014, 2017, 2019, 2021).

BMC® brings touch and movement together in an experiential process. In BMC®, we learn to be touched and thereby how to touch, and vice versa. The hands are sentient as is the body we touch. The language of touch is encountered as a rich palette of qualities or tones that adjust in response to the body that is touched. We may touch with an intention and then attend to the response in the body that is touched. We may engage in a dialogue, with deep listening and sensitivity – which implies that both are responding in a reciprocal sense to the quality that arrives in the bodies of both. Touch is then 'more than' tactile; it is a meeting place of consciousness – the body–mind unity encompassing both. The roles remain clear in terms of who is 'holding' whom, and the person touching remains responsible for the person in their hands, to accompany and attend to the changes appearing, holding the membrane that contains both, the relationship and the understanding, in a space–time in which the bodily self and sense of self reunite.

This moral sentiment arose from meeting and working closely with disabled, deaf, deafblind, blind and partially sighted people – our labels were only socially significant; in our work, we were performers and makers. I am eternally grateful to all those whose paths have crossed with mine and this work over the years. Whether through CI or BMC® my encounter with other practitioners has supported me and enabled all my professional roles, as BMC® teacher and program director, as Dance Movement Psychotherapist and Body Psychotherapist, as a dancer, workshop leader, dance filmmaker and researcher. It is odd to name them separately as they are all part of the 'one' unitary being, of my interrelated multiple 'selves'. I have tried to include as many as possible of the colleagues who have been, and still are, treasured. Many remain irrevocably attached to this work as it continues to be life-enhancing, a resource of methods and techniques, of practice and theory, of a way of being. The praxis is an invitation to be immersed in the present moment to find out more about what it is to be alive in an interrelationship with self or in the company of others.

Since childhood, I have considered my body as the first place of knowing, where 'knowing' refers to feeling complete, whether standing on the wet sand on the Isle of Wight, looking at the sea disappear on the horizon or swinging on a rope under a tall tree before falling into a pile of grass. Both of these childhood experiences are echoed in the internalised *small dance*[3] practice that Paxton has relayed to thousands of people; kinaesthesia is the root of the three-dimensional sphere of CI praxis – embodying the body's innate reflexive response to gravity.

I was glad to experience the early stages of Touchdown Dance, as this really made the ongoing journey less bumpy, as did working with the methods on my own in Denmark with blind and deafblind adults from 1989 to 1993. Difficulties accompanied successes, like finding visually impaired people to participate and instilling trust in them that this could be a valuable experience. Or finding sighted people who also felt it was less about a worthy cause than a genuinely rich learning experience, which is covered in the following chapter. When I took over as director of the Touchdown Dance Charity and company in 1994, I moved the centre of operations to the border of Yorkshire and Lancashire. From this new location, I brokered new relationships with the RNIB Leisure services setting up workshops with Simon Labbett, and with SENSE UK in Birmingham and London with deafblind adults and later children.

Other opportunities arose as I had young children in school and realised the importance of working with visually impaired children and young people, having heard from several adults how they wished they had such opportunities as a child. The RNIB had events around the United Kingdom at which I ran workshops and taster sessions, SENSE UK invited sessions on-site and also published an article in the SENSE International journal. Special RNIB schools and colleges were also places for workshop delivery and the application of the core principles succeeded in all these settings. Examples feature in Chapter 5 and we continue to work in these settings today. Careful consideration is given to the content of the workshops in each setting to ensure it is appropriate to the age and ability of the group. Cultural differences are significant – whether in this country or abroad, with children from ethnic backgrounds in mixed-gender classes, or a large class of children with different learning needs and mobility issues as well as visual impairment.

Creative Partnerships was a UK Government programme that placed artists in schools. Touchdown became a partner in Manchester, Salford and the surrounding Pennine area, up to Burnley, Nelson and village schools. Other partnerships arose through contacts in the dance sector who had heard about Touchdown Dance and wanted to support the training of local dancers to be able to offer the model locally. Other partnerships evolved with venues and arts organisations both at home and abroad, including invitations to present at conferences and to teach at Contact Improvisation Festivals. This included curating at the twenty-fifth and thirty-sixth-anniversary celebrations of CI; CI25 took place in 1997 in the studio at Oberlin College, Ohio, where 25 years earlier Paxton led CI research workshops. In his conference address at CI36 at Juniata College, Pennsylvania, Paxton spoke of the importance of Touchdown Dance and Dance Ability to the history of

CI and its significance to his career. Our conversations have continued over these years and parts are shared here and others elsewhere.

The integration of visually impaired and sighted dancers in the workshops continued in the context of performance with the establishment of Touchdown Dance integrated company. Company members became teachers as the praxis evolved idiosyncratically within each performance context. The core methods had been tried and tested; it was a question of application and adaption and artistic development.

THE CHAPTERS

Introduction: An ethno-historical overview is provided in three parts about the cultural climate in which Touchdown Dance was founded and the interconnections that grew between the community arts, disability arts and other minority cultures. Practitioners with shared beliefs became allies for inclusion and legitimacy as artists. The challenges that befell these communities were imposed by historical social and cultural exclusion on the one hand, and the lack of cultural diversity on the other. Providing opportunities for diversity in dance and theatre, companies like Touchdown Dance, Candoco Dance and Graeae generated new participatory arts practices (Paxton and Kilcoyne 1993; Benjamin 2002) and began the process of change.

Chapter 1 – Part 1. I write about the early years from 1986 to 1994 which has historical interest in terms of the contribution that Paxton and Kilcoyne made to the contemporary dance arena with this work. I present a brief history of how the initial idea to set up workshops for blind and visually impaired people arose. I consider the differences between these two groups and how they were met by the formulation of a pedagogic approach that evolved over time creating the right content and accommodating environment for this new type of learning to take place.

Part 2 gives more information on the workshop content and includes participant accounts to illustrate how in practice over the eight years of delivery, all their thinking and intention to involve visually impaired people in movement workshops did change people's lives. The impact experienced directly in the workshops continued in the day-to-day life of individuals and was transmitted further over time into society.

Part 3 is a brief account of the transition when I took over from Steve and Anne and the eventual relocation from Dartington to the North of England. Key to this was the provision of workshops for children and young people and the further application of BMC® principles to CI and the core pedagogical approach.

Chapter 2. The practice-led research methods arose within the context of Dartington College of Arts where the ethos of equality and dialogue in CI supported inclusion. Reflexive accounts of partnering from myself and a blind participant, Eleanor Tew, introduce the practice-led aspects that I brought to the work, such as discursive methods, reflexivity, embodiment and phenomenology.

Chapter 3 is focused on touch, introducing the intricacies of touch communication and the discourses that become entwined whenever touch is presented as a possibility – even though it is integral to the transmission of 'the work' touch can still be cautioned.

Chapter 4 looks at the modality-specific approaches to the workshop content, detailing the exercises, approach and caution when a specific exercise may need more careful attention. This is taken from the education pack we made as a company in 2015, and embellished with photographs.

Chapter 5 gives examples of workshops in the contexts in which we have worked and insights into the type of experience we have provided for children, young people and adults.

Chapter 6 presents the various performance projects that we have delivered as a company, here in the United Kingdom, and some that toured abroad (to Germany, India, Sri Lanka and Portugal, to name a few). This is with the view to illustrating the extraordinary nature of the work and the reason for our care and attention in nurturing the work, specifically for visually impaired audiences and the use of audio description or cross-modal perception.

Chapter 7. Some final words from me to link the book and the membrane theory together in three continua.

Chapter 8. Three long-established members of the Touchdown Dance Company have written 'point of view' chapters on aspects of their experience of being an artist in the company, as an individual and as part of the body of work. Holly Thomas and Jamus Wood are both visually impaired and Robert Anderson is sighted. These points of view are included to give further insight into the breadth and depth of the work as it is experienced and to show how it evolves from the direct participation of individuals in a process that nurtures originality and authenticity.

The book ends with a 'backword' from Steve Paxton originally written for the CDROM I produced in 2000.

My relationship with Touchdown Dance has been consistently one of being in awe of the experiential process. I cannot claim to be responsible for the impact, though I may take credit for the planning and delivery, for the holding that each project requires – all this requires organisational skills and conviction. What I come to write about in this book is a record of moments that had significance at the time, moments that seeded impressions and changed my view of the order of things. I am aware that many needs remain unmet and that there is much left to do, hence the urge to document and make the work more available.

The many moments referred to are sourced from the workshop experience as evidence of the ongoing dialogue between the dual roles of practice and research that Touchdown Dance has depended on for its growing epistemology. I go on to refer to related areas of research undertaken that influenced the language used, including consciousness studies, reflexivity and phenomenology. These served to flesh out the bodily sense that I have of CI praxis and to shape it into a three-dimensional 'animate form' (Sheets-Johnstone 1990: 307). All this I take with me at a time of increased restriction, isolation and fear of touch – the COVID-19 pandemic. Fear of touch has become a self-imposed restriction, one enforced by a coercive and divisive culture. Backed by medical science this fear threatens, whilst revealing, the vital need for touch. How to substitute the hand-on-hand gesture of kindness, the restorative impact of human-to-human connection that is mutually felt and sensed through touch? Other sensory modalities have been shown to be inadequate and many have suffered psychologically from touch deprivation, from loss of human-to-human contact. This we encounter still, as we return to face-to-face and 'in-person' sessions and notice the state of fear and withdrawal. We work with the membrane of containment, to find ease with the self and self-other relational membranes, to re-instigate the sense of trust that has been displaced by fear, in recognition of the need to receive or to reach out for support, but not all are ready. More pertinent than ever is the need to acknowledge the impact of 'social distancing', which is now a regularly used term of reference, and the state of restriction and control that so easily reinstates neglect and deprivation. As one visually impaired and disabled child recently stated: 'Don't touch me...sorry, I used to be OK with being touched but now I really don't like it'. I replied, 'I won't touch you, that's fine'. Then within a few minutes of the session she asked me to touch her leg, 'Feel my scar, there...' where she had been operated and took my hand to place it behind her knee as we moved. I wondered at this moment – had I been coerced into touching her or had she been 'coerced' into asking me? I realised my own vigilance; it was trust and need, an opportunity for her to venture back into a sense of normality, of human-to-human connection, and for me too.

June 2021

Acknowledgements

In e-mail correspondence with Steve Paxton back in the later 1990s, I had spoken of the idea of a book about the work. What followed instead was an exchange of e-mails and ponderings on the senses that Lisa Nelson included in a publication (Paxton and Dymoke in Nelson 2001). That was a moment in time that had been forgotten until I took a look at the archives and fell upon it. The book idea was shelved until 2016 when I started to think about telling the story. I spoke with my elderly mother and she came back with many questions – 'What does it mean for a visually impaired person to learn to dance? What do you do to teach them, for them to learn the movements? I mean, how on earth do you do that?!' Steve Paxton and my mother are two of many who have stood by with interest in this book about the Touchdown Dance journey.

I would like to thank Dick Matchett, Mr John Ireland (JP), Mrs Carolyn Steen, for their dedication in the early years, as chairs and vice presidents. Their mentoring and guidance assisted me to believe in the work – that however small our outreach was, it was incredibly important. More recently, Amanda Redvers-Rowe, Christine Yates, Lizzie Chappell, Shelly Owen, Natasha Stott and Stephanie Bairet have given unconditional support for the organisation and its sustainability.

I wish to acknowledge the company members, past and present, dancers who participated in our experimental productions – Patrick Beelaert, Alan Foster, John Hurley, Janee Hall, Holly Thomas, Jovair Longo, Rick Nodine, Mary Prestige, Jo Blowers, Robert Anderson, Stuart Jackson, Ray Chung, Indra Slavena, and collaborators, Scott Smith, Becky Edmunds, Katrina McPherson, Simon Fildes, Barnaby O'Rourke, Jim Parris, Julyen Hamilton and specifically Angus Balbernie, a genius choreographer and collaborator, who also co-created and performed with Stuart Jackson in more recent years.

Others still feeding the work and continuing this story and who have supported me to do this writing over the past years are the following artists; Toni Charlesworth who has worked as a dance teacher for Touchdown Dance community projects, delivering classes to the most hard-to-reach members of the community at Waterside Arts Centre in Sale and in more recent times, on the internet. Since

2016, she and Amanda Parsons-Hulse, Paula Hampson and Julie Williams have been working with me, locally in Greater Manchester and Merseyside, to deliver the work in schools, funded by BBC Children in Need. Together we have been standing by during the pandemic and jumping in at short notice to benefit children experiencing isolation and exclusion.

Part of this book was a major contribution to a Ph.D. by publication, at Manchester Metropolitan University, with a supervisory team who have questioned, admired and challenged me, and their belief helped to get the book to print; with thanks to Dr Patrick Campbell, Dr Anna MacDonald, and Professor Juliet Goldbart. Ben McEwen assisted with copy editing and as a reader for Holly Thomas, at a much-needed time.

Finally, I thank and consider all the children, young people and adults who have come to our projects and workshops and given us the privilege to share this work. It would be amiss not to mention Vicky Smith from LOOK UK, Dr John Patterson of St Vincent's School in Huyton for his belief and enthusiasm, likewise Naomi Harris at Manor Academy Sale, and Louise Preston at Newbridge Academy and College in Oldham.

Introduction

An ethno-historical overview of the origins of Touchdown Dance: A radical initiative in a radical climate

The story of a particular dance company and its movement practices would, one could assume, only be of interest to a small, particular, audience. In writing this book, I wish to tell the stories that account for the evolution of an inclusive and specialised movement practice based on a movement form called Contact Improvisation (CI), from the point of view of the visually impaired and sighted participants. The impact of the workshops and touring company productions, has been far-reaching and contentious, whilst connecting cultures and communities and challenging pre-conceptions. The work of Touchdown Dance evolved discretely within the arts sector but more vibrantly within the lives of participants; it exceeded expectations and shifted perceptions of what change is possible, wherever the work has been delivered. It is important to acknowledge the contribution made by each artist, participant and their community, who enabled this work.

This introduction is retrospective and historical. I locate the origin of Touchdown Dance within the contemporary cultural climate; a landscape in which those of 'like mind' sought to shake up old cultural constructs and make the world a better, more accessible and inclusive place. In this context, I consider the disability arts culture in the United Kingdom since the early 1980s, and the founding of Touchdown Dance in 1986 by Steve Paxton and Anne Kilcoyne at Dartington College of Arts (DCA) in Devon. Touchdown Dance was one of many new arts initiatives, but unique in seeking to establish an integrated approach to dance for visually impaired and sighted participants, which is the central topic. I also introduce the corresponding path that I took, to contextualise the evolving relationship between myself and Touchdown Dance and the aspects of my epistemology that I brought to the work. On moving to Denmark in 1988, I co-established another unique initiative called Performance Training Workshops (PTW) and got state funding to work with integrated groups of people with diverse needs, and then more specifically with disabled and sensory impaired people. The need for expertise led to the opportunity to bring Touchdown Dance to Denmark in 1989. The eventual meeting of these two projects early on in their development was significant for the continued collaboration and evolution of the praxis even after I took over as director in 1994 on my return to the United Kingdom.

The respective paths taken by Touchdown Dance and PTW were distant and distinct but intertwined in purposeful ways – they both sought to acknowledge difference and to achieve equality of opportunity. These two aims were not necessarily coincident or relative – such was the complexity of the endeavours. Whilst Touchdown Dance in the United Kingdom focused on dance with visually impaired and sighted people, the Danish project involved disabled, deaf and deafblind people, as well as visually impaired people and non-disabled people, the population being a tenth of the United Kingdom. In both projects, participants inevitably became trainers and/or performers, and grew the work further. Studio practice brought a diverse group of people together to collaborate as an ensemble so that inevitably, through inclusivity and practice-led research, new forms of pedagogy and performance projects arose and much learning was exchanged.

This chapter is in three parts:

Part 1. To start with, I look back at the social and cultural landscape within which Touchdown Dance was seeded, to identify the interrelationships that fed and warranted the wider societal change taking place beyond the studio and set the scene for potential new practice-led research. Touchdown Dance was of a particular time with a particular vision about learning by doing, and the aim is to show how this experiential approach, like a pebble in the pond, sent out waves of new thinking within individuals, institutions and further across society, where it thrives today.

Part 2. In this part, the field of dance in which Touchdown Dance was rooted is contextualised with consideration of the cultural changes taking place in mainstream dance and of the climate of dissent that gave rise to new perspectives, practices and discourses and to new forms such as CI. The growth of interest in embodiment and somatic practices influenced changes within the social and cultural climate which also became hospitable to the advent of integrated and disability arts sectors. All this paved the way for Touchdown Dance, for a more democratic arts scene, with participant-led processes and experiential practices.

Part 3. The climate of the time was one in which the potential of experimentation was a force for change and CI had a major role in this. Descriptions of some early experiences provide insight into what CI was as a practice at that time with reference to a close collaboration with a deafblind colleague, who experienced CI prior to the Touchdown Dance workshop and after for over a year. This set the terrain in which the membrane concept emerged – a concept that embodies the non-visual sense of containment, tactility, motility and mutability, that held our dialogue in a reciprocal realm of discovery (see also Chapter 2).

PART 1

Taking a stand for inclusivity in an exclusive society

To contextualise the story, I summarise some key influential features of the rise of community arts in the United Kingdom from the 1960s. I consider the commonality of these aspects with the contemporaneous rise of disability arts culture in the United Kingdom with reference to an anthology that I have kept on my bookshelf of the *Disability Arts in London* magazine (*DAIL* 1991) entitled *The First Five Years*. The anthology contains the voices of disabled artists and activists who contributed to the magazine between 1985 and 1991. Synchronous to this time was the rise of the community dance movement, which I consider in discussion with the views of Ken Bartlett, who was the founder and, for many years, director of the Foundation for Community Dance UK, taken from an article published in *MAiLOUT* magazine at this time (1996).

The book by Owen Kelly entitled *Community, Arts and the State: Storming the Citadels* (1984) was very influential for me and a seminal text for applied theatre and dance students at DCA in Totnes, Devon, where I spent two years mentoring students (1986–88). Kelly looks back at the beginnings of the community arts movement in the United Kingdom in the 1960s, at how it was started by a group of artists who collaborated and discussed common values and beliefs. 'Arts labs' were set up in different cities that were actively 'concerned with trying to break down the restrictions deemed to be inherent in "straight" social codes and structures: in the art world, the education system, the communications industry and the nuclear family' (Kelly 1984: 9). This early movement raised questions as to 'the role of art in society, and the relationship between cultural and political struggle' which led to the following tenets: that creativity was linked to empowerment and social presence; that arts practice enables learning from experience and gives back to the people the means to control media and communication (Kelly 1984: 9). Kelly's observations are evident in the evolution of disability arts in the United Kingdom, and on the thinking of non-disabled artists from community arts sectors, like Touchdown Dance and myself, who had similar beliefs.

Kelly describes the momentum of social agitation that had been building from the 1960s in demand for a change in 'occupancy' of the societal institutions, including their type of service provision. Such agitation became the means for community artists and disabled artists to act together and speak out for social justice. This battle was apparently won when community artists finally received grants for the arts from the Arts Council of England, but not for all. Kelly refers to the irony in this fortunate financial backing, as it led to a demise; it placed the artists in the hands of the very institutions they were fighting. There were some who wanted to be artists first, and so welcomed the funding as approval of this status, and others who saw statutory patronage as a renunciation of their right to determine their cultural identity, as well as divisive and perpetrating inequality (Morrison 1991). Rather than fight to defeat state control of the arts, Kelly advocates that such control 'must be reduced in order that activity in other spheres can have space to grow' (Kelly 1984: 46) and transformation would then, in his view, depend on how these sectors interacted and spoke to each other to make this space available.[4]

Kelly (1984) acknowledges that the gathering of energy and the formation of a convincing critical mass of new initiatives spawned from the agitation in other new sectors of society. These new 'spheres' of activity set the scene for the rise of community arts, disability arts and other culturally oppressed minorities (Benjamin 2002; Kuppers 2011). Magazines such as *MAiLOUT* and *DAIL* enabled conversation and debate, for these 'movements' to be mutually acknowledged for offering viable alternatives.

As a community and disability-focused initiative, Touchdown Dance accessed resources from DCA precisely as a space conducive to the research practice. The radical methods and new creative practices that Arts Labs and others like Touchdown Dance founded sought to return autonomy and control to the artist/practitioner, and away from statutory patronage. Efforts to spearhead these into the cultural arena depended on the commitment to activity and debate, firstly in local communities, then nationally and/or internationally, like Touchdown Dance, which travelled to Denmark and to the United States.

Back to the 1980s – The birth of disability arts culture in the United Kingdom

During the 1980s, I witnessed continued and widespread community arts and related political activity, undertaken by women, disabled people and minority communities seeking equality of access and social inclusion. The hub of activity was in London with funding sources from the Greater London Council and the London Arts Board. Artists came together to join the struggle with the National Disability Arts Forum founded in 1984, with the mission 'to create equality of opportunity for disabled people to participate (in) all aspects of the arts'.[5] As sub-cultures, disability arts (Morrison 1991: 2) and deaf arts found strategies to realise their legitimacy, with many non-disabled artists and companies standing by in solidarity.

As I was writing, I took a look through my archive and found copies of the journals I had subscribed to and kept. The *DAIL* magazine documents the politicised struggle that disabled people undertook to establish disability arts culture and is in line with Kelly's account of the community arts agenda. With the voices of disabled activists and thinkers, the *DAIL* anthology is a treasure of insights that indicate their resolve to take a stand for 'greater participation, and access to, the arts' (Morrison 1991: 2). Elspeth Morrison, the editor for this collection, determines that 'communication and information have been crucial in allowing the liberation of disabled people to take place' (Morrison 1991: 2). *DAIL* was written and published by disabled people for their cause; in it, we read the political resolve of arts campaigners in the ongoing debate and discourse about the inseparability of arts practice and politics. Disabled writers express outrage at the media's portrayal of disability, at the images used that rendered the dismal day-to-day life of a disabled person. Morrison introduces the social imperative – to admit and change the inter-causality of negative self-perceptions and societal prejudice:

> Disability Arts is making an impact on the way we disabled people see ourselves, strengthening our self-image and self-esteem. It has severely dented our own perceptions of ourselves and those of non-disabled people. Rather than being seen as passive and dependent, through Disability Arts we can explore ways of being creative and active.
>
> (Morrison 1991: 2)

Morrison sought to justify why the magazine couldn't just 'talk about art'. Disability arts was changing society, impacting the status of disabled people and this was inextricably political – I underline her emotive language:

> Call it 'protest art', call it 'political', whatever *waves* we create move us towards greater participation in, and access to arts activities.
>
> This Anthology is not so much a retrospective as a testimony of disabled people's journey to *unfettered* artistic and cultural expression. Here's to however long it takes to achieve that.
>
> (Morrison 1991: 2, emphasis added)

The resolve was to establish a culture in which disability was defined by disabled people, just as feminism was defined by women (see Finkelstein [1987] and Colleran [1988] in the *DAIL* anthology). As the first quote declares, disability arts was a means to reclaim an active role in society, to change the perception from being 'passive' and 'dependent' to one of being 'creative' and free to express. Morrison makes explicit how a negative 'self-image' can be a result of patronising, socially assigned attitudes, and a strategy of societal control: so a negative body percept can be a media construct. Disability arts, however, offers a transformative experience, a means to regain a positive body percept; it has given disabled people an ability to move 'unfettered', make 'waves' and be liberated from the backwaters of inactive segregation. A common view voiced in the anthology is the demand to redefine barriers and boundaries, to rule out being 'passive recipients of help and charity, needing to be helped by non-Disabled people' as Mandy Colleran writes in the disability arts and culture section (1988: 7). This implies that a positive self-image is defined by oneself and others through collective action and shared values; it expresses the realisation of a sense of self that has defined, to concur with Kelly, its 'space to grow' (Kelly 1984: 46) with values that engender self-esteem rather than diminish it (Barsoe 1991; Kuppers 2011). In other words, a sense of self is not static or passive, it is alive and participates in wider societal change and transformation; identity is a human right, not a social construct.

Like the community arts movement in the United Kingdom, the disability arts culture was formed by groups of artists who made new work, had shared values, established new cultural movements, and took a stand against a dominant and exclusive mainstream cultural arena (Brisenden 1991; Kelly 1984; Harvey 1989). By dint of being different and with resolute determination to participate in reshaping the social and cultural constructs that misrepresented them (Morrison 1989), the disability arts movement redefined their identity and place in society through making art and creating new cultural forms – hence the rise of Candoco Dance company in the United Kingdom by Celeste Dandeker and Adam Benjamin (Benjamin 2002) and of significant disabled commentators/practitioners such as activist and professor Petra Kuppers (2011, 2014). In his article 'What is disability culture?' Simon Brisenden (1991) has the view that chimes with Kelly (1984), that this mission was not straightforward or uncontentious. The aspiration to bring disabled people towards disability culture meant, in his view, moving away from the privileged, statutory-funded, non-disabled arts sector that hitherto provided disabled people with non-disabled role models to aspire to, i.e. to become part of a normality that oppressed them. Brisenden offers a way forward and argues for disabled people to tell their life stories, to admit

> that we are valuable people in ourselves, and that we need not avoid each other or hide behind a cloak of false integration [...] Disability culture is being built upon a ruthless honesty about the people we are and the role we play in society.
>
> (Brisenden 1991: 9)

The evocative analogy of the 'cloak' echoes Kelly's claim that disabled people must take their place alongside other 'cultures and lifestyles that are demanding a space to communicate and be themselves' and ends with a positive sense of celebration; 'Disability culture is about expressing ourselves in whatever way comes naturally and about realising that these expressions are valuable' (Brisenden 1991: 10). These ideas are further debated by Morrison for those disabled artists who refused the 'dis-' label. Others who did concur were empowered to feel valid and valued, despite, or in direct regard to having, and living with, a disability.

The voices of disabled people in *DAIL* seek to show society the way forward, demanding a debate and for non-disabled people to actively remove all the barriers causing exclusion. They expressed an aspiration that was shared – to belong to society, to be able to access and contribute to a culture that embraces difference. They voice and demonstrate that reality can change, and that participation in the Arts offers a means to debate, shift perceptions and create community. Many joined the charabanc, concurring with these voices, sharing their aspiration to create a cultural landscape, identify with others and belong to a collective consciousness, whilst being responsive to the wider political climate.

Fundamentally, I shared this belief in social responsibility (Becker 1994), in the co-responsibility of the artist and society to ensure that the 'cloak of false integration' is trampled on by all (Brisenden 1991: 9). Sian Vasey, also writing in the *DAIL* anthology, concurs that 'through the arts we can make discoveries about what we have in common' (Vasey 1989: 11) so that respect for individuality creates solidarity. This I experienced before heading to Denmark in 1988, in the lively community arts movement in both Cardiff and London. The same respect was voiced in Touchdown Dance when visually impaired and sighted people shared their stories. The journey involved 'discoveries' of new hybrid, cross-art forms like dance-theatre, arts and media. It was a time to stand up for a 'multicultural society', for diversity and against 'the dominant culture' by advocating values that served the rights of those it failed (Finkelstein 1987: 3), and community dance, an emergent sector within the community arts field, did just this.

Community dance, an assemblage of diverse cultures, its vision and 'coming of age'

In 1996, Ken Bartlet's article was published, in the cross-art journal *MAiLOUT* entitled 'Community dance; coming of age'. As director of the Foundation for Community Dance,[6] he celebrates a set of values that resonate with Morrison, Brisenden and others, values that have accompanied the advent of community dance. He writes avidly of the impact of the plurality of dance cultures on communities of all ages and abilities, to pitch positively for the future of community dance as we approached the new millennium. I look to this briefly as it illustrates how Touchdown Dance found its place alongside other inclusive arts projects and how new discourses became framed by some core values that connected the multiplicity of community arts practices into one large cultural movement. Bartlett, like Kelly (1984), points out how the community arts sector was susceptible to being sidelined by the nationally funded flagship companies that drop into communities, with funding and resources that community artists lack, only to move on and take these resources with them. Once more the 'cloak of false integration' (Brisenden 1991: 9) is imposed by an identified oppressor, the Statutory Arts sector, which, in this case, hides behind a benevolent intention to work with grass root community arts companies. Both Kelly (1984) and Bartlett (1996) witnessed how credit for arts provision was lost to the community artist and claimed, or expropriated, by the cultural economy to sustain their self-assigned jurisdiction.

Bartlett (1996) lists core values that he identified. First, he values *creativity*: community dance participants experience the true value of participatory arts practice, and discover the 'place of arts in their lives'; second, *empowerment*: community dance recognises 'the power and contribution of dance in transforming and empowering the lives of individuals and their communities' (Bartlett 1996: 7); third, *diversity*: community dance 'respects and values the diversity of form and practice in dance, the diversity of contexts in which dance happens and the diversity of reasons which led people to be involved in dance' (Bartlett 1996: 7).

These values, he claims, are politically and culturally liberating; they challenge the 'aesthetic priorities' that bolster the unjust cultural hierarchy. Inclusivity appears to be the outcome warranted as a means to embrace 'differences in cultures, traditions and aesthetics' (Bartlett 1996: 7). Bartlett questions what community dance does for all the multifarious participants and audiences, and his answer is a proposition, that collectively we consider all forms of dance equally, taking into consideration the diversity of their respective values. So inclusion is high on the agenda, for society, along with diversity, for everyone who dances to be valued along with all the kinds of dance that are available. The values assigned by Bartlett to community dance evolve from the shifting territory, the sea change taking place across society here in the United Kingdom. Today diversity and inclusivity remain in the strap line for the foundation; 'community dance is for anyone and everyone'; in other words, all bodies are welcome. This sadly is still not a value shared by all, including the Arts Council, as resources are not available to enable it.

The above viewpoints frame a cultural ecology in preparation for widespread change and set a mood, a climate in which values pertaining to the 'old' cultural hierarchies are challenged by calls for diversity and inclusivity across society. The old tenets and aesthetic priorities that held certain cultural activities in high regard had been out of reach of the majority of the population – by defending a certain type of 'body' in a certain type of activity – e.g. the ballet dancer and ballet. The societal construct that defined 'what dance is', and so who *can* dance, was a 'citadel' that would be 'stormed' from outside and from within its own walls (Kelly 1984) by artists who, like the disabled artists, sought autonomy rather than tutelage. The concept of 'body' was not a priority within the citadel, rather the dancer was bestowed a fictional character and her body, a trope of the classical form that she served. Yvonne Rainer shared this view in her stage and screen work (Rainer 2006), and others like Sally Banes (1987) and Mary Fulkerson (1996).

The concept of diverse or multifarious bodies was not well received – nor was the idea that artists should express themselves 'in which-ever ways came naturally' to them (Brisenden 1991: 4). Disability arts successfully proffered these alternatives and united in solidarity with others likewise seeking wider social change – such as community arts, Black arts and Feminism.

The cultural divide between mainstream and alternative arts was, and still remains, political as much as cultural. In dance, the idea of liberation was embodied in the reclaiming of the artist's main medium – the body – through embodied and pluralistic practices that experientially explored the science and philosophy of the 'Soma' (the body) (Sheets-Johnstone 1966, 1999, 2010; Kolk 2014; Bainbridge Cohen 2012). Just as the disabled artists voiced in *DAIL*, the non-disabled dancers also voiced by moving in whichever ways 'naturally' evolved from saying 'yes' to their subjective sense of having a body and dancing it (Banes 1987; Novack 1990; Fulkerson 1996). Not all were in revolt but simply seeking to identify with change and diversity, which allied those seeking liberation of the body with those seeking enfranchisement and representation, as would be the case for, and the roots of, Touchdown Dance.

PART 2

The body as the locus of liberation

In this part, I explore the 'sphere' of the body as a locus for the awakening of new concepts of the role of the body in dance, and for the origin of Touchdown Dance. Again historical in nature I connect the cultural landscapes of the 'new dance' world with the discrete evolution of Touchdown Dance.

The potential for new forms of dance had been established since the advent of modern dance and the ensuing genealogy of radical approaches to the body, from Isadora Duncan and Ruth St Denis, to Martha Graham, Merce Cunningham and then the likes of Trisha Brown, Yvonne Rainer and Steve Paxton. The continuity of interest in the sensation of movement and the subjectivity of the body created new genres within the field of dance, setting a trend of experimentation and subversion. For example, physical virtuosity and the corporeal expression of emotion varied according to the interest of each artist in their body and movement-focused research (Rainer 1965; Banes 1987; Fulkerson 1996). Whereas culture had been the province of the state, these artists reclaimed their place as makers, who determined where the dance took place, when and who it was for. No longer comfortably referred to as 'dance' the work was first and foremost of the artist's praxis, in which the body became the locus and the instrument of the creative process, and movement the subject of subjective interest, a praxis explored through research

(Rainer 1965; Banes 1987; Fulkerson 1996). Each artist had their own formula or interest in the body and so the art, like a form of auto sculpture or ritual (Boal 1993; Barba 2001), appears experiential and pre-cultural in nature. The artists provided a centre around which people with like minds could gravitate, to collaborate on new initiatives based on innovative or unusual ideas that came, paradoxically, from movement. This stimulated an interest in staging work that challenged stagnant societal values, changed the cultural map and altered perceptions of the body (Banes 1987; Novack 1990; Koteen and Smith 2008).

The roots of dissent and the interdisciplinary pathway to democratic change and CI

In the non-disabled arts scene, artists from the contemporary and modern dance sector, such as the Judson Church collective in New York, or Ballet Rambert and X6 in the United Kingdom (Prestidge 1997), had explored the cultural constructs of the body. The aim was to be liberated from the restrictions of classical ballet and modern dance; the gendered roles, the power of the gaze of the audience over the body of the dancer/mover, and the emotional and psychological control of the cultural economy (McDonagh [1970] 1990; Banes 1987; Novack 1990; Fulkerson 1996; Nelson 1996). Seen as dissent and initially perceived as the avant-garde, this work provided a path for disability arts to follow; it was now possible to say 'no' to the dominant ideology, directly, as Rainer's 'No manifesto' intended; as she writes, it was a 'provocation' in a live performance, that meant 'to clear the air at a particular cultural and historical moment' (Rainer 2006: 234). Artists discovered the potential of the arts to present new agendas, new aesthetics and to contribute to discourses, learning and change. At this time the dance artist could not remain untouched by media hype, or avoid becoming one of 'an undifferentiated mass' (McDonagh 1990: 93), or 'that abstracted virtual body which is the subject of so much postmodern discourse' (Stinson 2004: 154). It was time now for artists to own and proclaim the value of the work they made, to be the subject of their art through 'self-analysis' (McDonagh 1990: 203) and to define new corporeality, new language, new methods and creative practices.

To reclaim the body from abstraction resonates with the disability arts discourse, as does the need for this to be a collective endeavour – in which the right to self-identity is for each person to have but must be a collectively experienced right. Individual artists with all these 'spheres' undertook subjective experiential investigations of the body as a medium for exploration. This was contemporaneous with the evolution of the field of Somatics, founded by Thomas Hannah (Hannah 1988) and the field of embodiment that offered forms of learning about consciousness and the unitary being, like Body-Mind Centering (Bainbridge Cohen [1993] 2012). The value and meaning of 'body' were explored further in phenomenology, most noticeably in the work of Maxine Sheets-Johnstone (1990, 1999, 2008, 2015), the exponent of a pre-reflexive approach to understanding movement as the foundation for thinking.[7] This philosophy pre-empted the forthcoming role of the body in theory making, i.e. the idea that the body can voice, challenge and contest ideas, to formulate a 'soma critique' based on experiential or practice-led research (Kampe 2015: 77). Introspection and subjective, first-person approaches, were explored to find mindful ways of understanding embodied experience, and provided new methods of analysis, a means to examine pre-existing beliefs or perceptions (Bainbridge Cohen 2012; Whatley et al. 2018).

Reflexivity – A new domain for non-Cartesian discourse

A subsequent methodology emerged within the field of consciousness studies, applying diverse theories of mind and phenomenology to the self-reflexive study of lived experience. This first-person praxis bridged the polarities of an objective perspective (familiar to science) and an introspective and embodied one, in a mutually informative way of perceiving, analysing, and so theorising, experience. A key text edited by Francisco Varela and Jonathon Shear, *The View from Within; First Person Approaches to the Study of Consciousness* includes a chapter on psycho-phenomenology by Natalie Depraz (1999). Depraz contextualises her perspective as 'a return to self' (Depraz 1999: 95) to explore the process of perception and meaning-making using the 'duality of the subject, at once theoretical and existential' (Depraz 1999: 96). Depraz brings together contemplation and action, objectivity and 'intuitive access to internal experience' in an embodied praxis that enlarges the 'field of experience' and allows 'a dimension which precisely frees me from the ordinary pre-givenness of the world' (Depraz 1999: 98). In the act of perceiving an object is the simultaneous subjective act of witnessing self and what 'I' habitually bring to bear on the experience. As Depraz elucidates, this inter-subjective praxis is affective; a cycle begins with the suspension of an initial interpretation to enable another 'dimension' to arise or come to mind, which may displace or replace the former with a new insight. This approach to analysing subjective experience interconnects the fields of research taking place in the arts, philosophy (including phenomenology), psychology and the science of consciousness, with a hermeneutic and reflexive practice underpinned by theory – a new form of objectification (see Chapter 3).

These advances in body-oriented approaches determined a time that was ripe for change, and a research environment conducive for new concepts and ethical practices to arise and be tested. As will be seen in Chapters 2, 3 and 4, the interdisciplinary nature of CI was one such environment, and one that I consider reflexively as a temporal-spatial context for transformational experiential learning, and a locus of the membrane concept. The background to this follows.

CI: Going with and against the flow … from convention to liberation

As is well known to dance aficionados, CI was founded by Paxton and a group of dance artists in 1972 in the United States (Banes 1987; Novack 1990; Koteen and Smith 2008). CI evolved from the contributions of the many participants over time, each bringing insights and experiences, skills and techniques to the mix (Koteen and Smith 2008). CI was part of a wider cultural 'new wave', referred to historically as Post Modernism (Harvey 1989). Postmodern dance came to Europe from the United States in the late 1970s (Banes 1987; Novack 1990), led by artists such as Paxton, Mary Fulkerson and Simone Forti, 'who were instrumental in redefining dance and opening up new possibilities of the art form – the where, how and what dance could be' (Koteen and Smith 2008: xi). From its inception, CI was not 'dance' but a form of non-competitive sport (Paxton and Kilcoyne 1993). Primarily a dyad or group form, CI could not be practised alone; it required contact between movers and had other attributes such as weight exchange, mutual support around the body's centre of gravity and referenced the kinesphere (the surrounding sphere in which the body moves). As CI entered the dance arena, it evolved in spectacular fashion as a locus for movement research, revolutionising the dance experience and opening up dance culture to communities of previously excluded populations.[8]

The other aspect, dance improvisation, was a solo and collective endeavour, as Daniel Nagrin (1994) notes in the introduction to his book on the origins of dance improvisation in the United States that it was 'working against the current of that time' (Nagrin 1993: x). Nagrin discusses how dancers turned improvisers due to being 'overwhelmed by the intensity and awareness of existences beyond dance itself' (Nagrin 1993: x). As a response to societal pressures, improvised dance took place outside of conventional studio environments and performance arenas, it took up 'new challenges' such as striking a balance between structure and improvisation (Nagrin 1993) and reset conventional artistic roles through direct revolt (Banes 1987; Nelson 1996). Improvisation provided a creative outlet to free up creative potential, spontaneity and independence. It reclaimed the body as the dancer's instrument – which the more 'aristocratic' (Pallant 2006) and formalised movement vocabulary offered by classical and modern dance training restricted (Banes 1987; Blom and Chaplin 1988; Novack 1990; Paxton and Kilcoyne 1993).

Paxton and others took CI into groups of like-minded practitioners and improvisers, such as the Judson Church collective in New York. CI, with its emphasis on play, embodiment and experiential approaches to learning, found its own flow and place alongside other new dance practices within a cultural movement that was for liberation and was borne on a wave of increasingly interdisciplinary practice (Novack 1990; Koteen and Smith 2008). Fundamentally the artists defined a new sphere within the dance ecology and created a research space to develop new forms of movement making such as CI. Through offering public workshops and delivering CI to dance students and professional dancers in 'labs' and 'jams', it became an open form, for exploration, for new questions to be asked and new understandings to be learned. CI was a non-dualist ontology in action and an emergent practice as a research methodology that adapted and accommodated different agendas. CI was used politically by many of those involved, as a form of dissent; it was accessible (and permeable). The CI movement grew in correspondence with other arts movements such as the disability arts movement and inevitably, the two movements found a means to share, cohere and grow stronger together within a new, independent arts scene (Ptashek 1992; Nelson 1996; Dymoke 2000, 2014).

Nancy Stark-Smith (2008), originally a gymnast and one of Paxton's performance partners in CI over many years, quotes from their personal correspondence of his acknowledgement of the 'co-development' of CI: 'It is barely clear to my memory who brought what to the table...I did my part, but in any group endeavour even a small suggestion may grow to large importance', he continues: 'We look around and see firm fabric but have no choice but to project our own strand, since we are all midstrand, as it were, and continuing the weave [...] I imagine all the strands are important, else no fabric' (Paxton in Koteen and Smith 2008: xiv).

Paxton presents the ethos of CI – owned by no one; it could be open to all to partake in its creation, inherently interpersonal and interdisciplinary. As a participatory and improvisational activity, CI was accessible and as a dyad or group form, it was readily applied as a means to explore movement, ideally divorced from other social or political agendas that could expropriate, be divisive or exclude. CI offered a subjective experience to be had by individuals within a reciprocal and collective relationship – a type of liberation from convention that required a particular consensual environment, and empathic engagement, with space and time to explore. CI enabled self-determination through self-directive practice and its transformatory impact, evident bodily, permeated membranes between self and others, shifting perceptions and bridging societal division.

CI – A liberatory praxis

In the 1980s, CI had become a form of inquiry into the physical responses of the human body to external forces such as gravity and to the body of another dancer. The interest was to establish relational interaction between dancers that did not happen in the mainstream dance context: CI served to defy the 'cult' of the body and the oppression that dancers experienced outlined above, as well as the non-touch agenda and taboo (Nelson 1996; Prestidge 1997). The fundamentals of CI are defined and analysed in detail in all the following chapters, but suffice it to say for now that it requires at least two dancers in a mutually dependent relationship, both of whom set aside traditional dance technique to explore the possibilities that arise as they improvise, moving around each other in ongoing physical contact. CI was a praxis in which the unexpected was immanent, a place of suspension and discovery, an enhanced field of experiential research – like a hermeneutic cycle in movement and on bodily terms.

In CI, the 'natural' ability of the body to move was explored and broke with traditional assumptions of gender, role and training. This did not deter groups of very experienced movers exploring the extremes of physicality in a specific 'closed' environment, nor, either the possibility for people with very limited mobility from participating in others. CI continued to evolve from further research and became situated beyond the constructs of culture, gender, and to some extent age and ability, within arts, health and therapy contexts. As a practice, CI is disposed to differences in body size, type and mobility (Novack 1990; Paxton and Kilcoyne 1993; Fulkerson 1996) and as an 'art-sport' is accessible to non-dancers. However, many dance artists practice and teach it discretely, or as part of general dance/movement programme. Dance has always been the art of the body in motion, a form of non-verbal communication that 'speaks' kinaesthetically, through gesture, to reveal the unspoken hidden aspects of self. With CI the purpose was also about dwelling in the subjective, lived experience of moving, and different for being in the here and now with others committed to the challenge of its immediacy and to a community-oriented ethos (McDonagh 1990). In her book on *Disability Culture and Community Performance*, Petra Kuppers discusses how naming oneself as part of a larger group can help to focus energy, and to understand how solidarity can be found, even if precariously, through improvisation (Kuppers 2011: 109). CI provided an opportunity for those previously excluded from physical activity to mix with others who were not, with mutual benefit (Koteen and Smith 2008; Albright 2018).

As Paxton states, the 'form was developing us' (Koteen and Smith, 2008: xiv) as each 'strand' that wove into the work brought new attributes, without compromising its primary ethos of natural movement and democratic relationships. CI was a unique aesthetic experience for being pluralistic (Kuypers 1918; 12), integrating 'gymnastics, eastern martial arts and dance' (Paxton and Kilcoyne 1993: 3–4, original emphasis) and pragmatic. CI integrated the arts, with collective play/social interaction and sport (Koteen and Smith 2008). With these attributes CI opened the dance floor to a multitude of people with and without dance training. It was just a matter of time before CI opened the door to many excluded from dance – in the 1980s in the United States physically disabled people participated in

CI and the wheelchair became integral to the dyad (Ptashek 1992; Kuypers 2018). Evidently cultural norms and boundaries were being challenged and redefined – some could run with this, others held on fast to the membrane of convention and stayed away, or kept to other cultural identities. Such was the opportunity of CI – to go with the flow, with a 'new frontier' and the avant-garde, and participate in innovation, renewal and change.

CI comes to the United Kingdom – DCA, Totness, Devon

In the 1980s, Paxton gave classes in CI and other movement explorations at DCA in Devon, in a faculty of contemporaries such as Mary Fulkerson. He shared his ideas and beliefs with other artists, like in the art labs, and was a huge influence and presence at the Dartington Dance festival over many years. In 1987, I saw him perform: he surprised the audience, by standing still for several minutes, causing consternation. I recall he spoke, he commented on the fact there had been too much movement in other pieces he had seen. In Paxton's piece, I saw the liberation of the mind in movement – the physical voice of the body came across as powerful, defining and radically direct – all the more so for being improvised. It appeared to be a way to confront contemporary practice, to position the 'old' ideas of dance making in a precarious position, undermined and challenged. It was time for a rethink, Paxton gave permission to reclaim dance from the citadel, for the dancer to explore the subjectivity of their dancing body, and for diverse bodies, with less movement, to participate.

The pedagogic ethos of DCA provided a conducive setting and appropriate context for Paxton's vision of CI to take root. Contrary to the contemporary mainstream vocational dance training institutions, DCA introduced CI as a form of 'detraining', a practice that provided students with a 'natural resource' to work in contemporary social contexts (Hulton 2018: n.pag.).[9] This was a deliberate move away from the kind of training expected of formal, monolithic forms of dance education such as Graham, or Cunningham. At DCA, the philosophy of 'detraining' sought to provide the perceptual, imaginative and creative concepts required by the social context that the student-artists worked in. Aligned with the educationalist Paolo Freire (1970) this meant a pedagogy embedded in contemporary social and cultural forms, in the current appropriate methods of communication, and in dialogue with the participants (Hulton 2018). With this 'de-schooling' experience (Hartley 2018) the students had the resources, and the 'know-how' to work with a whole range of people. DCA was an ideal environment for practice-based research projects that championed community participation.

At DCA, CI was embraced like a new contemporary language within the radical culture of independent artistry, of hybrid and pluralistic practice, across dance, theatre, music and visual arts (Novack 1990; Fulkerson 1996; Hartley 2018). In the 1980s, CI suited the applied arts study and community-based practice, as a methodology that would work within disaffected or disadvantaged communities, where formal training that was in keeping with 'societal diktats' would not (Hulton 2018: n.pag.). CI training prepared the students for the Touchdown Dance experience and featured in interdisciplinary performances as a practice and research method, that was new to academic settings at this time (Nelson 2014[10]). The impromptu Touchdown Dance performances at DCA with Paxton, Julyen Hamilton and Gerry Overington (visually impaired) in 1989, illustrated this self-generative process (see Chapter 1).

Paxton's ambition, from a seed to a body of work

DCA was an obvious home for Touchdown Dance; Steve Paxton and Anne Kilcoyne were involved in teaching and research, Paxton in the dance department and Kilcoyne, a clinical psychologist working in applied theatre. In 1986, Paxton and Kilcoyne set up the 'Skipping Visual Impairment' workshops, with students who remained involved and dedicated to the research, such as Caroline Waters and Charlie Morrisey. As I recall from my time there from 1986 to 1988 as a student mentor, it was a period where seeds could be planted and then grown. Paxton spoke to me directly of his germinating idea in our online conversation in 2019; 'It was an ambition of mine, it seemed like CI and blind people had something to share, a touch based dance form and people who are touch based' (Paxton 2019). This idea de-emphasised the need for a cultural form, by emphasising instead the appropriate forms of communication required by the context. This 'ambition' became a research programme in which sighted and visually impaired people could move together in a carefully facilitated CI practice. A brief account of how this idea became a coherent research project will establish the corporeal territory in which this particular seed would grow, as I come to in subsequent chapters.

The forming and reforming of the social body – Bodism or barbarism?

Looking out onto the wider social and cultural landscape mentioned earlier in the chapter, I realise that the *body* is conspicuously omitted in Kelly's accounts of the counter-cultural relationships in the societal 'body' (Kelly 1984) and remains an abstraction in the disability arts anthology (Morrison 1991). However, Bartlett (1996) celebrates the advent of dance for all and pre-empts the concurrent liberation of the participant body with societal liberation through dance/

movement – and the intrinsic expression and creative representation of the self, by the self, and to the idea of 'self-making' through arts practice.[11]

As the accounts will tell, this was to be the experience of the majority of people coming to Touchdown Dance workshops. Moving together with another person in a safe space brought many undisclosed beliefs to the surface, beliefs that had been conditioned in a society in which aspiration was absent. Touchdown Dance evolved into a new culture, a form of body-ism, one in which the body is liberated from previous inhibitive experiences and becomes re-experienced as mutable. CI transformed and unified body, mind and memory, in a 'sphere' for safe exploration where fear of difference was relinquished.

Evidently, DCA was the space for new dance development in the United Kingdom in the 1980s and in 1986 the place where Paxton and Kilcoyne established a research project that became Touchdown Dance in 1987. Their journey is the matter of the first three chapters of this book and their legacy unfolds in subsequent chapters. Touchdown Dance, like other community dance endeavours, started by bringing people together and being prepared to engage with the unknown. The vision to involve visually impaired people in CI reflected the untapped potential of CI as a touch-based approach to movement practice (Figures I.1 and I.2).

FIGURE I.1: Aller Park Studio, opening circle wearing blindfolds. Photographer Kate Mount.

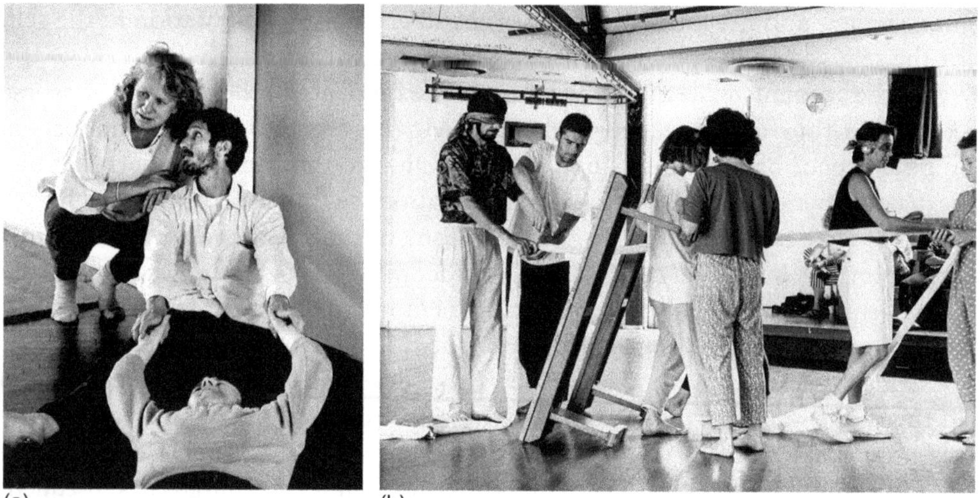

(a) (b)

FIGURE I.2: (a) Steve Paxton and Anne Kilcoyne in conversation. (b) The workshop group exploring space and form through touch and movement. Photographer Kate Mount.

PART 3

Bringing CI and Touchdown Dance to Denmark

Having moved to Denmark in 1988, I discovered the potential for CI in diverse communities and contexts. Starting at Aarhus Theatre Academy, I then taught at a street theatre school and delivered community-oriented projects. Significantly, the grotesque buffoon work raised particular interest from some physically disabled performers who saw the initial performance project and believed it would suit their physiques, their ideas to challenge deformity and their personalities. They were competent with writing and text but believed they lacked physical performance training. As a non-national, the question arose as to why this was the case in a culture which prided itself on social integration. An investigation into why vocational training programmes in performing arts excluded disabled people provided answers such as: 'Well our staff are not trained to teach them.' Based on these facts and the intention to bring in expertise from Europe, in 1990 the Danish Cultural Ministry granted funding to deliver a year-long project initially called *Hidden Bodies* (*Skjulte Kroppe*). After a research and development phase of skills sharing through laboratory-type sessions, we formed a group of artists interested in collaboration. Existing community theatre groups were found by contacting the associations for the blind, the deaf and other disability organisations, followed up by meetings with local groups and individuals. Preliminary sessions and exchanges were set up to establish contact, to explore what could be of interest, any possible access needs and the most appropriate methods of communication. I contacted institutions and their 'gate keepers'. I learned to use the 'writing phone' which translated spoken words into Braille and enabled me to speak independently to deafblind participants.

The blank page

I later recognised the challenges I faced to connect with the Danish blind and deafblind community to publicise the Touchdown Dance workshops. I had no idea at this stage about their work, but only about CI and Steve Paxton, and I had not met Anne Kilcoyne.

I contacted the Deaf House in Arhus and went to meet the deafblind theatre group. I couldn't speak Danish sign language and felt the barrier and distance straightaway. As there were interpreters, I offered a barter; for them to teach me sign language for deaf and deafblind people and I would provide some taster workshops in CI. In addition to learning sign language, I also learned to speak Danish, and the expressive physicality of signing came relatively easily, having learnt other languages and training in mime. Before sharing any CI material, I had to start with communicating through touch and movement, as voice, vision and hearing were redundant. I used the interpreters at first, or wrote on paper in large writing for those with residual sight and then sign into the hand of a deafblind person. I had to start with a 'blank page', as I was often to describe it, with simple movements, using physical contact in small doses in the least intrusive body-on-body positions, such as standing back to back with a partner. I learned later how hard it was to move backwards without sight and hearing, and so having support from behind freed up the body to move in any direction. I found out how to start by demonstrating with one person, to feel the connection between us. We each then taught another person and so on till everyone had a go and were given time to say what they felt. The time to share was revelatory as there was no other way to know what impact this gentle start was having, there was a lot of inhibition and encouragement needed to continue beyond a few seconds.

It was a gentle process, slow and considered but it became meaningful and the outcomes for the deafblind group were significant – they had inspiration for their own theatre work and showed me short scenes the group had devised. They used tactile cues, tapping and signing on the body, and used any sight or hearing that was available to navigate the space and keep the scene moving. The scene was very theatrical, the idea was that everyone co-wrote the text and knew it, each spoke to one person in sign language and sat attentively till the next cue came their way. Comic interest was also important, they wanted to embrace the banalities of being sensory impaired and how this condition was of course one of difficulty and discomfort but also of comic relief.

Their skills base showed imaginative and creative ways to deal with practicalities, how to get on stage, in a chain, hand on the shoulder of the one in front, to find the set and how to know when to start. I was impressed by the artistry and honesty and realised the potential to nurture further what was already working. Within our short time together they learned the possibility to move together without sign language, how to indicate and communicate in space, and how to work with the body as a natural development of sign language itself. I witnessed a widening of the sense of space, a hunger to explore and a need for reassurance that what they wanted to be seen and heard, was in fact received. Without sight and hearing, sign language on the hand and body was the normal touch-based language of communication, and CI became a natural sequel to expand on this. From signing into the underside of the hand to meeting arm on arm or back to back, touch and movement were the common languages, providing a realm of mutual awareness and dialogue, the realm of the membrane mechanism.

The real deal

When Touchdown Dance arrived as part of the workshop training series, some of the group had a basic understanding of CI but came really open-minded. We had very few participants during the first weekend and then a much larger group for the second and third; we had people expecting to 'dance', dressed in skirts and shirts and ties on the first day, but it was a magical experience. The transmission from English, to Danish, to Danish sign language to the deafblind person, and then back through the same interpreters became an essential preoccupation. Listening to each person's experience was an important part of the learning so that the time this took was a key example of the communication issues, limiting the amount of 'learning' that could be done through movement. Finally doing the work, taking simple steps directly through moving in partners, was the best way to communicate – with simple verbal and sign-supported prompts, and feedback such as 'Feels OK?', 'yes', 'no', 'Again?' and 'good'. The sighted, Danish-speaking participants, took on the transmission of information with the help of an interpreter speaking hand-in-hand with the deafblind mover when required. Paxton and Kilcoyne marvelled at this process and the fluidity of transmission.

Paxton and Kilcoyne had to accept that hardly any time was spent on dancing, but the experience of partner work was positive. I recall two deafblind men sitting back to back on the floor, responding to instructions to move together forwards and back bearing each other's weight and stopping to comment. How would they know what was ahead? How to show them the potential of this initial exploration? It became evident to us that this was just the beginning and that the work would need to continue intensively to have an impact. It was following this weekend that I continued to visit the deafblind group and to work with Torben[12] who asked for regular sessions on his own. This was the start of a practice-based research project for both of us and the beginning of a working relationship that culminated in performances, presentations and many travels over the next four years; he relished performing in public.

The changes manifested over time as we moved more freely together. By videoing the sessions I could see the relationship evolve. I taught him Paxton's yoga stretches, and the importance of a long out-breath. I learned to sign by watching him sign with others who spoke at the same time. I learned more signs as we paused to discuss the experience; it was up to me to give positive feedback by signing into his hands and by expressing my feelings in thumbs-up gestures when I saw him smile or become animated. I saw him grow more and more adept with his body in movement with mine, taking on the basic journey Paxton and Kilcoyne had taught us; we worked from lying on the floor upwards to standing, working with bodily coordination exercises, such as rolling in different formations. Such exercises I discovered were essential to the educational purpose of a workshop, where movement that is limited due to habitual lack opens up space.

Torben discovered his body by moving in a way that was remedial and reminiscent of the sport he did in his youth in the deaf school. PTW became an integrated 'arts lab' (Kelly 1984) for praxis and creativity. Intensives were given by other performers, including the duo Jag Plagh and Jack Klaff, and Common Ground dance for the deaf from the United Kingdom. We then rehearsed and produced a touring production that took an extended Nordic tour. We gave workshops and performances in the Faeroe Islands, Iceland, Sweden, Aaland and then back to Denmark. In these locations we gave workshops to blind and deafblind people; a young woman I name Ida features in Chapter 3. We gave training workshops to institutions such as the Swedish National Ballet in Stockholm who found the first morning painfully slow, but over lunch the 'penny dropped' and they valued the process as inclusive, embodied inquiry. I recall travelling on buses, ships and small propeller planes – with a fully integrated company, musician and crew.

After eighteen months, Torben became more agile, his walk more confident, striding out rather than following the carer's lead. He told me; 'I want to do this for the rest of my life. This is my way of being in the world', now that he had found an outlet for isolation and an activity that aligned with his senses, he no longer experienced touch deprivation – in a world in which contact with others was through the hands and rarely on the body. We continued our sessions for nearly two years before heading to Dartington for one of the intensive workshops in 1992 and discovered how much we had achieved (see Figures I.3 and I.4). Torben went on to get a job teaching the 'writing phone' and our sessions stopped when I returned to the United Kingdom (Figures I.3 and I.4).

FIGURE I.3: Katy Dymoke and Torben Hansen dancing CI at Dartington College of Arts, 1992. Photographer Kate Mount.

I had learned so much from this experience – following this, I would struggle to find any other teaching experience daunting. I had had to build something on trust, to work together with the group, following their lead. Danish sign language became a natural part of the dance partnership in lieu of the visual teaching methods I was used to, I learnt from each single moment of movement that we made – each initiation was a point of communication and each movement response a new possible outcome that shed light on a path leading to somewhere new. Each step was a proposition, an initiation of a dialogue that was painstaking and careful, giving space and time to consider how, when and where to move next. In all, these PTW days were preparation for the Touchdown Dance experience that was to come.

FIGURE I.4: Katy Dymoke and Torben Hansen dancing CI at Dartington College of Arts, 1992. Photographer Kate Mount.

Serendipity ... finding a common ground

Both Touchdown Dance and PTW started off in a non-disability arts and non-Statutory Arts culture, with an ethos and understanding that the training and creative process would be inclusive and research-oriented. Paxton and Kilcoyne introduced the practice of CI to visually impaired and sighted participants whom they embraced as researchers and contributors to the development of the methods and processes. PTW was also the initiative of non-disabled people and the research project was in response to the request of disabled actors to receive physical training for performance and was undertaken in partnership with them.

In both instances, the work survived by maintaining a community and a body-oriented, holistic approach and by embracing values that emerged outside of the mainstream arts scene that was progressive advocating social change. Both initiatives were situated independent of mainstream disability services and emphasised the idea of integrated cultural exchange, promising new types of learning experiences, and a new aesthetic to achieve this. New initiatives like Touchdown Dance and PTW applied the revolutionary teaching practices of postmodern dance theatre, to precipitate change; it was time for a new participative landscape, for disabled people and other minority groups to determine their place in a new inclusive cultural ecology, through direct action.

The demand for access and equal opportunity for *all* was not easily met across the Arts and Culture in the 1980s. Inclusivity and diversity were values and aspirations that were current at the time and shared by the wider community arts movement. Actions such as setting up integrated arts research projects could have been seen as a form of cultural agitation and politically confrontational. According to Elspeth Morrison, writer and activist (Morrison 1991), the culture of oppression and control appeared as a norm to many, including disabled people themselves. The climate of cultural change would require determination but also a reflexive approach and a more accommodating pluralistic view, to ensure that the 'old', tired and exclusive perspectives were edged out (Kelly 1984). This process gave space for a shift towards a revitalised and more inclusive cultural landscape, spawning a new set of values including diversity, accessibility, social injustice and equality, which were eventually enshrined in government legislation in the United Kingdom; the Disability Discrimination Act in 1995 and the Equality Act in 2010.

A personal note – Finding the self-membrane

Rather like Rainer's *No manifesto*, writing this book has brought many things 'home' to me. There is both provocation, and an intention to 'clear away' the debris caused by injustice, to provide a space for something new. It became clear to me, as a member of society, that this same society was failing many of its community, and worse, depriving people of the opportunity to participate, of a right to be as alive as possible and to fulfil potential. I have pictured society as a place that segregates types of people in a carefully contained box, carefully labelled, with no chance to shed or reframe this given construct – literally body-less and there is no body to be liberated. Being alive is not compatible with such constructs, as aliveness requires space for change and transformation. Liberation from social constructs can only take place if these are deconstructed as false; the 'box' can then reconstruct its animate form, into a permeable and motile membrane that enables the self to receive and respond to what other animate forms are there, in recognition from both inside and out of the right to determine its path. Change is 'decided' with the membrane in reciprocal support, as is the right time to break free or transform. If not there is inevitably a break-down between the self and society and rifts form in a landscape of separated spheres cloaked in false beliefs.

A sense of self-contained within a sphere for growth is, conversely, motile, free to transform, seen and heard, and in mutual recognition of a sense of justice takes a rightful place in the world. If the right to dance is a proclamation of a need to be met, then, if deprived of this need, society has abused this right, or, if addressed, the proclaimer can participate in defining the world as it is lived by others; so individual and social responsibility gravitate together and movement practice becomes a political activity (Albright 2018).

This thinking was matched by lived experience as I moved with visually impaired people; initially very afraid to move they start to perceive a 'new word' and then dive into the movement experience as if 'hatching' from a shell, compulsively, perceiving freedom and life on the other side. Later, from the concept of breaking through the shell, into the light of the world beyond, I formulated the idea of a motile membrane that responds relative to the experience and supports choice-making and change. The membrane enables communication with others, shifts in states of consciousness and its animate form is in constant movement and contact with 'other', which I return to below and more specifically in Chapter 2.

1

Returning to the Origins: The Journey Taken by the Founders

PART 1

A chance encounter – Where it all started

This chapter gives a retrospective overview of the journey taken by the founders, Steve Paxton and Anne Kilcoyne, to set up and run the first workshop series, with participant case studies to illustrate. The story of the early days of Touchdown Dance is sketched out, with reference to the documentation and archives passed on by the founders. The discrete development of the work is contextualised within the wider culture to flesh out the discourses outlined in the introduction, with a reflexive commentary that draws on insights I make in the process of writing, based on my experience of the work. Contact Improvisation (CI) and other influential practices that informed the foundational purpose are explored, mostly taught by Paxton, and other attributive touch and movement-based practices that were researched further in the following years by Kilcoyne.

In our recent online conversation, Paxton recalled an invitation to an institution for visually impaired children in Boston (USA), where he demonstrated CI with a visually impaired boy who actively sought to connect with him (Paxton 2019: n.pag.).[13] This direct interaction had surprised staff and touched Paxton, so he believes his interest in working with visually impaired people was kindled. He shared this idea with Anne Kilcoyne at Dartington College of Arts (DCA) and having discovered the local rehabilitation centre for people experiencing sight loss, they undertook visits with the view of introducing CI to the residents.

As I re-read the history of the work, it became apparent to me that this first phase established the ethnographic basis of a long-term practice-led research project. Accounts of the early days are found in short articles in the *New Beacon, the Journal of Blind Welfare* (Paxton and Kilcoyne 1989, 1990a, 1990b) and in a long article entitled 'On the braille in the body: An account of the Touchdown Dance integrated workshops with the visually impaired and the sighted', published in *Dance Research*, the journal of the Society for Dance Research (Paxton and Kilcoyne 1993). I draw on these sources and interweave them with reflections taken from a transcription of a recent online interview with Paxton. Evidently, the work was located within their respective professional fields of dance (Paxton), and theatre and psychological therapies (Kilcoyne), creating an interdisciplinary praxis rooted in movement (re)education and potential rehabilitation (Paxton and Kilcoyne 1993). The focus of these early workshops was on inclusion between the visually impaired and sighted dance students from DCA, as if in replication of the Boston experience. The key aims were first to meet the challenges faced by each of the participants with sight loss and second to manage the complexities presented by the meeting between visually impaired and sighted participants.

The first seven years of the seasonal workshop programme established an ethos of accessibility and trust. It becomes evident that the touch and movement methods of CI were baseline from the start and continued to underlie a foundational pedagogic approach. The following ethnography contextualises the value of CI to the contemporary dance arena and other arts-based, body-oriented disciplines that enabled CI to evolve. These methods continue to imbue the work today and are covered further in Chapters 2, 3 and 4.

The first years of Touchdown Dance 1986–88 – Finding a common way of seeing using CI

CI became the heart of the Touchdown Dance praxis and served the inquiry into non-visual learning methods appropriate to the inclusion of visually impaired people in an otherwise sight-dominant culture. After several workshops, Paxton and Kilcoyne identified three attributes in support of access and opportunity for the visually impaired participants: 'Contact Improvisation seemed an ideal activity for the blind, based as it is in the natural physics of movement, in human communication, and focused on a touch-based learning path' (Paxton and Kilcoyne 1992: 47).

Paxton had noticed that in practising CI, sighted people tended to close their eyes, to sense and experience more consciously the movement that they are making. This indicated to Paxton that the practice of CI brought into focus the kinaesthetic sense, at the expense of visual perception, i.e. it could be available to all including those without sight (Paxton and Kilcoyne 1993). Paxton had become aware of kinaesthesia as the perceptual process of sensing the body whilst moving, and indicative of the 'natural physics of movement' assisted by touch, the tactile sense. Primary to vision, kinaesthesia brings attention to the present position of the body relative to gravity and to its corresponding movement. Sight, if available, supports spatial orientation and without sight, an internal optical sense can be experienced as a commonality for all. This became apparent in one of the influencing and founding kinaesthetic practices that Paxton explored alongside CI; a practice of standing still, called 'the small dance'.[14] Whilst standing for minutes at a time with eyes closed, the mover registers subtle shifts in weight transmitting through the legs to the feet that manifest the body's innate responsiveness to move with or against gravity. This conscious attention to the subtle reflexive movements that maintain verticality and prevent falling provides a means to pre-tune to the kinaesthetic sense and an amplified awareness of each movement response, however small or short.

Paxton perceived how the experience of the small dance and the attributes of touch that bring awareness to the inner landscape of the body in CI were non-visual practices. By closing the eyes the external distractions disappear from 'view'. As CI was not dependent on sight,[15] the hypothesis appeared logical – that it would be an accessible form for visually impaired people to do with sighted and interesting to explore what use CI could be in that context. As the title of their seminal article implies, *On the Braille in the Body*, Paxton and Kilcoyne (1993) perceived CI as a form of movement-based communication, the main form of transmission being through touch and so 'instinctively available to those who can see and those who cannot, alike – a shared language learned through the body' (Paxton and Kilcoyne 1993: 6).

Touchdown Dance would provide a new domain for Paxton to deepen his research into movement pedagogies. In the small dance and more generally in the practice of CI, Paxton explored the innate responsiveness of the body to gravity and other external forces, the tone and innate motility of muscular tissue, and discovered the potential for ease and fluidity in improvised movement (Paxton 2018). His tools were interdisciplinary; dance improvisation, experiential anatomy, somatic practices like yoga, and influences from tai chi, Aikido and meditative practices. With Anne Kilcoyne's psychotherapeutic background and her shared interest in research, these attributes would be integrated within a group process. Together they developed an interdisciplinary praxis rooted in movement (re)education and potential rehabilitation (Paxton and Kilcoyne 1993). What was new to both of them was the prospect of teaching to a group of both visually impaired and sighted people; this was unknown territory and various challenges, with both individual and societal implications, arose.

Bringing visually impaired and sighted people together through CI

In 1986, the practice of CI in the United Kingdom was mostly local to DCA and was, as yet, more of a participatory activity than a performance modality. CI was relatively unknown outside of the contemporary dance world and particularly unknown to the very separate world of visually impaired people, who were at that time, the most excluded from Dance (Dymoke 2014).[16] One of the first challenges was how to define the CI experience for visually impaired people in a way that indicated a movement practice and non-dependence on sight. The first workshop series called *Skipping Visual Impairment* appeared, through this title, to offer a means to literally hop, skip and jump back into a moving body (Kilcoyne 1989). The title implied that movement and sight loss were at the heart of the experience, which was the case for the local participants, and instilled a certain spirit of fun and risk-taking that characterised their pragmatic and reflexive approach going forward.

The first visually impaired participants were very diverse; some had an interest in movement, sport or physical activity and others not. In our conversation, Paxton recalled the tentative start as no one had ventured here before:

> Somehow we got a car full of people to come over, once a week or something and do sessions, and then I would endeavour to teach and you know what that is like. If you don't know what you are doing and haven't been around visually impaired people much, I mean there are just all kinds of protocols.
>
> (Paxton 2019: n.pag.)

Paxton experienced the reality of the unknown as I did in Denmark, and the prospect seemed fraught with having to manage the potential barriers caused by societal segregation. His resolve is apparent, as is his belief that learning how to 'do sessions' would evolve, and that 'protocols' would address the magnitude of this 'endeavour to teach'. In these initial meetings, Paxton and Kilcoyne explored how to engage a diverse group of visually impaired people in movement with some basic CI principles and witnessed the loss of ease and confidence in some. Naturally concerned for the safety of all in the room, 'protocols' were put in place to enable familiarity with the space. Each person's response was subjective, contingent on for example, how much input could be given in the time, on the level of independence and mobility strategies each person had, or on whether the sight loss was recent or from birth. The idea to involve sighted dancers provided a potential way to mediate these singular aspects – to assist access to the space and to the information in a one-to-one interaction. However, much work needed to be done prior to any first meeting, so preliminary research was undertaken with both participant groups.

The visually impaired group – Managing the challenges of sight loss

The visually impaired participants were adults who had lost their sight due to accident or illness, or genetic reasons. Still in a process of rehabilitation this was the key consideration for Paxton and Kilcoyne as they planned the initial workshop content. Being in rehabilitation indicated that the visually impaired participants were likely to have experienced more recent sight loss. From consultation with the centre, the implications of possible trauma, along with the challenges of adapting to the loss of sight, led to a conscious decision to not involve people with other disabilities at this early stage (Paxton and Kilcoyne 1993). This was out of respect for the range of complex communication and learning needs that the visually impaired participants experienced, to focus and give space for these first. The challenges of sight loss for the visually impaired group were, evidently, way beyond the type of challenges faced by participating blind staff members, who were accustomed to being visually impaired, and the sighted dance students. For those with sight loss the world that they had previously navigated with sight was now partially, or totally unseen, full of obstacles, steps, doorways and elusive paths ahead. Many questions arose that were addressed in the studio practice and research writing. How to encounter one's own body, now inhibited and reluctant to move, when one's day-to-day life has changed so dramatically? How to respond with ease when faced with the challenges, the impact of the loss and the affect of ensuing grief on both interoceptive[17] and proprioceptive systems? The process and protocols would need to be aware of such bodily states, to not coerce or push, but rather relieve and reassure as required, and how this was achieved will unfold in the many case study accounts that follow.

Group 2, sighted students adjust to a non-visual pedagogy

Before bringing a few visually impaired and sighted people together for the first time, another question arose as to how the sighted students would respond to this situation: what challenges would befall them in this new *detraining* experience? Located in the familiar environment of DCA, their day-to-day life in the dominant visual culture was privileged but limited. Paxton and Kilcoyne understood the dichotomy facing the students who volunteered; they were used to a predominantly visual mode of communication and could find difficulty in adjusting to predominantly non-visual learning methods. As already noted some sighted dancers had explored moving and practising CI at times 'unsighted', by closing their eyes – however this voluntary adjustment was a natural response in the more sensate moments and a far cry from living with the trauma of permanent sight loss. As a result, there would be some degree of a divide between their expectations of what could be achieved and their availability to communicate, at least initially.

Paxton and Kilcoyne focused on how the prior training of the students could be useful and raise awareness of the barriers that exclude visually impaired participants as well as acknowledge the contribution that each individual's lived experience makes. A preliminary training would help Paxton and Kilcoyne to become more familiar with working without sight and to explore how to transmit the methods and process of CI in ways that facilitated non-visual learning, bearing in mind the barriers that would arise. As this was unknown territory for all, further preliminary research was warranted to identify protocols within a pedagogic approach that would embrace diversity of need. Paxton and Kilcoyne could not assume a specific constellation of visually impaired or sighted participants but could predict that CI was inherently accessible and accommodating of diverse preconceptions.

Working towards a collective learning experience, establishing protocols, managing obstacles – Establishing the 'membrane' of inclusion

Prior to their meeting with the visually impaired participants, it seemed appropriate for the sighted students to explore working with blindfolds, to gain experience of being unsighted for a longer period and to sense the limitations imposed on movement. The blindfold prevented peeping but enabled the eyes to remain open, and of course, it was optional and could be removed at any time. In the context of CI, the focus is on experimentation and self-directive play or inquiry, so blindfolds were a choice for each person, to use as and when they wished. Blindfolds served, to some extent, to displace the fear projected, and so expected, of being unsighted. In a subtle way the realm of non-visual communication started to open up, to become more familiar to the sighted students; they learned to rely on touch and movement, the languages that visually impaired people shared. This was the only way ahead for the research programme, it could not happen the other way round – the sighted group in the community had to adapt their territory to include the visually impaired, for a mutual space to be established. With the blindfolds, the sighted students were then less 'in the dark' about sight loss, and were able to move together with the visually impaired as 'unsighted'. As will be seen, this experience unlocked the potential for a paradigm shift, for further questions and for shared learning, as the research endeavour intended. CI became useful to break down the aforementioned barriers as Paxton and Kilcoyne note:

> Cultural barriers which often constrain or even impede healthy human communication disappear as touch, movement blindness, sight and equality within the work combine to provide a radically different climate for one person to meet another.
> (Paxton and Kilcoyne 1993: 8)

This quote defines the transformative space in which Paxton and Kilcoyne evolved their teaching, the protocols to support the ongoing process of both intra- and interpersonal learning. In this 'climate' the 'fabric' of CI formed a realm in which further concepts evolved, including the concepts of individual and social body.

During this embodied experience certain negative preconceptions of blindness, or visual impairment, were brought to conscious awareness. For example, the presumption that a visually impaired person is dependent on mobility aids (as some blind people don't use them), or unable to move in the world for fear of collision (as it depends on the environment, and other compensatory senses such as echo location) and of the idea that visual impairment means living in permanent darkness (as many blind people don't define their world in this way). Another misunderstanding was the belief that visual impairment causes the exclusion of visually impaired people from society and culture; that the disability is the cause rather than the inability of sighted people to accommodate and integrate them (Paxton and Kilcoyne 1993; Kuppers 2014). Paxton and Kilcoyne anticipated that the direct experience of visually impaired people would inform these societally imposed perceptions differently.

As the medical model of disability prescribed,[18] the sighted students typically preconceived blind people in terms of their disability and didn't have any idea as to the range of visual impairment or the ways these factors impacted the rights of a visually impaired person (Paxton and Kilcoyne 1992, 1993). It became clear to Paxton and Kilcoyne that the fear that some visually impaired people experienced of moving, was due to 'physical and psychological ramifications' that were 'compounded by the attitude of many of the able-bodied towards disability' (Paxton and Kilcoyne 1993: 9). These projections were 'the elephant in the room' that could undermine transparency and inhibit trust if left to wander. To some degree, the Touchdown Dance experience would dispel these 'myths', or at least bring them into question and other 'truths' would emerge in the workshop experience. Touchdown Dance proposed a pedagogy that would shift such perceptions, in a safe attentive space that enabled equality of access for all to move and learn.

Later, as a participant in this 'radically different climate', I conceived the permeable membrane, initially as a metaphor for the meeting with a visually impaired partner, and the inter-subjectivity between ourselves as subjects that rarely meet, then as a locus for analysing the learning process. The membrane represents the ethos of empathic engagement that Paxton and Kilcoyne shared between them; they held a safe space for the encounters to occur, for engagement between unknown people, and for the therapeutic benefit to be realised. The success of the workshops would determine a collective belief in the possibility of relating through touch, movement and other common senses – a radical idea which I consider next.

The first encounter – A mini revolution

The first meeting of the two groups was planned to commence with the arrival of the visually impaired participants at the studio, an annexe on the Dartington Estate called Aller Park. The Hexagon studio had a reception, toilets and residential accommodation just across the road. There were large full-length windows looking out onto grass, trees and other buildings on one side and walls with wooden rostra on the other. It was accessible and the movable rostra provided opportunities to change the space, to create a three-dimensional sculptural environment. When participants arrived for the first time, they experienced the long drive through the Dartington Estate, the bumpy road to the studio, and then the studio itself, an open unusual space with obstacles to be encountered as well as the people. Orientation to this new space enabled opening to self and each other – to a journey of discovery.

Paxton recalls the moment of arrival in our conversation:

> I know we took them all in that little studio, we would take them all around it, and let them feel the space and go outside [...] and let them know that we were at a little 'dead end' and so there weren't many other people around and they were safe to be outside.
>
> (Paxton 2019: n.pag.)

Spatial orientation was foundational to the activities on the first day. Carefully considered, the group process was instigated outside the building, where the sighted students met the visually impaired participants from their transport and guided them into the studio.

This 'ice-breaker' activity became a protocol, a ritual mapping of the space and the transmission of information from those who see to those who don't in which both discover what each other finds. Once oriented, the visually impaired partner may then return the favour and guide the sighted partner, who had their eyes closed if they wished, so both were unsighted. Evidently, it took more time to 'see' the space through movement and touch but this was effective as it was also with the view, more specifically, to experience how it is to encounter hazards, however minimal, to locate structures like windows, exit doors, and how they open, and to experience the textures of the environment. This empathic and respectful relationship brought to consciousness the otherwise unspoken differences in perception and levelled the assumed hierarchy of sight in the guide role.[19] In this way the guide modulates their pace in response to the guided, to work compassionately with their partner rather than detached and nonchalantly. Then with a pause to discuss, to chat, came the acknowledgement of the discoveries in common and any different interpretations of these.[20] This exercise was based on the idea of arriving, for all to feel welcome, to feel met and the space became familiar. Paxton referred to these moments as a form of 'socialisation' for the group; it was also part of the group process model that Anne Kilcoyne held, and that from this sense of containment, of safety, then it became possible for each person to begin to explore their own limits, little by little, and to find ease in terms of orienting to the material. At times nervous and tested this required careful reading and facilitating for each to feel their own process whilst being one of a group.

Much was to be discovered for the instructors too – by improvising and responding intuitively Paxton found solutions to events as they occurred. His understanding of how to teach grew in response to what was needed in each moment. He recalled how the protocol established itself as a principle for the work:

> so one thing we did was of course to have sighted people and blind people paired ... so that we could ... transmit to the body of the blind person ... you could teach to a whole room and then each sighted individual in that room would convey the, the *work* to the visually impaired person ... so many ways to try to be considerate ... to try to fill in the picture for somebody who is visually impaired.
>
> (Paxton 2019: n.pag., emphasis added)

This was really what the 'work' sought to do; to find ways and means, principles and methods, without creating a 'one-way street', in which the visually impaired participants were always guided and not reciprocating. The point being that by changing roles having already had the experience, the visually impaired person then 'instructed' the sighted with the aim to 'fill in the picture' of the movement as they made it. This would be a non-visual 'picture' using touch facilitation methods like tracing the pathway on the body, and tracking the body as it moved.[21] At times, and voluntarily, both could be unsighted, by use of a blindfold or closing their eyes. In this way the assumed role of the sighted guide was turned on its head, as was the assumption of disability.[22] Being led, the sighted student experienced the passive role experienced by a disabled person in society, the loss of control and assumption of dependency, whilst also being in a position of trusting in the visually impaired person to lead and to 'see' the space together, without sight.

To illustrate this in action, Paxton recalled the onerous and slightly comical reality of the 'blind leading the blind' in a variation of this exercise outdoors, when the visually impaired participants arrived outside the studio and were met by a blindfolded sighted partner:

> SP: You know ... everybody could get blindfolded and conducted across the landscape by somebody using a cane ... I must say, it was thrilling and scary.

KD: Outdoors?

SP: Yes outdoors and setting off across the landscape and … not on a path necessarily as somebody setting off with a cane doesn't necessarily know where the path is …

KD: Yes … Brilliant!

SP: … anyway lots and lots of adventures and mutual learning.
(2019: n.pag.)

This was the climate of play and demystification. Each exercise was an adventure, and the work was to realise at each point of the journey how each person held their own piece of the fabric and allowed it to interweave in the wider landscape.

This protocol revealed how the assumption of ability resolved quite quickly, as, in general, both partners adapted to accommodate and reciprocally support each other. Both could respond to the present moment and find ways to meet the obstacles, so both became *able* to find their way and often went so far as to explore outside. A sense of liberation arose as this experience affirmed different ways of being. A mini revolution was an effective means of challenging individual/personal limits or expectations and socially imposed if well-meaning convention turned stereotype.

In our conversation, Paxton acknowledged how the role change emerged as a moment of magnitude and with surprising significance for both partners. When guided, some of the sighted were challenged by their own unsighted state (whether blindfolded or with eyes closed) and their partners, and so felt momentarily doubly 'disabled'. However, this changed with the process of accommodation that took place when, for example, the visually impaired guide started to navigate the space with hearing and touch, the senses that both could use to locate activity and to orient through sound and movement till the first encounter with a surface such as a wall or a chair. The role change was generally a conundrum, at least to begin with, as the 'blind' led the sighted, so 'who' identified as disabled was now a paradox – it was not about who could or couldn't see, nor about control, it was about cooperation and dialogue.

The environment provided a third factor, whether outdoors or in the studio. The nature of the location had to be indicated and mapped out for the visually impaired people to feel independent or at least able to recognise where they were. At times on a residential weekend, the sighted students explored the Dartington Hall estate with a visually impaired partner, walking through the gardens, along pathways through the picturesque and sensory environment and repeatedly around the studio where the workshops took place if required. For someone in rehabilitation for sight loss, an unknown environment added a significant challenge, but in this co-learner relationship it became normal to discuss and respond together, to learn from each other.

In view of what Paxton and Kilcoyne had observed on a preliminary visit to the rehabilitation centre, the ability of individual visually impaired people to engage with CI in the workshop was likely to depend on factors such as feeling at ease, on whether they had an antecedent experience of movement practice or on how well they were recovering from sight loss. As the later examples reveal, it required confidence and depended on how ready they felt to re-engage with aspects of their previous life, alongside sighted people, and recover (even if partially) from this loss (Paxton and Kilcoyne 1992). It was not known at this initial stage how or to what extent, the pairings of visually impaired and sighted participants would establish the territory for mutual gain, growth and development of the self. It required space and time for review and the continual accommodation of individual needs.

The circle – Being seen and heard, the membrane of unity (see Figure 1.1a)

Forming a circle was intended for each person to say their name and where they were from. Being visually impaired and unable to see where the circle was presented another moment for facilitation. Step by step the membrane of separation was met and brought awareness to the possibility of assumed inability, to the need to indicate, offer support, to listen and communicate for each person to state a preference or choice. It was a delicate matter for Paxton and Kilcoyne to attend to avoid over-intentionality of a sighted stranger seeking to 'help'. It meant taking time and space and subtlety became key to enabling participants to be self-directive and fully present.

It took sensitivity and attention to relational dynamics in order to pitch each planned activity, Determining how to start meant consideration of the familiarity of the task (such as walking in the space), so the pathways of communication were introduced with a spirit of lightness. For example, standing up in the circle, holding hands and passing a squeeze from one to the next till it comes back to the start. This opened up another kind of listening, not the outer ear or its inner hearing mechanism, but the kinaesthetic sense that attends to the body in movement.[23] It was an interactive game too, stimulating discussion and information exchange as to whether or not a squeeze was a real squeeze to pass on or just an unintended haphazard one. There were usually far more squeezes going both ways around the circle, creating surprise and instigating the possibility of responding to physical cues.

FIGURE 1.1: Personal and social acceptance – the body-on-body approach.

Moving out of the circle into space followed, by getting into pairs (a visually impaired person partnered with a sighted), the first experience of partnering and meeting in movement in the space. Then a 'roll down' may be offered, in pairs, from standing, one does a slow forward bend by leading with the head and the partner walks their fingers from the top of the spine down to the sacrum, with a focus on each vertebra. This was not assumed to be a new experience, but in this context, it introduced touch differently and the way it worked had ramifications for the direction of the workshop. On the one hand, it indicated to Paxton and Kilcoyne if anyone was resistant, felt discomfort or withdrawn; on the other, it was a means to introduce movement language and terms. Through observing they were then informed of individual limitations or difficulties and able to make changes and adaptions to accommodate any variations in need – such as slowing down the time given and giving smaller steps towards completing the full thing. Again, partnering was the key form of transmission, through sharing with each other in the doing of the exercise, the two partners noticed their differences as part of the experience, so if one needed more time this was accepted.

In our call Paxton spoke to me of his instructor role:

> It relies not so much on teaching or being taught [...] the visually impaired were put in a situation in which they were supposed to be teaching themselves and learning [...] so that my job was to provide a way for a learning experience to occur.
>
> (2019: n.pag.)

Paxton depicts the essence of the 'protocol' underlying the intended learning process – he wasn't to know what was being experienced by each person in the workshop, at any moment (2019); he wanted to trust in each person's capacity to self-direct and collaborate, to enter a new phenomenological realm of learning. To do this required trust in himself and in the moment of movement making, which he transmitted to the room in the permission he gave to all to explore. Further protocols will follow but the key one to state now was the nature of the experience that he expected to take place, he had to trust that the experience on offer was engaging. Paxton notes how his attention was directed on the room:

> As instructor, I had to track everything that was happening in the room and if anyone seemed upset, or disoriented, or anything by what was going on, for instance the facings or configurations of the body [...] for somebody who is visually disabled, they might become disabled, they might not know which way they are facing.
>
> (2019: n.pag.)

Paxton picked up the disorientation issue as it proved very variable and individuals could become quite disturbed if it wasn't addressed. Having lost sight (the primary sense of orientation to space), it could be challenging to be asked to 'find a partner' and to feel unable to do so, and worse, to have to wait to be selected. Paxton reflected on this disabling effect and how he managed disorientation:

> I guess they would know which way my voice is coming from but then they don't know where I am in the room and if [...] the 'if' question being for me – if that disorientation is disabling for them [...] as it would be you know if a person with vision is disoriented, they're sort of dizzy, not knowing where they are [...] a feeling of insecurity, you know, all of that?
>
> (2019: n.pag.)

Disorientation, as I discovered, was not a uniquely negative experience, it could also be a requisite for re-orientation to a new learning experience. However, feeling lost could be a potential disabling experience for either visually impaired or sighted people, until ways to reorient were shared.[24] This emphasised the mutual relationship that generally did evolve but which could be more or less effective. It became evident, whilst tracking the room, that changing partners gave new information and a wider range of movement responses to learn from, as well as changing the use and sense of space. These became further established protocols.

The significance of Paxton's observation is that he admits the reality of difference; homogeneity was not considered nor was any standard preconceived learning outcome expected to validate his research ambition. The sphere for the work had to evolve in a way that was safe, by establishing relational attunement and finding ways for very different individuals to become familiar with each other. The workshop provided an experience full of unknowns and the unexpected so that everyone was at a disadvantage in some way. To normalise the unique nature of this very new experience Paxton sought to build collaborative, respectful and reciprocal relationships. The following exercise was one physical way that achieved this, a way to literally bring heads together!

In our conversation, Paxton enjoyed recalling the atmosphere from over 30 years ago, and how directly the participants moved into partnering. He gave the example of working 'directly head to head [...] which is a kind of non-threatening part of the body to touch some other person' (2019: n.pag.). Once the initial position is found, standing face to face, forearm distance apart, the partners lean forwards to connect forehead to forehead, then shift gently to determine the comfortable amount of weight that could be supported there (see Figure 1.1). Following Paxton's commentary, they feel the connection through into each others' feet[25] and begin to follow any slight movement into rolling across and around the forehead. Paxton recalled this experience:

> it's a really amusing part of the body you know [...] compared to other parts and from there it was a model for the whole body, so the movement of the body and the touch on the head would be co-ordinated so that people could roll their heads around so they were facing the opposite direction but still touching heads, or whatever direction you ended up in [...] you are just allowing it to happen, everybody's going very slowly, exploring the touch you know, just the words 'exploring the touch' set you off on a kind of journey, and then moving the touch down onto the body onto the back, the shoulders, other non-threatening areas, until everybody is apparently alright with it.
>
> (2019: n.pag.)

Paxton emphasises the need to go slowly to maintain the contact and prevent disengagement. The head is a rounded surface and peripheral to the body, and moving head to head gives limited physical contact but introduces the basic principle of CI which will follow. This rolling touch is essential to CI praxis and required a normative approach, but was admittedly not a normal thing to do (Koteen and Smith 2008); no doubt this was an unusual experience or 'weird' to some. Paxton comments on the nature of this journey that had no known, fixed or presumed destination, and the pairing of blind and sighted participants marked the start of the collective adventure, of 'exploring the touch' together:

> so just trying to keep the spirits up [...] I always felt that the instruction should be amusing, like I comment on how weird we are being and everybody agrees that we're being weird rubbing heads together and [...] you know just pointing out that we're on a trip, and the trip is a little bit strange, and we'll get to where we are going soon, and is everybody comfortable? And all of that!
>
> (2019: n.pag.)

The partnering experience was essential to Paxton, for the CI experience to have any significance – the 'touch-based dance' with 'touch-based' learners did not preclude the sighted, or any sense of discomfort. Paxton reflected on how radical it was; he recognised the unknown aspect for visually impaired people and then realised it was not unique to them; 'they're not used to, you know, *we're* not used to it; it's not a usual thing to do by anybody's standards' – it was important to take time to realise what had changed. He also pondered the accommodating process and the transformative space:

> It does make time go by differently, you know, it does slow time down or speed it up or cause alterations of your normal perceptions, of all kinds; time and space [...] and the socialising – you are doing this with someone you don't necessarily know, maybe some did know their partners, but some didn't, and that's OK, it's a kind of social acceptance of whoever it was that you were working with.
>
> (Paxton 2019: n.pag.)

For Paxton, the key to this endeavour was to acknowledge collectively what was being learned through touch – the essential form of communication: 'I thought of it as fundamental to learning, you know, what touch could tell you, which is the differences between partners, so having partners change was another part of the program that I felt was important' (2019: n.pag.).

The idea that touch tells us something is not so radical after this experience. It makes sense of the process; being in touch with a partner is a reciprocal state; we are both being changed by the touch we share but only feel our own altered perception. In changing partners, each partnering was a different experience and gave different insights; it avoided unnecessary 'personalising' caused by staying with one partner (Paxton and Kilcoyne 1993: 25) and the consequential negative impact this may have on the learning experience.

The ramifications of what touch revealed ran deep into the personal and group process. CI involved a willingness to partner with others; each partnering opened up a different movement quality, a different touch experience and a different response. Each stage or phase was set in a time and space and opened the body/mind to a new possibility or altered existing perceptions. The purpose of partnering was to experience full-bodied movement, in response to being in contact with a partner. The partnering taught us about our bodies in a way that could not happen alone; touch revealed me to myself and in parallel did the same for my partner, and our collective experience revealed collective changes taking place which transmitted into the beyond, within the social body – a realm in which the 'work' was working some magic?

Very soon questions arose in support of the collective experience, from encounters experienced in the process, and were responded to pragmatically through the activity, as is innate to the iterative nature of CI. Questions inevitably came from the perceptual challenges that seeped into the partnering, for example, from the sighted, such as not knowing how much sight the visually impaired person had, or from the visually impaired person not knowing if the sighted person would allow space and time for their needs. Paxton and Kilcoyne took an approach that would catch these thoughts before they fell, using the 'physical plane' of movement and touch; this meant that any negative assumptions were negated (so made positive) by the experience of the potential for collective learning that no one had foreseen. Other aspects held in common became evident to both the visually impaired and sighted participants so that the outcomes for both had some collective territory and not to mention some fun! I continue the chronology as to how the CI praxis contained the pedagogic process and as to what this learning experience was like for the participants in the next chapter sections.

PART 2

Touchdown Dance (1988–94), Breaking new ground, new discourses, new science, new praxis: Re-inhabiting the body brought into question the perception of the visible and invisible

This second chapter section covers the continuing programme of seasonal summer workshops at DCA, told historically with reference to the archives and from participant case studies found there. From the first workshop in 1987 to the last

Dartington workshops in 1993, the collaboration with Manor House residential and rehabilitation centre for people experiencing sight loss (Kilcoyne 1989) continued to support the research. The Aller Park annexe continued to be used, and it eventually became possible to offer residential facilities there so that other participants could attend from further afield and 'from all walks of life' (Paxton and Kilcoyne 1992: 47). The majority of the participants who came from afar were sighted, some visually impaired participants did travel long distances, independently by train. Others, like myself with my Danish deafblind colleague, came from abroad.

The pattern of a week-long, non-residential workshop was repeated in the spring of 1988 and the following year this increased to a series of weekends and a week-long workshop over the spring/summer. In November 1989, Paxton and Kilcoyne offered an autumn research and review session with the aim to provide continuity in provision and to gather feedback from participants (Kilcoyne 1990a). The Skipping Visual Impairment Programme continued till 1991 when under the new name of Touchdown Dance, Kilcoyne made an application for Charitable Trust status that was completed in October 1992.

The workshops were publicised in the New Beacon journal of the Royal National Institute for the Blind, and more widely through talking newspapers (Kilcoyne 1989, 1990a, 1990b). I refer to these and other articles in which participant insights and reflections document the research program experience, including one by Kilcoyne that specifically considered the topic of touch taboo, entitled 'Common sense' (Kilcoyne 1991). Another, similarly entitled 'A common sense' (Paxton and Kilcoyne 1992), published in the Contact Quarterly journal a year later, discusses the role of touch in the transition from the visual to the non-visual mode of moving. The already referred article 'On the Braille in the Body' (Paxton and Kilcoyne 1993) is a source of information on the workshop content which is revisited here.

A reflexive approach – Establishing the mind of movement

The ongoing evolution of the research reveals further roots of the practice in the embodied, non-dualist ontology of CI (Paxton and Kilcoyne 1993). Paxton and Kilcoyne bear witness to how each experience revealed subsequent realisation and learning, and I note the underlying circular idea that just as thought is the *mind* in movement, so movement is the *bodily* foundation for thought (Sheets-Johnstone 1990). Movement, the kinaesthetic sense, is the realm of bodily experience, the sensory realm of being in action that the mind brings to conscious awareness. In the continuum of this unconscious–conscious dialogue, the mind is *reminded* of its mutuality with the body and this knowledge is referred to as experience (Bainbridge Cohen 2012). Alternatively, the mind assumes a dualistic disembodied 'attitude' and ideas define an a priori predefined reality separate from the phenomena of affective, bodily experience. Between these polarities lies the space–time continuum worthy of excavation or 'research'. It is a realm for the former metaphysical and phenomenological perspective that enables the emergence of reflexive and discursive commentaries that appear in this book. The practice-led inquiry is a form of corporeal re-experiencing and serendipitous ideation that has become a fertile ground for the membrane construct.

Significantly the various articles referenced above provide brief participant self-reports and observational accounts of the workshop practice. These accounts capture the 'interworld' (Daly 2016) of the participant experience, of the dyad and of the overall session. In the first-person descriptions of their experiences, each participant case bears witness to both the accomplishments of the early stages of the research and the limitations. Interlaced with observations by Paxton and Kilcoyne, this material resonates with layers of perceptive understanding and interpretation that reveal the personae in action, in dialogue with themselves and their partner. The process of interpretation and meaning-making became a form of discursive analysis (Harre and Gillet 1994) like an improvisation of the mind in movement, as I am doing often in writing this book.

Later in this chapter, I introduce Gerry, Peter and Bonnie[26] who speak to the embodied and transformative nature of the workshop experience and reveal a process of self-steering, of opening to a possibility for change in a dialogical process with self and/or another person. Over time the accumulation of experience leads to insight and new information that informs the mind of the body's instinctive intelligence expressed in movement, uniting and deepening the participant's sense of self. Fundamentally the movement research undertaken in the workshop experience sought to provide participants with a greater understanding of the universal human attributes assigned to movement and touch, as well as the value of subjective, and idiosyncratic ones, to self-perception, healing and growth.

I now recapture and discuss the fundamental attributes of CI, the ontology that provided a framework for a body-oriented pedagogy and for the reassembling of body–mind unity.

The seasonal workshops (1987–93) – finding the body

Following the first workshop in 1987, the continued collaboration with Manor House rehabilitation centre enabled Paxton and Kilcoyne to learn from individual participant experiences and ascertain effective pedagogic principles to use going forward for individuals experiencing sight loss. They noted how the 'loss of sight contributes to loss of movement' and how, more generally, a resident visually impaired person 'engages in little physical movement' (Paxton and Kilcoyne 1993: 8).

They interpreted the 'range of fears and inhibitions that affect everyday movement quite specifically' (Paxton and Kilcoyne 1993: 6) as a form of bodily retention, 'the body and its own particular intelligence becomes dormant' (Paxton and Kilcoyne 1993: 13). In this account, much is written about the impact of sight loss on the body, a condition that was worsened, they believed, by further inhibition due to the 'physical imprints' on the body caused by fear of collision or falling. The negative impact of lack of movement upon the body was evident, and how this had compounded the 'chronically tense' or the 'apathetic and flaccid' states in muscular tone (Paxton and Kilcoyne 1993: 10). Their combined expertise in movement and psychotherapy led to a decision to address the state of loss before any expectation to move *more* could be realised.

From these observations, Paxton and Kilcoyne trusted in the shared experience of CI that this could support the visually impaired participants to feel re-enlivened in the body and to rediscover movement by moving with others, specifically with people who didn't have such trauma. Evidently, for this participant group, finding ease and letting go of tension were key to addressing the mutually symptomatic loss of sight and loss of movement and also to the protocol of inclusion. Indeed, inclusion felt contingent on whether this positive impact on visually impaired participants could mutually benefit sighted participants who come with their own challenges; 'relaxation is something that each person has to discover and rediscover' (Kilcoyne 1989: 20–21). Paxton and Kilcoyne anticipated that moving in physical contact with another person could serve as a self-regulatory activity for the symptoms of stress and the immobilising states that endure due to trauma, even the more subtle traces in the body.

Having considered these issues, movement re-education and improving mobility became the central purpose of the Skipping Visual Impairment Programme (Kilcoyne 1989a, 1990) using fundamentals from CI to encourage more movement, through play and confidence building. This could not happen quickly, a sensitive and subtle approach meant that much was learned for everyone to carry forward (Paxton and Kilcoyne 1993). The focus on 'specific research work with the visually impaired' (Paxton and Kilcoyne 1993: 7) meant that sight loss became a condition integral to the learning process, a variable rather than a barrier to learning, and required a predominantly non-visual approach. Paxton and Kilcoyne considered carefully how to set the mood and a contained, safe, atmosphere for the interactions to be undertaken with ease and comfort. The content could then adjust to the participant responses and be directed by the need of the moment rather than some imposed standardised goal for all to meet.

This positive, pragmatic, 'can-do' attitude to learning proved to be effective, as they reflect; 'the condition is easily improved. Use it. Move it. Have it moved. Feel the range' (Paxton and Kilcoyne 1993: 7). As it transpires, it was not doing the movement per se that changed lives but the way of finding the movement, the methods that enabled the movement experience regardless of having sight or not, and this was the aim of the research – finding solutions to the challenges that emerged in this unchartered territory was the navigation task that time and the CI pedagogic ethos would steer.

Paxton admitted how not knowing what would work or how an idea would materialise, had led him to discover how confusion was innate to problem-solving. It took these years of exploration to start to understand the broad potential of the work, as Paxton puts it: 'There were a lot of levels for us to figure out. The pay-off was a long time arriving' (Paxton 2019: n.pag.). The difficulty they faced was to verify how effective CI methods were in creating a level-playing field for all to self-direct and still feel part of the collective experience. Paxton stated how he had experienced this:

> my job was to provide a way for a learning experience to occur, keep it light, keep everybody you know, the spirit up, and guide them through it but what I'm guiding them through I don't know, I don't know what they're finding, I don't what they are experiencing. They talk about it or note it as we go by it, [...] but basically I really trusted that everyone was having an experience of space and socialisation, and time slowing [...] and speeding up [...] and all of that [...] and yeah it is part of the experience, I expected that was what was happening.
>
> (2019: n.pag.)

Paxton and Kilcoyne gave time for each person to experience bodily and then to observe consciously the movement responses taking place, taking time to give subtle cues or insights into what else was possible for them to explore further. It was a laboratory for movement, for research into the body's natural responsiveness, for self-discovery, self-perception and ideation – a sphere that put into action and demonstrated the value of the sensory being and of nurturing self, creatively.

Establishing the basic rules of CI: The drive for equality – defining a new 'normal'

'Skipping visual impairment' is the first article in the archive to be published by Kilcoyne in the New Beacon journal in January 1989. It looks back on the first two seasons of workshops and names the 'basic rules' that defined the four fundamental aspects of CI:

> 'Contact improvisation' starts with a couple of basic rules: Movement arises through the sense of touch; each partner follows the other in a 50/50 balance. A dance of

> responses and reflexes then ensures. If neither partner leads, then each person is following movement generated unpredictably by a two-way accommodation.
>
> (Kilcoyne 1989: 18)

In practice, these 'rules' were modelled in the teaching methods and created clarity for the participant's learning experience. The 'rules' established the ethos of 'a level playing field' in which differences and issues were met as they arose: such as the pre-expectation of a mutual 'accommodation' to bridge the gaps caused by difference and to open up new possibilities for integration.[27] The principle of 'a two-way accommodation' was not a new feature of the CI experience where sighted people were concerned but was initially impeded in the combined experience with visually impaired participants, due to the known leading role of a sighted guide, in which the visually impaired participants were usually not so well accommodated (Halliley 1991; Heller and Gentaz 2014). These rules were reiterated over the workshop period and implemented in working towards learning more specific CI practices. The concept of 50/50 balance becomes a metaphor for the consideration of all forms of difference and ability, as a baseline for the workshop process and for reference to the membrane of inclusion in this chapter section.[28]

The intention to balance the teaching input for all participants was an inclusive attitude towards learning that Paxton and Kilcoyne sought to model. It meant attending to the impact on visually impaired people of their restricted mobility, of the challenges to orient to space and relate physically with other people, which impede their CI experience. Just as sighted people had learned to slow down and close their eyes to benefit from the sensory, touch-based and improvisatory attributes of CI, so too, it was justly assumed, would be the visually impaired participant experience. To what extent CI would reduce states of tension and alter perceptions that underlie such states was to be discovered. In due course, Paxton and Kilcoyne learned how much to do to assist, or how little to do to facilitate movement. It was contingent to understand the circumstances in which the 50/50 balance with a sighted person was impeded, and how to mediate this by engaging empathically in the bodily experience and accompanying this with mindful exchange. Inclusion was key to the research endeavour, requiring a balance of focus, method and execution and success enabled four basic principles to evolve from the four basic rules, to provide a coherent foundation going forward.

The four basic principles

The first basic principle underlying the touch-based pedagogic process is 'learning from and through the body', defined more explicitly as: 'The body has its own intelligence. It is that intelligence which we hope to awaken, revitalise and use. In taking this approach we diverge from the dominant learning culture which places mind over body' (Paxton and Kilcoyne 1993: 12–13).[29] Paxton and Kilcoyne perceived the loss of movement caused by the trauma of sight loss, as a manifestation of the dissociation of mind from body. As the case studies tell, the impact of sight loss diminished as more movement and body awareness were gained. This became a win–win situation, as by regaining movement within a supportive touch-based relationship, the sense of self that had been lost through lack of social engagement, trauma and distress was also regained (Paxton and Kilcoyne 1993; Dymoke 2014, 2017). In reclaiming the body in movement, the healing experience of the work served to heal both the trauma and the body–mind split as the cases of Gerry and Peter will confirm later in this chapter section. The workshops proposed a physical process in a space where the inner, silent murmurings of the body–mind disconnection were listened to, where hidden, latent or suppressed aspects of self were revealed, and more so in the socialised sphere of the dyad.

The *second* basic principle evolved from the second rule of 50/50 exchange; an 'integrated exchange' was fundamental to the dyad (Paxton and Kilcoyne 1993: 15). The shared movement experience was a means to seek balance, in more ways than one, in the exchange between the visually impaired and sighted partners. *Integration* took place at multiple levels, where an imbalance pre-existed; of body and mind, of relaxation and alertness, of sensing and moving, of inner and outer, of ease and tension, of mobility and immobility, and of resistance and availability. CI could accommodate difference and the repatterning of experience; it provided a sphere within a mutable membrane – conducive to self-regulation and balance. Further attributes of CI contributed to a sense of stasis, but not stuck-ness, a sense of progressive expansiveness, or continuum, in which these potential polarities and differences become threads that interweave in the experience, to nurture a substantially strengthened persona who moves differently in the world – i.e. integrated.

The *third* basic principle is defined as '(a)n attitude towards Learning' in which '(c)onscious observations of the physical experience here and now is the basis for a study' (Paxton and Kilcoyne 1993: 17). The participant cases below will account for these principles in the context of the experience of the working with the body, in an integrated exchange, a process that Paxton and Kilcoyne interpret or comment further on. Their observations and reflections consider how emotions arise in the 'dance of the reflexes' and significant social interactions occur in the movement relationship (Paxton and Kilcoyne 1989, 1992, 1993). In these interactions, Paxton and Kilcoyne put the focus back on the shared learning that takes place through moving with others, to the commonality of sensation, to learning to trust the body. By finding solutions to make movements that are at first unknown, then known, the fear and mistrust of the body's potential were replaced with a more positive perception and renewed awareness of how 'muscular states are intrinsically linked to feeling states or moods'. The combination of 'physical experience, physical sensations, and emotional feelings' underlies the core attitude to this learning process: 'The lesson works in the psyche and in the soma, but our work focus remains centred on the body' (Paxton and Kilcoyne 1993: 17).[30]

The *fourth* principle considers the use of touch communication to establish the non-visual learning environment and the various protocols required for all participants to be included, and on equal footing. In relating through touch and the skin, vision and the perception of the world peripheral to the CI partnership took second place (Paxton and Kilcoyne 1992, 1993). The focus away from sight lessened the amount of movement in space and placed the attention on the detail of the sensation of movement, and upon the body itself, its inner spaces and the space close to the skin. These are the spaces and gaps which were discovered in the sphere of the moving partnership, by being listened to and inhabited by the partners and then shared.[31]

As the dimensions of touch perception and communication are slow then time was taken with the transmission of information and in the approach to movement instruction. This enabled both visually impaired and sighted participants to find gaps in movement awareness and to discover more subtlety and new qualities. By moving slowly new patterns became attainable. The active partaking in movement gave a 'direct experience in the body of the student' which meant that the teaching process was a 'hands-on or body-on approach' (Paxton and Kilcoyne 1993: 14).

Paxton and Kilcoyne sought to share their practice and also their insights into the experiences they had with visually impaired participants, to tell the story and give a glimpse into this discrete world. The revelations appear to me, the reader, as real as any epiphany or sudden realisation of what the world can bring to my attention that my conditioned mind prevents. Evidently, the root of this process is bodily initiated; just as the small dance awakens the body, so does sensed experience, relative to the external world. The reciprocal process of sensory adjustment and enrichment is the next topic to revisit. The following section considers the key areas of input that evolved and were further developed with the ongoing caretaking required to facilitate an environment for the experiential praxis to take place.

Establishing ease and a communal atmosphere – The pedagogic protocols to begin the workshop

To create a positive, inclusive environment for all to feel at ease, '50/50 balance' was the target to be reached. Paxton and Kilcoyne devised starting points for the first day to enable all the participants to buy into the idea that everyone was on equal footing in the workshop environment. Social rituals and community-focused interactions established a congenial mood and atmosphere for the group to feel accommodated, open and available to participate. This reiterated the pro-body and pro-touch discourses, with a focus on play and socialising. A state of ease and collective self-regulation established 'an energy with a subtle nature to replace such energies as tension, flaccidity and wilful straining to move' (Paxton and Kilcoyne 1993: 25).

By attending to subtle energy states, a 'tone' is set in the space, a state of collective consciousness referred to in Body-Mind Centering® (BMC®) practice as 'the mind of the room'. This 'mind' is a form of perceptive presence that imbues the room with the collective energy of the bodies present – like water in a pond that responds to the wind. Comparatively, in psychotherapy, this is experienced as a form of empathic resonance; as in the mother–infant context of Daniel Stern (1985) and Allan Schore (1993, 1994). It is a state of resonance in which embodied learning takes place, a form of conscious–unconscious dialogue, primarily sensory, expressed reciprocally in movement (Bainbridge Cohen 2012). This benign energy field established an environment that invited and gave permission to open up to something new, in a space that could otherwise be feared and perceived as unknown. This would be predicated in other day-to-day social environments where those with sight loss and sedentary habits remained passive or resistant, or in anticipation of being directed (Paxton and Kilcoyne 1993). We now look at the workshop activities, or rituals, to see how the proposed energy states and empathy were realised.

Reinventing the CIRCLE – A community ritual, the membrane defining a 'common place'

In the previous chapter section, the rationale for starting in a circle was given. It served to establish a ritual for the community to sense and feel itself and to communicate equally, first orally, each person says their name and where they live to the whole group. The circle constellation developed further with other interactive ideas, such as communal sounding and passing around a hand squeeze, vocal chants and 'clapping rhythms'. Clapping marked the transition into movement in space with 'stamping, walking and running rhythms, and stretching exercises' (Kilcoyne 1989: 19). These activities introduced who was present, the size of the group and the idea of the 'ability to act cooperatively' (Paxton and Kilcoyne 1993: 28). The playful atmosphere created warmth and security and helped people to relax, to laugh and be ready to partner, or conversely, it was perceivable how committed the participants felt and both informed the process. Once the environment was contained, greater motility was possible – for both individuals and the group. The socially exclusive hierarchy of blind and sighted people in society seeps into the room, as does the manner in which this premise shapes into socially constructed binaries. Assuming a circle invited discrimination of any kind to be levelled out, without compromising the diversity of a group, and for the 50/50 balance to be effective as a continuum. The circle embodied a protocol of inclusion, the membrane that developed over time to establish a common understanding of the ethos of CI that Paxton and Kilcoyne were offering (Paxton and Kilcoyne 1993).

Finding a relaxed body but alert 'mind' – Introducing kinaesthesia, the 'role' of Introspection – The space of deep listening

These exercises focused on a mindful sense of engaging with the body's physiological tone, as a means to access associated negative states of discomfort, lack of confidence and retained emotional distress. For example, simple introductory partnering works such as the spinal roll down and rolling point of contact (already introduced in the first section of this chapter) were found to work directly with the positivity of 'sensations and physical experience' (Paxton and Kilcoyne 1993: 13). I introduce more detail to reveal the layers that appeared as the process deepened. By taking a long time to work through the exercise, the revelation of change came from sensing the movement as it occurs – kinaesthetically. By giving each person their own time to do the roll down, Paxton invited voluntary engagement to sense and feel the body's limit in the movement. By taking time to release the body forwards, the potential release of tension could be felt on the return to standing. Named 'The Dive into the Centre of the Earth' the initiation was in feeling the pull of gravity in the fingertips, then following the pull from the fingers 'the spine allows itself to follow through on the sensation of weight and the pull of gravity in the hands' till the spine, bone by bone, is flexed forwards as far as is comfortable (Paxton and Kilcoyne 1993: 28). This method ascertained the mobility of each person and contributed to a state of physical readiness to do further pedestrian movements such as navigating the space, running side by side, and then lying on the floor for guided relaxation. These preliminary interactions established the group, and developed an emergent understanding of how to adapt the attributes of CI in this context, which I italicize: 'In the act of releasing tension into a state of readiness and allowing the senses to expand first, we are *making a space* for the physical movement to follow. We see that the *body-learning* is faster when one is *slow*' (Paxton and Kilcoyne 1993: 11, emphasis added).

In 'making a space' for the discovery of movement, Paxton and Kilcoyne's endeavour was to show how learning new movement led to the un-learning of old habits and worked best when the body–mind attended to internal sensation, reciprocally to the feeling of the movement response. To teach CI in this context 'requires a fundamental approach which has to be slow and delicate' (Paxton and Kilcoyne 1993: 10). This body-time approach was essential in this context and determined the cool, regulatory 'climate' of the teaching and direct respectful use of touch. Sensitivity to detail gave the possibility to analyse the movement methods traditionally taught in CI, to focus less on virtuosic athleticism and more on the intrinsic state of the body. CI became a praxis of listening, of pre-tuning away from the external world to the interoceptive climate of the body's sense of itself – to the self-regulatory interrelationship between the visceral and somatic systems (Bainbridge Cohen 2012). In this way positive responses to the experience were openly shared, feeling states such as the vitality effects on the body (Stern 2010), or the softening of tone, transcended the disassociated states of disuse or dormancy and the body–mind moved as one again – summarised here: 'Ease does the heart good. The overall goal is to give people heart (with metaphorical connection intact) via physical effort, ease of touch, empathy and sensible attitude' (Paxton and Kilcoyne 1993: 7).

From this gentle starting place, the results became predictable and observable, indicated by factors such as 'improved posture, ease of gait, range and extent of movement, and physical and mental self-confidence' (Paxton and Kilcoyne 1993: 7). Each of these aspects could be observed, each one in itself carries insight and potential transformation at a level that can impact greatly on an individual's sense of self and wellbeing and so on their ability to relate to others with trust in themselves to carry it off. Self-trust appeared as a significant gain, often identified positively in this introspective and interactive environment, one in which both sensitivity and sensibility were given time and space. In the realm of this workshop process and forthcoming to the practice of CI, human contact was normalised as an exchange in which any power play is mediated, and the word *trust* became a common descriptor of the felt sense in the room, and almost synonymous with *ease*.

The 50/50 use of touch and movement – To touch to move with ease

It has been established that in day-to-day life, visually impaired people are more dependent on touch for communication with the world than sighted (Dymoke 2000; Heller and Gentaz 2014). Paxton and Kilcoyne realised that this was a provision of CI, but that touch as a topic would be key to the research and would need further exploration within the workshops. In the touch and movement practices of CI, they sought to maximise the use of touch as a primary sense, to underline the normal use of touch in interpersonal relationships, to remove the obsession with the 'non-erotic' label and advocate against non-touch. This also served, directly, to balance awareness of the role of the body, to enforce the unitary concept of the body–mind, its application to learning, and to both personal and community wellbeing (Paxton and Kilcoyne 1993; Sheets-Johnstone 2008, 2009, 2010).

In the archived research publications, a 'common sense' attitude towards touch became increasingly significant. The value of touch was consistently questioned, researched and clarified, inviting reflection and assertion of the ethical non-erotic use of touch in CI, as in sport (Dymoke 2014).

> Touch in everyday life is normally restricted to ritual greetings, sexual relations and aggressive encounters. Within 'contact improvisation' touch is an intrinsic and necessary aspect of the movement experience. Normal taboos and associations have to be humorously noted and then left behind.
>
> (Kilcoyne 1989: 19)

The usual awkward feelings that arise in touch could be dropped when the atmosphere was relaxed around it, and then they found the groundwork for CI practice could start.

> Once this has happened the more profound benefits of touch that have been noted by many researchers in the field of tactile research can begin to occur at the same time as the basic building blocks of 'contact improvisation' are introduced. Many of the initial exercises in the workshop move everyone gently towards this position.
>
> (Kilcoyne 1989: 19)

A new 'normal' pro-touch discourse is established in place of the deviant touch taboo and socially prescribed non-touch agenda. However, objective definitions of the benefits of touch were not always useful. The affective touch experience could be masked by 'social preconceptions' and by 'personal feelings being centrally involved', so the resolve was to set aside touch classifications as touch 'is not always felt by the body', and to begin with 'subjective sensing'. Touch becomes a common good, a means to embody, to connect, empathise and communicate bodily, as summarised by Kilcoyne: 'The learning that happens is located in the body rather than the head' (Paxton and Kiloyne 1993: 23–24). With everyone barefoot and comfortably dressed in loose clothing, they could attend to sensation, to the tactile feedback from the floor to bring awareness to *weight* (or *mass*), to the pull of *gravity* earthwards and once the body started to shift over the feet, to the sense of a *centre of mass*. These three aspects became more clearly indicated and explored in the workshop programme as time went on. In the absence of non-touch discrimination, the physical experience took precedence over that social construct and established a new 'truth'.[32] I return to these aspects in more detail after the participant anecdotes and more detail on touch comes in Chapter 3.

The building blocks

This section documents the deepening of the practice of CI, through admitting new attributes, protocols and a fuller body-oriented experience. In developing the practice with a soft and clear approach, participants experienced the importance of space and time. As will be seen in more detail in Chapters 2, 4 and 5, the work evolved specific exercises to introduce key movement principles that operated on different spatial levels. These were the 'building blocks' of CI movement-based practice (Paxton and Kilcoyne 1993: 31) and involved specific partnering and touch-based methods. A range of partnering exercises started the journey into the land of CI, in which one person was the support for the other to move around, like a 'human apparatus' that 'senses, adjusts, encourages and learns whilst being used', they continue:

> The building blocks are filling in on some basic positions and possibilities within the contact form. They allow a controlled investigation into trust, reliability, balance, the giving and taking of weight, falling, leaping up a partner, sliding downwards from the leap, lifting, the timing of movement, the quality of the contact.
>
> (Paxton and Kilcoyne 1993: 31)

Paxton introduced other full-bodied movements and applied some basic principles to the existing ones. For example, to connect the limbs to the torso and spine; to reposition the body's weight distribution through the spine in the roll down; to shift the centre of gravity over both feet in the small dance or to turn around the centre and axis of the spine in a spiral twist, in either standing or rolling. In each movement, the participants traced the pathway from standing to the point of completed movement and then returned back to the start, working both sides of the body to balance. Floor work became significant too, to pattern full-bodied movements close to the floor, to prevent falling and to establish connections that would then enable ease in transition from lying, to sitting, to standing. Paxton ensured ease and comfort in each orientation before finding the pathway to the next.

This developmental, step-by-step, approach worked through any gaps appearing to complete a movement or lack of flow along a pathway into space. Using the attributes of CI such as the hands-on and body-on approach, touch supported each person's sense of fulfilling a level, such as rolling which precedes sitting, before moving on to the next. Learning to orient to one's own body landscape with the immediate surroundings provided the basis for moving out further into space. Paxton and Kilcoyne simplified movements into patterns and developed ways to teach and explore them with partners using incremental steps, pausing to gauge the sensory input, to allow choice, to permit and encourage a revitalising of the body's physicality, for all participants regardless of sight.[33]

> Developing body-patterns; the extension of physical experiences beyond the usual range of movement; and the language of instruction itself, all reinforce other sensations within the group [...] The sensory material – sensing the body's state in the present moment- ultimately allows the body to move more freely, more appropriately and more safely. That is the goal of this journey.
>
> (Paxton and Kilcoyne 1993: 25)

However, CI in itself would not address the issue of being unable to transition from one habitual state of tension to a more relaxed state. To address this, Kilcoyne introduced head massage at a strategic point in the day such as after lunch. Facilitated with a light touch and repetitive circles on the skull bones, this massage impacts the whole body, releasing the breath and physical holding (Paxton 2019). Limb manipulation was also introduced, as an exchange between partners to release tension or joint compression. These methods reemphasised body–mind integration, normalised touch-based interactions and served again as means to build trusting relationships, which I return to below with the case of Bonnie. With these regulatory practices, the work succeeded to rebuild aspects of the self that are lost with sight loss, or that are awakened with the emotional affect in the present moment of letting go. For the sighted and non-traumatised visually impaired people, the benefits of such healing interactions were part of the general learning about human-to-human connection and the shifting of stereotypical perceptions.

Kilcoyne's clinical background brought a person-centred focus to support self-regulation for those experiencing other forms of psychological distress symptoms. Many, whether visually impaired or sighted, experienced psychotherapeutic benefits from the work. The realm of touch through massage and other hands-on practices calmed the activated autonomic system whilst releasing the effects of stress on the body. Breathing exercises were also vital to this process, which Paxton led in conjunction with yoga postures, extending the out-breath to allow the body to yield and soften, for the stretch to deepen: everyone experiences a physiological change, a state of 'alert relaxation' (Kilcoyne 1989: 19). The influence of touch research is evident in the use of touch and movement to provide a wide range of body-oriented possibilities to draw from, in the necessary analysis of individual needs:

> For those who have no movement training and poor muscle tone, the physical approach has to be clear and soft. This has called for an analysis of exercises to try to determine what exactly is suitable in this situation. The proportions and durations

> of things have to be considered. Relaxation and massage are brought in frequently. Exercises requiring just one part of the body to relax in isolation, such as 'floating the head' or arm drop' are very useful.
>
> (Paxton and Kilcoyne 1992: 50)

Together with Paxton's embodied perspective, the combination enabled the work to 'work' for each individual, contained within the limits of the time and space available – as the intensives were not regular enough to function as psychotherapeutic interventions. Once the workshop ended, participants returned home and it was not possible to track them further. These limitations led to the idea of more regular workshops and to the intention for ongoing participation, rather than a one-off experience and many did return (Paxton 2019b).

Further on into the Skipping Visual Impairment programme, with the increasingly wide difference in individual participant needs, the therapeutic benefit that was inherent to the general experience was evident. Paxton and Kilcoyne became more aware of this sense of purpose when looking back on their progress in 1993; 'our explicit goal is movement and health, implicit in the work are goals of health, fitness and the integration of mind and body' (Paxton and Kilcoyne 1993: 29). With this shift in context, the practice of CI was also tested differently and became a more expansive and holistic endeavour, with touch and movement methods deployed to affect more deeply and directly the sense of self-identity, and self–other relationships. The explicit benefits felt bodily would then seep into intellectual insight with motivating potential and mostly positive emotional affect, as the experience of Gerry and Peter indicate in the cases that follow. Such benefits would be implicit in the positive nature of the experiential, relational process (Paxton and Kilcoyne 1993; Dymoke 2000, 2014).

Paxton and Kilcoyne consistently altered the content of the workshops to accommodate participant needs, demonstrating the user-led focus of their programme. If the movement had been the initial focus of the work, it was to be contextualised more widely, to include influences from other movement disciplines and embodiment practices. I consider these changes as they indicate the nature of the transition that took place and continued to add value to the experience that I picked up and developed further. This was very much in the flow of the times, as the dance world was changing and endeavours such as this would force this process further.[34]

The experience of participants indicated two main gains; first that the range of movement experienced in the workshops through CI contributed to improved fitness and confidence; secondly that empathic engagement was needed for the dyad to be effective, and this led to the introduction of other practices that supported all-around wellbeing. These practices involved other reciprocal touch and movement methods (such as massage), embodied approaches to sensing and feeling whilst moving, and playful interactions to shift negative self-perceptions or to balance and unify the body–mind. It became apparent, through the experience of working with the trauma of sight loss, that imbalanced physical states manifest disturbed mental states and difficulty to move indicates difficulty to self-regulate (Paxton and Kilcoyne 1993; Dymoke 2000). It also became apparent that touch and movement had a beneficial impact beyond the direct experience of an individual participant. Where levels of stress and distressed states of being were experienced, these were generally relative to levels of touch deprivation, or abuse. Lack and involuntary loss of movement were relative to lack of access to movement education and were accompanied by loss of confidence – again symptomatic of isolation and a culturally exclusive society (Kilcoyne 1989a; Fulkerson 1996; Nelson 1996). Touchdown Dance discovered ways to serve all these needs effectively, with Paxton attending to the body by inviting ease and spontaneity and Anne Kilcoyne attending to the emotional and psychological aspects through affective touch and empathic relational interactions. Both would engage in the educational and performance strands as they evolved.

The accounts of the visually impaired participants that follow give their insights and comments on their experience, together with witness statements and reflections from Paxton and/or Kilcoyne. These cases illustrate the continuum that is created between the two socially estranged worlds of visually impaired and sighted people, embracing their mutual discomfort and dilemmas in a space of personal transformation.

Case study 1. The late Gerry Overington

The experience had by Gerry Overington was the first significant success story in the research programme. Gerry's story illustrates the inherent functionality of the embodiment practices in CI and the self-creative realm in which transformation is 'just that thin membrane away' (Bainbridge Cohen 2012: 75). In Gerry's case the 'thin membrane' multiplied three-fold to contain three relational spheres I have identified as realms in which change occurred; firstly in movement, secondly in movement with sighted partners and thirdly in public performance.

Gerry was a blind participant and member of staff at Manor House rehabilitation centre who, we discover, lost his sight twelve years previously. He attended the first workshop in April/May 1987 and returned in 1988 and 1989 as it nurtured aspects of his past. He established a strong relationship with Paxton and went on to perform at DCA, which I come back to below (Kilcoyne 1990b). Kilcoyne (1989) describes how Gerry took to the work 'like a fish to water':

> Feeling that he might be bored with repeating the basic exercises of 'contact', he was encouraged in the partnering work to move towards more complete 'contact improvisation'. His freed movement and his ability to engage full and confidently with the contact form made it clear that the potential of this approach was far beyond the most optimistic projections that Steve and I had made.
>
> (1989: 21)

Gerry's comments are quoted by Kilcoyne:

> As a totally blind person I was originally rather sceptical but interested. I no longer have any doubts. I have progressed from simple exercises which help in confidence, fitness and deportment to performing an improvised dance which enabled me to express myself artistically and allowed me to move with complete confidence and freedom. This is something I have been unable to experience for many years. It would be nice to see the project developed in the future and adopted by groups working with and catering for the visually impaired.
>
> (1989: 21)

Gerry's story appears across three different articles, as a workshop participant he improves physically in a short time, to the extent that he applies himself to improvise in front of an audience with Paxton and Julyen Hamilton – two internationally influential and independent[35] dancers. Kilcoyne writes about their performance '*Blind Faith*' a few months later in the New Beacon:

> The performance, which was totally improvised combined a sense of poetic human movement with moments of humour and generosity. Images of fast athleticism between the men were juxtaposed with still, held images evoking human vulnerability and brotherhood [...] The audience warmed to the unique event of which they were a part.
>
> (1990b: 56)

Gerry is the first case of a participant who takes up the gauntlet to perform – this was thrilling to Paxton and Kilcoyne. It was also thrilling for the community of young and professional artists and the audience. As Gerry records for himself, quoted by Kilcoyne:

> By the end the audience were flying as high as me. It was very exciting and surprisingly easy. I was apprehensive when it was first suggested to me. I thought: I'll never be able to dance for a whole hour. But the pressure came off during the performance. I easily did things I hadn't done before, without any fear. I'm not the normal build

> for throwing myself around a stage, but it's excellent to think that I've got to a point I can really do it. I've got a lot of pleasure from being pushed that far.
>
> (1990b: 56)

There are five photos of Gerry with either Hamilton or Paxton, actively moved and moving, lifted or lifting, moving over or around their support or actively supporting them. There is one of Paxton and Gerry, talking and laughing about something. The sense of achievement after just three intensive trainings over a three-year period is a clear accomplishment that inspired others, all be it there were no visually impaired audiences at this time; it was a discrete event. Gerry's reflection on the difference between dancing with the two men indicates the level of his ability to tune into the uniqueness of each partner and himself:

> One of the most interesting things was how different it was to improvise with Steve and Julyen. Julyen is as if he is made of India rubber he seems to bend around every which way-whereas Steve is much more powerful and dominant. But then they are two different people, and in contact improvisation you seem to truly meet the person through the dance. It was great, the culmination of the work we'd been doing in the workshops.
>
> (cited in Kilcoyne 1990b: 56–57)

There was a clear bond between the three men, likely helped by the visual memory that Gerry had. Kilcoyne comments on the change she witnessed as 'Gerry's new found physical sensibility, confidence and agility' (Kilcoyne 1990b). A year later Mark Halliley writes in the Independent newspaper (22 October 1991) having interviewed visually impaired participants in the spring 1991 workshop in which Gerry's epiphany is shared:

> Some of Touchdown's work culminates in performance. Steve Paxton danced [...] with one workshop member Gerry Overington, a former steel- worker from Sheffield, who has been totally blind for the last 12 years. 'Gerry had never danced before, and he just couldn't get it. Then one day, all the pennies dropped and he said: 'Bloody hell, it's an aesthetic form isn't it?' recalls Ms Kilcoyne.
>
> Mr Overington said it was the first form of exercise he had found where he was not led and controlled by a sighted person, on whom he was totally dependent.
>
> (Halliley 1991: n.pag.)

To end her article on the *Blind Faith* performance, Kilcoyne writes a definition of CI that helps to make explicit the activity that the photos capture but rend static, the activity that is in practice, a fluid and continuous movement experience. It wraps up and summarises the direction of travel, that the work was pitching the 'health-enhancing' and so therapeutic application of CI.

> Here blind people and professionals from the fields of blind welfare join together to action-research the possibilities of extending mobility through the relaxed and health-enhancing principles of 'contact improvisation', a dance form developed by Steve Paxton in the early 70s. Often called an 'art sport', contact improvisation is a two person dance form based in naturally sensed movements which arise when the two bodies move together and around one another whilst maintaining a continuous but changing point of contact.
>
> (Kilcoyne 1990b: 57)

Participating in a workshop lead inevitably to performing, and far more than simply 'health-enhancing', and it was a personal achievement for Gerry. Chapters 6 and 7 consider performance and the advent of audio description in more depth as visually impaired audiences were reached. I now revisit and consider these threads and how they intertwine for another participant, Peter.

Case study 2: Peter – The interdisciplinary potential of CI as a therapeutic modality

The therapeutic benefits of touch and movement modalities became apparent to Paxton and Kilcoyne through their empathic engagement with the participant experience. Their written texts express the intent to get to grips with what the nature of this experience actually achieved, as well as what it meant, to each person and in the wider ecology in which it was operating. Halliley documents comments from Tony Rahim on the accommodating and liberating atmosphere of the workshop:

> Although not officially a therapy, it seems Contact Improvisation can work as one. On the most basic level, simply sharing time with other human beings is a rare experience for many people who are blind. 'Most of those who are officially (visually impaired) are depressed and lonely', says Tony Rahim, [...] who has lost all sight. 'In the workshops you can find other people and interact with them – you cast away all your shyness'.
>
> (1991: n.pag.)

Peter's experience illustrates Tony's point and signposted new directions for the research to take.

Peter – Finding a new vocation

In the article entitled 'Common sense: Making use of the sense of touch', Kilcoyne (1991) discusses her research into the benefits of touch and tactile stimulation with reference to a BBC documentary featuring research studies undertaken at Universities in the United States.[36] She applies her findings to two participants named Peter and Bonnie (who featured later); for now, I quote her portrait of Peter and his experience.

> Peter is big and burly and in his fifties. He has virtually no sight, having been badly injured in a traffic accident. His sight loss was instantaneous and irreversible. He could no longer work at his trade, he no longer had an income and he felt very depressed. He came to the movement/dance workshop with some trepidation – he was not 'arty crafty' and thought dancers were a peculiar and probably effeminate breed. However, he thought he would give it a chance as a friend of his had spoken very warmly about the workshop. Having been athletic and interested in sports when he was sighted, he experienced no difficulty with the work. He began to enjoy himself, giving and receiving massage and learning contact-dance.
>
> (Kilcoyne 1991: 48)

This portrait indicates the evident benefit of the work for a man like Peter. He reclaims his physicality and regains lost movement. Kilcoyne relates his 'psychological gains' to 'the primacy of touch' (Kilcoyne 1991) which her research revealed as unique to human-to-human contact, the foundation of strong relational bonds, and effective in trauma healing.[37]

Peter's participation in the workshop features in another article entitled 'A common sense' (1992) published in the Contact Quarterly journal (CQ) for a very different readership. There is a photo of Peter sitting on a rostra in the studio with Kilcoyne behind him. The description of the photograph states that she 'traces the pathway on Peter, from sacrum, under shoulders, underneath arms to the little finger'. Paxton and Kilcoyne appreciated Peter's sense of humour and give an insightful account of how Peter responded to their request for him to participate in their video-based research by walking across the gym to see if there had been any gain in 'posture and gait'. Peter realised the sincerity of the idea but true to character, 'treated us to his version of a "bad walk" as he put it'. He then gave a solo 'Chaplinesque performance' which they describe in detail:

> He stumbled his way forward through the vast dark territory of the unseen gym, the blind cane then became a fencing sword and he fought his way through a duel with an imaginary foe. Victorious, his stick transformed into horse which he sat astride and

> whipped into a gallop to the far end of the gym. Turning, he lurched drunkenly back towards the chair which he could not see but sensed the position of. Having reached it, he failed to sit on it, somersaulting his way through falls: and ultimately, threading his cane through the back of the chair and hoisting it aloft over his shoulders, he trudged off in the proverbial sunset. Peter had found improvisation and performance.
>
> (Paxton and Kilcoyne 1992: 46–47)

In this very memorable moment of self-assertion and abandonment, Peter reclaimed his lost autonomy. He affirmed for himself that despite losing his sight that his body can and will move and that his body will show him the way. In sharing this outcome publically, Kilcoyne celebrates his journey of rehabilitation and recovery. Evidently, the workshop had been a life-changing experience for Peter and his story is an example of how CI impacts the life world of visually impaired people – illustrating the basic principles in practice.

By engaging in the workshop, Peter regained a lost world and aspects of his former active and expressive self-return. His ensuing emotional catharsis was instrumental in healing the trauma he had sustained due to sight loss and his renewed identity took form. His display marked the culmination of a positive experience that had been contained and held. It was a moment in time, of transformation and reconnection, of a return to a coherent sense of self, from a discordant one. The workshops were not intending psychotherapy (Paxton and Kilcoyne 1993), rather an awakening or re-awakening of corporeality that had been erased. Kilcoyne reveals the moment of catharsis that Peter experienced.

> When the workshop finished Peter broke down and wept copiously and deeply. It was the first time he had cried since his accident. His tears were not only tears of grief for the ending of a workshop, they were also tears for his loss of sight and tears of appreciation at having found a new level of experience which opened up the future for him.
>
> (Kilcoyne 1992: 47)

This opening to a new outlook provided space and motivation to move on to pastures new. Following the intensive five days, Peter took a course in massage to work in the field of sports injury. From the outside looking in, the overall story tells of the body's innate ability for movement, even without sight. Visual impairment does not prevent movement, and movement is possible without being dependent on others. Peter's experience is a reminder of the uniqueness of each person, of the importance of autonomy, and of rebuilding self-trust and confidence. Peter had an opportunity to say yes to change – to movement, touch and a return to work – the same three membranes of body–mind unity, 50/50 balance and inclusion.

Case study 3: Bonnie; touch – A healing, self-regulatory experience

Embodiment practices provided the calming CI climate of the Touchdown Dance setting. The emphasis on the sensed state of the body became more consciously experienced as it settled inwardly, and the sense of calm arose deeply, creating a more meditative and receptive state. By calming the interoceptive state of the inner visceral body, a state of 'ease' supported a more responsive somatic nervous system so that the body became available to resume natural movements that have been halted or forgotten (Paxton and Kilcoyne 1993). With the re-use of the body in other positions than sitting, the natural reflexive responses that underlie human movement are re-awakened, motility returns to the self-membrane, as does confidence in being more of oneself whilst being in the company of others (Bainbridge Cohen 2012).

In her article 'Common sense: Making use of the sense of touch' (1991), Kilcoyne reiterates the importance of 'the primacy of touch' to healing and transformation in the workshops. The workshop experience provided an environment in which 'non-erotic touch' is primary and persistently used as a baseline to any relational dance – whether with myself and the floor (which is the first partner) or with others. A participant, renamed Bonnie, who had experienced relational abuse, found this environment significantly challenging.

In this article, Kilcoyne writes a report of Bonnie's experience as she recalled it. Bonnie was a partially sighted young woman. She disclosed her anxiety about working with men and before being in a situation of partnering with anyone. A workshop exercise brought the past trauma to the room. The exercise involved assuming positions that were associated with her trauma in 'embodied' memory. Here 'she' is Bonnie:

> At the beginning of a movement class she talked to another participant about the fact that she did not want to work with any of the men in the group. She had suffered violently in the past at the hands of men – she gave no further details. When the class was asked to lie on their backs in order to focus on muscular relaxation, Bonnie found this impossible; rolling over on her side she assumed a foetal position with her hands clenched.
>
> (Kilcoyne 1991: 50)

The characteristics of trauma are described here, and then, as Kilcoyne observes Bonnie's discomfort, she is able to offer support, and after further verbal disclosure the shift in Bonnie is then evident; she has been heard, and she can then trust the space as it is now understood that her physical resistance is grounded in her emotional distress. With the ebb and flow of her body in the space calm starts to resume.

> In private conversation, Bonnie revealed that she had been incestuously abused in her childhood, and had been raped by an acquaintance as a young adult. As the workshop progressed Bonnie became more relaxed; this showed in her body and in the quality of her communication with other people in the group. She could dance beautifully in contact improvisation, her movement having a vibrant and extremely flexible quality.

The emotional healing arises from the safety of the workshop environment, and it becomes a haven in which she can quietly find her way. Bonnie had decided to attend the workshop due to her interest in dance, and possibly subconsciously feeling that this opening to an experience could help her 'move on'.

> By the end of the fourth day of a five-day movement workshop Bonnie was working happily in physical contact with each of the men in the group. She was able to lie on her back without anxiety and her hands were no longer clenched. She was at ease within herself and with the group.
>
> After the workshop had disbanded, Bonnie felt well, no longer depressed nor overwhelmed by the images from her past.
>
> (Kilcoyne 1991: 48)

Bonnie continued to dance after the workshop and started to access counselling services – which shows that from this experience she was resourced and able to take decisions to meet her needs, and had the presence of body and mind to do so. Kilcoyne was able to give space and time to assess the situation in a way that de-escalated Bonnie's anxiety. The held space accommodated her needs and helped Bonnie discharge the affect of her fear, which was initially the lens through which she perceived those in the room (Levine 2010). Her pre-understanding could then change and she was given an opportunity to feel receptive to the nature of the touch interactions, which mediated the usual triggers (Bainbridge Cohen 2012). Bonnie's story, like Peter's, brings Kilcoyne to a positive conclusion about the impact of the work, even at this early stage, and that more research needs to be done:

> Even very small amounts of tactile contact in everyday life seem to have a dramatically positive effect.
>
> The Touchdown Dance project for the visually impaired cuts right across the established systems of touch taboo. The issues raised are discussed and worked through so that the workshop experience can be understood within a cultural context. Clearly more detailed research is needed to establish the relevance and permanence of the psychotherapeutic effects of this programme. The pilot studies are encouraging; they seem to indicate that the place of human touch in the rehabilitation

> and learning environments of blind people should be looked at very carefully and utilised more freely and imaginatively than has hitherto been the case.
>
> (1991: 49)

This undertaking was followed up and the primacy of touch continued to be key to the work and has been looked at very carefully. Touch is covered in more detail in Chapter 3.

Evaluation and observation of impact – The research protocol of care

I complete this chapter with reference to documented feedback from visually impaired participants with reference to an article by Kilcoyne, 'Physical conversations' (1990).

Professionals working in 'blind welfare' were targeted by Paxton and Kilcoyne to provide insight into potential ways to change the culture of the institutionalisation of visually impaired people. The participants had indicated issues caused by lack of movement, lack of touch, lack of access to movement education and lack of human contact. The imbalance caused by these lacks is also shared with sighted people who led very inactive lives and may have experienced trauma themselves; however, contributing to 'blind welfare' remained a priority of the work going forward. The Mission Statement was to 'provide opportunities for visually impaired people to participate in dance' so that through social and cultural inclusion and active participation, this may improve quality of life.

Whilst the workshop experience had evident therapeutic benefit, the spirit of self-development and inquiry remained the primary aim and emotional catharsis was more formally contained:

> There are deeply felt difficult emotions and painful social configurations which accompany the loss of a person's sight: however, our working contract is physical rather than psychotherapeutic.
>
> In the teaching we acknowledge connections between emotions and muscular tensions, but formal space for the expression of emotions is reserved for the closing process of the workshops.
>
> (Paxton and Kilcoyne 1992: 50)

Having disclaimed a psychotherapeutic purpose, there is much evidence of therapeutic impact, body–mind integration, somatic and cultural psychology, and so of a plurality of practices that participant feedback and accounts have validated. There was evidence of an improvement in physical appearance, which reflects the state of inner ease and postural re-alignment.[38] Often the closing circle witnessed 'tears' and feelings expressed that were 'overwhelmingly positive', and this was a point at which Paxton and Kilcoyne more consciously assessed a shift or change:

> Objectively we can see that the emotional landscape is accompanied by a changed physicality – improved complexions, relaxed features, lowered shoulders, alert posture, more confident gait and increased animation. The participants' own improvisations have provided the means whereby these changes have taken place.
>
> (1992: 50)

Paxton and Kilcoyne took measures to evidence if physical improvements did occur, such as taking a video before and after of participants doing the same sequence of natural movement 'walking the length of a gymnasium, finding a chair and sitting on it, and rising from it' (Paxton and Kilcoyne 1992: 46). As with Peter, this was effective. The videos showed 'visible improvement in all areas of movement' (Kilcoyne 1990: 100). However, the credit goes to the touch-based process, to the autonomy of each person's 'own improvisations', to the moments of natural human movement in the company of others.

In her paper 'Physical conversations: Research into movement for visually impaired people', published in the *New Beacon* (1990), Kilcoyne reiterates the key findings.

> The results of the workshops were very positive. Participants found the skipping visual impairment work life enhancing, creative, and spontaneous.
>
> The research discussions focussed on three questions: The physical effects of the work [...] The psychological effects [...] Life after the workshops in the sense of lasting effects.
>
> (Kilcoyne 1990: 100)

In the hands of Paxton and Kilcoyne, CI proved itself to be accessible and malleable, by enabling the participants to determine their own discoveries which were often profound and existential in nature. Despite the intention to improve agility and motility, the work impacted more fully on the self-world of the mover, as through moving the body the mind moves too, capturing memories and transcending habitual self-perceptions. Kilcoyne shares some of the words and comments she had gathered from the closing circle:

> Feelings of joy emanated from the physical conversations that are at the heart of the work. One of the major points made was that people had rediscovered their bodies, and in particular the capacity of their own body to be a wise teacher.

> On the emotional level people variously described the workshop experience as 'a stimulating journey', 'exciting', 'egoless play', 'humorous', 'an opening of emotional channels'.
>
> (1990: 100)

The negative impact of sight loss, of the trauma of loss itself, is displaced by a sense of being able to participate in an activity in which I meet myself anew, move with myself and become more of myself – I regain a sense of self-trust, as Kilcoyne concurs:

> Working in physical contact with a partner had enabled the re-awakening of trust to find a new freedom of movement, and the start of a level of physical awareness that improved mobility and self-confidence. One participant described the workshops as an 'MOT garage for the body, where I could recharge my batteries and learn to spend time and attention on improving my health and my body's capacity to move freely and confidently'.
>
> (1990: 100)

In CI, self-trust is a re-discovery that emerges from the internal dialogue of the moving body re-informing the mind of sensation. New movements, or those that have been forgotten, are awakened and become choices or possibilities to explore. In some cases this process, as with Gerry and later others, the possibility arose to explore the realm of performance.

> It is clear that the contact improvisation training allows for two levels of participation. Primarily it imparts building blocks for a healthy and enjoyable approach to the body and its capacity to move, and the other where the release into full bodied movement can lead to a level of dancing that can and has been publicly performed.
>
> (Kilcoyne 1990: 100)

Kilcoyne mentions the concern that Paxton and she had of the need for follow-up sessions for some participants who had profound experiences, such as Peter, to avoid the possibility of feelings of 'grief' on returning home after a workshop; 'Some members had experienced real grief as the workshops closed and had wondered how they would cope with a return to everyday life' (Kilcoyne 1990: 100). Follow-up workshops provided a place to express and share these feelings and avoided the possibility of a sense of loss and abandonment that may be felt on returning to an environment in which the process of self-growth could be curtailed. Kilcoyne returns to Peter's story to illustrate the point:

> Peter reported to the follow-up workshops that the tears that he had shed at the close of the May 1989 workshop were the first that he had shed since he had lost his sight. His feelings of loss had lasted through the next day. He had then felt stronger than before, and this sense of being strengthened has continued to grow.
>
> (1990: 100)

Evidently, the practice and process of the workshops served each person's needs in some way that had not been foreseeable. The ability of Paxton and Kilcoyne to steer this research was due to the dyad that they created professionally; Paxton's ability to dive into the unknown and Kilcoyne's ability to discern the psycho-physical holding in which each participant felt supported on their way. The self-world of each mover changed in this experience, through encounters with the body, with different selves that arise or re-emerge relative to the affect that the practice effected. Through the practice of CI each person came to recognise their potential, in an experience that revealed the next step, the path to follow. In the dyad, combined with the self-world of a partner, a space materialised for another dialogue to take place, in which a plurality of possible selves emerges in moments of movement. The movers allow these possible selves to communicate with each other, resonate and become more of who they are. CI did not only rehabilitate disuse of the body but it allowed unspoken voices to be heard and touch deprived bodies to be touched, so there would be an array of sensory, motile and mutable experiences to recall, to use to support changes in lifestyle or altered life choices into the future. This is where Paxton's statement about being attentive to what touch 'tells you' makes more sense; touch creates an existential realm in which we transcend who we are and encounter what we have yet to learn.

PART 3

Finding my place

In this part, I discuss my early experience of Touchdown Dance with reference to specific themes. These reflections reiterate points already made and shed light on the process from a sighted participant point of view, particularly with regards to the sense of space and time, to embodiment as a form of inclusive inquiry and to my growing vested interest in the process.

After the Touchdown Dance workshop in Denmark in 1989, I continued the project with a deafblind colleague and came to DCA to attend a workshop in the late spring of 1991. We were not the only ones coming from abroad; others came from the United States, some returning from previous years. Over time

I experienced how new lines of inquiry opened up for us in response to this new environment and how affirming and, at times, questioning feedback resourced an ever-deepening experiential research process.[39] Like others, I brought my own lines of inquiry to the praxis, informed by BMC® I took an embodied perceptual position in the continuum that embraced both introspection and interpersonal exchange. In being part of the process and living the open and permeable nature of the exchange, I engaged in the facilitative methods and approaches and established the membrane concept as both a hermeneutic device and a resource for embodied reflection – for ongoing practice-led research.

A natural atmosphere – The time–space for inclusive inquiry

In my first meeting with Paxton and Kilcoyne, I encountered their complementary relationship and the integration of somatically informed and mindful approaches within the workshop process. As a dancer and psychotherapist, I relate to both of these positions – to the sense of space and time, to the self-directive position of each person relative to their experience, whilst taking us all on their journey, we were invited to take up a proposal and 'run with it'.

From my interest in embodiment, I recall the characteristic starting points intended to create an experiential realm of the senses. By levelling down the emphasis on sight, the use of touch and movement created a space in which each person aligned to their other sensory affinities, whether sight or hearing. This generous space of discovery opened up new possible responses and potential activity, and reminded me of the richness of the available sensory-perceptive realms. It was refreshing to cohabit in a realm of being in movement without an imposed 'quality control', and each exercise felt unique and was shared by each partner and the group.

I recall the spacious atmosphere Paxton and Kilcoyne created, both in terms of an attitude to time and to the range of movement in space. Much time was spent on exploring a simple idea, such as rolling around all sides of the body, exploring the three-dimensional sphere of movement on different levels. I recall the breath cycle underlying a movement, a sense of body time as I felt my movement response, sensing weight, the assistance of gravity, embodying tissue motility and intersystemic tensegrity. I recall corporeality being the focus of attention, underlying the pedagogic intention to nurture individual differences towards a common goal – i.e. more movement. I witnessed each person enabled to experience growth and fullness of being, engaged in moving with others, bringing all that experience and bodily information to mutual effect.

These threads are interwoven in the fabric of the work and substantiated the research. Success required careful holding of the space, to prevent getting lost in the unfamiliar or unknown, or of one aspect becoming too contingent for all and the process becoming exclusive. Success came from giving equal weight to the discoveries of each person and equal value to even the smallest of insights or impressions, creating colours and new shades for the palette of potential possibilities that represented and defined the collectively determined experience.

The active participant position

I recall the handover from Paxton and Kilcoyne to the group, instilling an active, self-directive participant position starting in the opening circle, and then the seminal familiarisation with the space in pairs. As there were fewer visually impaired participants than sighted, some sighted partnered with each other and worked 'unsighted' to experience the non-visual realm. I was familiar with being a sighted guide but appreciated being corrected by my visually impaired partner; I stood one step in front of them to the side they preferred, with their inside hand holding my flexed arm at the angle of the elbow. Having already 'seen' the space I felt at an advantage, and initially awkward as I felt how easy it was to place my visually impaired partner in a subservient position. It soon became apparent that the purpose was to begin to relate to each other's limitations. Rather than let sighted participants off the hook, I recall feeling relief at being able to assist visually impaired participants correctly when needed and noted that some had the means to find their way autonomously, having mapped the space.

This active collaboration was reaffirmed when the roles changed and I was led, eyes closed if I wished, by my visually impaired partner around the space. Together we encountered obstacles so that collision became an inevitable reality. As the concern to guard against colliding was a shared responsibility, it reduced the sense of anticipation and fear, that otherwise imposed a sense of restraint on the freedom to move. I encountered my own distrust impacting on how far I could go before I reflexively opened my eyes, only to discover that I had intuitively sensed an obstacle. Trust grew when I realised my visually impaired partner did the same – having mapped the space they had measured and mapped it in their own intersensory way. This was an instance of learning to not take control based on fear or assumption, but to dialogue and listen in order to attune to my 'inner ear' and intuition, and to remain oriented through touch with my visually impaired partner. It meant letting go of preconceived ideas of blindness and accepting how much we gained together. As I had experienced in Denmark, the gap between the blind and the sighted disappeared and in its place was an understanding that the differences were a symptom of social and cultural inequality, inaccessibility and elitist attitudes, that at that time excluded many from dance. In either case, whether blind or sighted, our corporealities reflect the impact of experience and socio-cultural constructs that become imprinted, defining how we move and relate to self and/or others. Through these positive experiences, the commonality of touch taboo transformed into a common area of inquiry, and a locus for establishing trust and attunement when the partnership felt consensual and compatible.

Touch – From the void to mutual understanding; the positivity of difference

From the start touch taboo was one potential barrier to success but Paxton and Kilcoyne remained uncompromising. Learning through the body necessarily included touch and admitted its absence from the lived experience of both sighted and visually impaired people. I noticed the differences between one person's sense of touch connection and my own, and this initiated thinking as to how to manage this lack of awareness, as a shared learning point. For example, I experienced how touch taboo is expressed bodily as self-restraint and how associated tension states manifest in the whole body, impeding the ability to perceive the point of initiation of a movement exercise. In turn the impact on an ability to kinaesthetically sense the movement and accomplish it in an effective, self-directive way (Paxton and Kilcoyne 1993). As a sighted dancer I noticed how the kinaesthetic sense of doing a movement could be over-ridden by mirroring the movement as seen, thereby missing the point of initiation through touch. I learned to sense and see kinaesthetically when touch was veiled by vision.

Touch is the bridge between the visual and kinetic worlds, to bring attention to the transmission of information at very subtle moments of movement initiation. Paxton painstakingly facilitated the visually impaired participant's awareness of their joint position, of their range of movement, using his hands to simultaneously mark specific articulations or muscle actions. Paxton knew that, without sight, learning a specific movement would take more precise input and time to perceive and become conscious; first, body segments have to learn their initial position, to find their mutual relationship, and second, having initiated the movement, to find where the movement ends. Touch enables the mover to measure how far to move in space and in a space that is not seen, in incremental small steps. Observing this as a sighted participant I learnt in abundance, absorbed in this intention, from the quality of the exchange and the conversation that took place 'somatically' and reiterated verbally. I experienced this differently with different partners, each one providing new insights, a new sense of my moving body and a new sense of what CI is about.

Encountering the unknown – Discovering trust

Having never encountered a visually impaired dancer since I began dancing at 3 years old, like the majority of sighted participants, I considered this normal.

Sharing this realm with a visually impaired person inevitably raised concern and dilemma. As a sighted participant I became aware of my assumptions and beliefs, unprepared as I was for how each visually impaired person had a different degree of impairment influencing how they accessed this experience. I was to be surprised at how soon these feelings and emotions found resolution over time, and new understandings emerged. For example, initial feelings of confusion such as 'how should I be with someone who can't see me – yet I can see them?' This fear of not being able to work together sat deeply within and was inhibitive and apparently unresolvable as it is not possible to know what it is like to be blind when one is sighted, and vice versa if one is born blind and never had sight. Fear of sight loss influences my idea of being blind, and commonly, I question how it must be for blind people who confront projection, pity and patronising assumptions for help. When getting to know them it becomes apparent that they see the world in other ways, and like my blind friend do not consider themselves to be 'in the dark'. Touchdown Dance addresses this estrangement, with reciprocal trust-based work, in which the sighted learn self-trust from moving without vision with a visually impaired person, and the visually impaired participants learn self-trust in relating to the sighted with touch, movement and sound (Heller and Gentaz 2014). An implicit empathy is evident in both cases – and so a trusting connection is possible.

Natural vision

Due to having a mild visual impairment from birth, curiosity about sight led me to research natural vision techniques based on BMC® and later the Bates method. In 2001, I invited Lisa Nelson and Karen Nelson to deliver a two-week experiential workshop for dancers on the visual sense and dance. In BMC® I discovered how sight relies on light to illuminate surfaces to show the colours, shape and form of things and how binocular, stereoscopic vision provides a depth of field, perspective and a wide peripheral visual field. I can try to do without sight and use other senses to perceive space, but this will never equate to being visually impaired. I do this to experience how I may otherwise be when without sight, to gain perceptual and empathic understanding and respect for other lived worlds.

This kind of reflection reiterates the limited understanding and misperceptions imposed in the absence and exclusion of visually impaired people from the sighted world. These feelings have evolved with the discovery of touch and movement as primary ways of 'seeing'. Whereas sight only 'sees' the frontal surface that is in view, and perceives objects in movement, an object has to be in reach to be touched and to perceive what is hidden from view (Noe 2002). By working non-visually, awareness of the substantiality of objects in the world comes through touch – by touching something and feeling it move it is possible to navigate with this, or if it doesn't move, to navigate around it – this tactile way of reading the world is relative to touch-time, not visual-time (Gibson 1966). This is pertinent to CI, when my partner is in direct contact and behind me, I can't see them but their body can be further perceived through moving. So in CI, being without sight is not, paradoxically, sight-less, we have other ways of seeing in common and other senses fill in the gaps cross-modally.

Time

I recall, as a sighted dancer, how I had to slow down my movement responses to stay in relationship with my visually impaired partner. We needed more time to sense and feel the movement we co-created, to track our bodies shifting from one place to the next, to 'know' where to go next and to trust in the approach. I learned to adjust and discover how *not* to disempower and deny the visually impaired person their own path of discovery. I learned how to step into the world of the visually impaired person without drawing on visual information which, as I had learned with Torben, was too fast and unhelpfully denying the more fundamental kinaesthetic focus. I learned that to attend to sensation enabled me to follow the inner landscape of the body as it changed, the inner eye and movement complemented each other, whereas having sight or learning visually did not necessarily guarantee ease in moving. This finding is born out in the way that over the years of research, many blind people were able to absorb a lot of information without any visual reference, yet with some 'optic' sense of their body gained from being touched as they moved (Schilder 1999: 17; Dymoke 2014).[40]

For myself and other sighted dancers, the adjustment away from vision to a new 'touch-based' learning was essential; it required a sense of trust in an ability to defer to touch, to communicate effectively through touch and receptivity to the language of touch. The detraining aspect was at times too great a challenge, at others complimentary to the process. It was perhaps the awakening to these challenges and encountering them together with others, that supported Paxton and Kilcoyne to establish the type of experience they then offered, as it did for me. CI provided the modality that was adaptable and accessible – it offered them the macro and micro realms of corporeal research – a breadth that accommodated diversity but could also be overwhelming.

I experienced the vastness and immediacy of sensory information at the point of contact and the simultaneous outflow in movement that was protensive of the unexpected. In this space is the continual unforeseen, the continual surprise of the micro-movements, the micro-events, of the movement of mass, of the possibility of taking weight, of giving weight, of being able to give support without collapsing into it as we may do in other situations. In this space I discovered the scope of movement, more than I consciously engaged in with sight, I oriented to image and shaped thought and meaning. I sensed movements inwardly as if for the first time. As Paxton stated, admitting to the marvel of this experience in a presentation in 2015: 'The human body has been my personal frontier' and there is too much to understand but then as we all have bodies 'there will continue to be one to find' (Paxton 2015: n.pag.).

Somatic-oriented techniques and embodied approaches to movement

In the context of a Touchdown Dance workshop, I witnessed the influence of the 'small dance' as a starting point for the initiation of movement from any point of contact that is then transmitted through the whole structure or through the centre of gravity. I witnessed the ability to breathe more deeply and to reciprocate in micro-movements. Awareness of reflexive responses provided insight into the innate responsiveness of the body, its ability to protect itself from falling, to prepare and learn how to fall and the possibility of small adjustments to avoid doing so. Fundamentally, the attributes of CI were demonstrated and experienced bodily, with attention to subtle shifts in sensation and movement initiation. As the following chapters will tell, the usual visual language served to describe the placing, positioning and orientation of bodies in space, particularly when movement instructions were given to the sighted to clarify or provide information to their visually impaired partner. This protocol enabled the development of ideas into exercises and later into categorised modalities that are put in greater detail in Chapter 4.

CI provided an experience that was not sight-dependent or contingent (other than to ensure against collision). The facilitator–learner roles were easily reversed and the visually impaired person could step into the place of doing rather than just receiving. The role reversal became, by default, another protocol and feature of the praxis, one which required time and patience. These two aspects led to the new idea that finding commonality was possible in a space that accommodated change, a space that became embodied for me in the physicality of the membrane concept. CI gave the possibility for all to move and for a new perception of self in the world to emerge from the experience of being 'on the other side' of the usual social perspective we hold. The membrane concept became a means to explain this shift, to explain the phenomena of movement, the mutability of the 'unitary' body–mind, in a way that pertinently honours disunity and the elusiveness of truth.

My early experience with Paxton and Kilcoyne indicated how well a situated research endeavour can improve perceptual and moral understanding, and how ethical engagement builds community. Touchdown Dance instilled trust in the corporeal experience, and a change in attitude towards other ways of moving with conscious awareness of the new impulse, sensation and appreciation for the self-directed experience. It was satisfying to be fully engaged in one's own movement, to define one's own movement experience, and the practice helped to nurture this new intelligence as a form of self-control, a skill. This skill established new ways of thinking, noticing connections, accepting differences and an ability to examine in detail. As a person who always sought resolution to any conflict, I felt at home, in a place where the 'membrane' moved in a way that supported this preference, and the space in which it took place.

2

Methodology: Undertaking Research That Is Practice-Led

Contact Improvisation – Sowing the seeds of self-determination through touch and movement

This chapter looks more specifically at the evolution of Touchdown Dance's inclusive practice-led research methodology and its concurrent interrelationship with a unique, experiential and body-oriented approach to pedagogy. The interdisciplinary aspects of contact improvisation (CI) are revisited, to reveal the underlying ethos of inclusion and to locate this praxis within a mixed methodological research paradigm. Reference is made to ethnographic and participatory research (Walmsley 2001;[41] Totton 2018), practice as research (Barrett and Bolt 2007) and practice-led research (Hazel and Dean 2014). Appropriate to the founding 'ambition' to integrate visually impaired and sighted people through CI, this discussion returns to the ethnographic context that determined the original basic principles for the Touchdown Dance pedagogic approach as introduced in Chapter 1.

True to participant and practice-led research,[42] Paxton and Killcoyne positioned themselves in multiple roles: as the 'professionals' who undertook the project, as co-learners alongside the participants and as co-researchers witnessing the participant experience. Prior to starting, they had little idea of what this learning would be like, of how to teach movement to visually impaired people. Their aim was to be accessible, by working in partnership, to provide the tools and the space in which to use these, and to involve everyone in the research activity. They created an inclusive pedagogy based in the body, a membranous realm, in which motility (movement) and tactility (touch) form a continuum, which was ideal for inclusive research into the diversity of subjective experience and realms of consciousness. The process of sharing these emergent understandings changed bodies, minds and lives. By embracing diversity and inclusion, the uniqueness and potential of each individual was honoured, in an environment in which autonomy was respected and valued. I explore the iterative nature of this process (Hazel and Dean 2014) in which preconceived outcomes were avoided, and new knowledge captured as it emerged (Redmond 2005; Nind 2008; McClimens and Allmark 2011).

Touchdown Dance opened the door to visually impaired learners who experienced exclusion from 'traditional' dance training contexts, due to non-adaptive visual learning methods and the demand for precise positioning and timing.[43] As expected, CI offered a predominantly non-visual model, where touch and movement practices formed the foundation of an experience in which visually impaired and sighted co-participants found new ways to 'see' and 'be seen'. Paxton and Kilcoyne discovered 'how effective non-visual communication can be when movement is the subject being taught' (Paxton and Kilcoyne 1993: 16) and consequently used touch- and movement-based communication methods, as the senses available, if variably experienced, to all. Essentially CI offered an environment in which habitual perceptions of learning and types of provision changed. Inevitably letting go of the dominant role of sight caused some discomfort, but a gain in the emancipation of touch and those who touch. Everyone stood to benefit in this new experiential realm where new potential for integration and social inclusion was unveiled. On the one hand, the dialogical touch and movement methods 'liberated' the teacher from having to know all the answers, as they could defer to the participant experience, and, on the other, liberated the participant from pedagogic control, and touch from the hinterland of taboo.[44]

CI – A practice-led approach to learning

The Touchdown Dance pedagogy evolved from a practice-led[45] approach to learning through research (Candy 2006; Nelson 2013; Hazel and Dean 2014), with a focus on the experiential dialogue taking place between participants in the praxis of CI (Paxton and Kilcoyne 1993). The practice was both a pedagogy, supporting participants to explore bodily movement, corporeal awareness and verbal representations of these, and a studio practice for devising performance work with the same methods underlying the creative process. In both instances, visually impaired and sighted people were equally involved in a collective ethos of co-participation, in a form of research that Harre and Gillett articulate as an inherently discursive learning and practice-led research process (Harre and Gillett 1994). In the Touchdown Dance studio practice for improvised performance and film, CI was implemented by Touchdown Dance as a practice as research (PaR) method (Barrett and Bolt 2007; Nelson 2014), which is the focus of Chapters 6 and 7.

This chapter continues in three parts. I argue that CI is not just an 'art-sport' in this context, but a framework for an inquiry into corporeality. Corporeality is more aptly referred to here as the *body*, and the inquiry into its perceptual processes brings into focus the interrelationship of perception and consciousness (Gallagher 2008; Bainbridge Cohen 2012). Herein, the membrane concept appears and operates to provide a system and a means to capture this realm of being, iteratively and metaphorically.

In Part 1, the cultural context in which Touchdown Dance's pedagogic methods originated is revisited, and more broadly, how the research contributed to socio-political discourses on inclusion and diversity. The workshops endeavoured to provide a supportive space for self-discovery and growth, inclusive to participants and teachers alike. The aim to teach CI to a socially and culturally diverse group required a careful, creative and iterative process of inquiry, to be receptive to, and perceptive of, the variable and often unexpected responses. The two experiential realms of the visually impaired and sighted participants met in the process of learning, and perceptual change took place at the membrane interconnecting them lessening the potential risk of reinforcing exclusion and inequality created by societal control.

In Part 2, the four basic principles of CI identified by Paxton and Kilcoyne (1993) introduced in Chapter 1 are considered in more methodological detail. Learning points that arose from the pedagogic process are animated in two short vignettes. These were sourced from the workshop experience and evidence the epistemic depth of CI as a pedagogy and research practice, influencing attitudes towards learning through the body, and serving as a basis for personal and interpersonal research (Dymoke 2000, 2014, 2017; Bannon and Holt 2012; Sarco Thomas and Zammit 2020).

In Part 3, an account from a visually impaired participant is analysed to authenticate the principles inherent to CI: as a two-way process of accommodation, an unconscious-consciousness dialogue which supports transformation of the self (Albright 1997, 2019; Bainbridge Cohen 2012; Paxton 2019).

A phenomenological and reflexive approach to practice and research

There has been much learned and shared that has been directly accessed in the workshop practice as it is experienced (Barrett 2007) and warrants illustration. As a participant–researcher and workshop facilitator, I have selected key aspects of the practice for descriptive accounts in which I identify strategies and flesh out the fundamental principles. I explore the space of transition and change, where membranes move between practice and research, between the physical and verbal, the implicit and explicit ways of learning, between teacher and learner, and the blind and sighted. I discuss the methods of movement that apply to CI that are adaptive, forming, reforming and transformative, creating a space–time environment for the early development and growth of the work. This is grounded in my experience of CI as a unique form of interpersonal scholarship that provides strategies for a practice-based research framework. CI is:

- an approach that explores perceptual experience through movement;
- a method in which movement becomes a 'semantic' corporeal account of that bodily experience;
- a means to gain greater corporeal awareness and learn the significance of this awareness relative to a unitary and singular sense of self;
- a means to learn how perceptual experience is influenced by the social and the cultural context in which it takes place;
- a means to experience how new perceptual experience is a place for change or transformation, a permeative membrane for acceptance or rejection, a place where past perceptions are revoked or affirmed – a locus for revolt or reclamation – where boundaries can be redrawn.

I start with the context in which CI evolved from an art-sport and established its duality as an inclusive praxis and somatic movement research method.

PART 1

CI – The inter-relationship of pedagogy and practice-led research – The advent of an integrated and inclusive approach

The historical overview chapter (Chapter 1, Part 2) locates the Touchdown Dance initiative at a time for change, characterised by revolutionary methods and philosophical perspectives in both Education (Freire 1970, 1976; Foucault 1972, 1991; Stronach 2010) and Health and Social care contexts (Cooper 2008; Wiiliams et al. 2008; Totton 2005a, 2005b, 2015). Inevitably new interdisciplinary research initiatives arose, implementing mixed methodologies appropriate to the phenomenological complexity of subjective human experience that quantitative, systematic methods alone could not do justice to (Harre and Gillett 1994; Meekums 2010; Walmsley 2001, 2004; Gallagher 2008). A concomitant increased awareness of social deprivation and inequality led to pluralism, in the recognition of the need for a greater diversity of approach (Cooper and McCleod 2007; Meekums 1993, 2014; Smith and Dean 2014). CI found its place in this landscape as an interdisciplinary practice and research method with a remit to self-define (Keogh 2002; Cooper Albright 2018). It was not an accident that CI, as a form that liberated dancers from the restraints of classicism, would liberate ourselves and others from the restraints of taboo, or at least question prejudice and social injustice.

As a 'body' of work, CI was 'an evolving encompassing and changing entity' (Fulkerson 1996: 41) that was admitted into higher education vocational performing arts programs. CI found its way into community arts contexts, situated within a wider social 'body' and permeable to the many systems and discourses alive therein (Novack 1990; Albright 1997, 2019; Koteen and Smith 2008). CI impacted upon and was influenced by the wider social context of exclusion and prejudice; it was a response to discourses against touch (Nelson 1996[46]), to abusive attitudes towards, and objectification of the body in dance and sport (Paxton and Kilcoyne 1993; Fulkerson 1996), to the subjugation of the female role in dance (Fulkerson 1996[47]) and, by being inclusive of disabled bodies, a critique of ablism (Ptashek 1992; Albright 1997, 2019). Paxton's 'social charter' was to involve the 'whole body' (Fulkerson 1996: 40) to promote an attachment to bodily life as he had subjectively experienced it in CI; it was 'that intelligence which we hope to awaken, revitalize and use' (Paxton and Kilcoyne 1993: 11). The diversity of this 'intelligence' was the remit that changed the contemporary dance scene (Fulkerson 1996) and gave rise to methodological challenges that are explored further here and more intricately, in practical and theoretical ways, in later chapters.

CI – A pedagogy in process

As documented in Chapter 1, the social and cultural climate of the 1980s was rife with discourses that confronted societal values and advocated for diversity, disability and human rights (Foucault 1977, [1995] 1961; Harvey 1989; Kristeva 2002). It was a time to re-examine traditional ways of teaching and learning, in both formal and informal educational settings, as well as the teacher–learner relationship. One such discourse was the 'liberatory praxis' of Paulo Freire, the educationalist and philosopher whose critique of social injustice and inequality opposes the unilateral role of the teacher and proclaims, in its stead, a reciprocal relationship and learner-led perspective – a pedagogy in process (Freire 2016). Freire wrote that *what* is learnt and *how* it is taught is not only embedded in societal values but also contingent on their implementation (Freire 1970). For example, autocratic societal values are embedded in autocratic pedagogies and democratic learning through dialogue is not only antithetical to such autocratic diktat but also undermines its implementation.

By implementing a non-hierarchical model of teaching and learning, using CI, Touchdown Dance's praxis, like Freire's, necessitated an ethos of co-operation and dialogical relationships. It intended to democratise dance to accommodate diversity and be inclusive of visually impaired participants (Paxton and Kilcoyne 1993). As a new and radical pedagogy, the research endeavour required Paxton and Kilcoyne to adopt a methodological framework (CI) to find principles against which to evaluate the effectiveness of the work, and critically, the means to modify the activity according to the emergent needs of both participants and they themselves (Smith and Dean 2014). The process evolved around a core organising principle that CI could be of use to visually impaired people. For this to unfold, a context was required that would enable best practice and monitoring mechanisms to manage when something did not work. Observation, reflection and participant feedback were key, so that the critical mass of the work was effective and the practice remained cautious (Paxton and Kilcoyne 1993). Evidently, the participant's experience led the research, as the way each person responded to the tasks was unique in some way. This is the nature of improvisation; the form is, in itself, a research method (Dowler 2019; Sarco-Thomas and Zammit 2020). It was not only what was taught but how it was experienced and valued that was the source and force for change and where meaning resided (Keogh 2002; Dymoke 2014, 2017).

CI formed the basis of Paxton and Kilcoyne's pedagogical approach and their respective backgrounds provided the theoretical underpinning and evaluative skills needed for ethnographic and practice-led research. Touchdown Dance was initially an inquiry in which the bodily realm of CI and the restricted lived world of visual impairment collided. But as they explored this uncharted territory, using their vast past experience, more ideas permeated in, or out, to determine what would be of use in each 'collision' – such as the building blocks of trust and tolerance and the focus on 'personal development' that were latent to CI, as Mary Fulkerson, herself one of the original researchers of the form reflects:

> The possibility of using the work to contribute to one's own personal development was implicit in the work structure and movement dialogues. It was important to be present within a specific world of experience with specific demands of intention and concentration. Sexuality and expression were not central to that world.
>
> (Fulkerson 1996: 41)

As the founder of Contact Improvisation, Paxton was familiar with practicing, 'teaching' and performing CI with trained athletic professionals such as Fulkerson (Novack 1990; Paxton and Kilcoyne 1993). Similarly, Kilcoyne was assured at holding groups in a humanistic therapeutic or creative process. In this context, the challenge lay in accommodating two distinct participant groups, habitually estranged socially, whose reciprocity was contingent on equality of access and inclusive approaches and who presented an array of dissimilar participant needs, experiences, movement range (mobility) and expectations.

Complementary to the specific demands of this remit, the participant perspective also had to adjust; they had to concede to a new norm, a shared territory with non-visual cues for the sighted and unaccustomed open space for the visually impaired participant. CI required acceptance of touch and movement methods and for all participants to engage in a self-directive, experiential process, from which new perceptual understanding arose. These subsequently impacted on the 'worlds' that they return to as changed, or even transformed individuals. As Guilo d'Anna puts it (in Smith and Dean 2014: 184): 'What has been given us from society disappears and what is saved is the reality of our selves together.' Paxton and Kilcoyne came to define CI in similar, radical, terms – as an opportunity to shed the cultural barriers that polarise the lived worlds of visually impaired and sighted people, for one in which they thrive and learn from each other:

> The integration of blind and sighted people into a common learning process is basic to our methodology. There is no good reason to perpetuate the *schism* between the seeing and the unsighted worlds. This is especially so when the material to be learning can be mastered with equal ease by both groups; both groups gain an enormous amount from sharing the dance with each other.
>
> (1993: 15, emphasis added)

Rather than position itself *against* the mainstream statutory culture, or against an idea of an exclusive, predominantly sighted culture, the Touchdown approach of 'sharing the dance' (Novack 1990) introduced new thinking, broke new ground and bridged the 'schism' that caused exclusion, as mentioned in the above quote. The lived world of the visually impaired participants provided the formula for the CI process to follow and, as a result, their voices have articulated changes to that world, changes brought about through access to a creative and inclusive, reciprocal process:

> Contact improvisation itself is based on a shared partnership, where the process of mutual co-operation is more important than any one person in the partnership. Because of this it is easy to structure the workshop learning process in a way which reflects an equal and democratic partnership.
>
> (Paxton and Kilcoyne 1993: 15)

Such changes in self-perception led to many life-changing insights and to questioning why opportunities to participate were not more generally available to accommodate them; some perceived an historical injustice of deprivation and ultimately, the exclusion of visually impaired people from the wider 'monoculture' (Smith and Dean 2009: 184).

With this pragmatic approach, Paxton and Kilcoyne established an essentially inclusive methodology. Time would tell how well the experiential 'reality of our selves together' would work in practice to interlink the two worlds (d'Anna in

Smith and Dean 2014: 184). CI provided a structure in which mutual co-operation was inherent and further principles iterated, as I come to below. However, the democratic process did not proceed without issues, as it took place in a society in which the exclusion of minority groups from participation in research subsisted (Walmsley 2001, 2004; Heller and Gentaz 2014).[48]

Questions naturally arose as differences in attitude were encountered by participants and teachers and explored by them, for example:

- As a visually impaired person, how do I work with a sighted person when, I don't know how they perceive movement, space, touch, their body? And vice versa?
- In a touch-averse and sight-dominant society, how to create a conducive environment in which learning takes place within the perceptual realm of visual impairment, i.e. in the domain of touch and deep listening?
- How to open up the possibility for alternative methods that apply to visually impaired people, (who may respond very differently to how I, as a sighted person, operate) without disengaging sighted participants or without them losing agency?
- How to 'language' the movement as it is made conscious and lived, using description as a way of telling what the experience is like, in a way that the visually impaired person can perceive it, i.e. 'see' it non-visually?
- What other ways can be used to represent the body in movement as we intellectualise and theorise? How to acknowledge the bodily sense verbally, without detaching from it or losing it?

These kinds of questions arose within the experience and opened up awareness rather than pegging the ground too soon, or putting theory before practice. In reality, visually impaired participants got to experience movement practices that dismissed their habitual way of being excluded due to lack of sight, and sighted participants were challenged to step sideways, to adjust and accept integrative and non-visual methods (see Chapters 3 and 4). This reciprocal relationship underpins the learning culture and attitudes inherent to the praxis of CI. As the questions were elicited, the answers arose from within the self-membrane of each person's bodily experience; their dialogical reflexive process takes place at the space/time membrane between self/other, between states of consciousness, allowing adjustment and precipitating change.[49]

In summary, the questions express the implicit or observed societally imposed schisms, or perceptual polarities that were confronted by a force for change – the cyclic and iterative process of embodied inquiry. Using the body-oriented focus of CI praxis, Paxton and Kilcoyne's research identified divisions, conflicts and their potential solutions. Their newly defined creative approach and methodology of movement for wellbeing rather than vocation, followed in the footsteps of others committed to the same values of applied arts practice for inclusion and social change, contributing to an interdisciplinary cultural ethos (Hulton 2018).

CI has a substantial role in this chapter for its emerging dual identity as a pedagogy and as practice-led research – an identity that has unified into a body of work, a three-dimensional form that is both permeable and mutable within a relational membrane that foregrounds the diversity of participant experience. I return to this embodied concept later, having retraced and clarified further foundational principles of CI as an interdisciplinary research framework.

PART 2

The foundational principles in practice

CI – An inclusive pedagogy in praxis and interdisciplinary research project

With the attributes outlined above, CI provided a foundation for learning and a framework within which to establish a research protocol. The four 'basic principles' (Paxton and Kilcoyne 1993: 11) provided a rationale for reflection on participant progress and anchors for the pedagogic process, which I consider in more depth, with examples. The principles predicate new domains of movement research and initiate subjective interest which those of us who have continued the work have since applied to the research and development of modality-specific approaches (see Chapter 4).

Paxton and Kilcoyne observed the habitual inequality of sighted and visually impaired people relative to the bodily sense and sought to fill the gap separating their lived experiences of the world. From early on Paxton noted how the 'physical and psychological ramifications' of negative attitudes towards visually impaired people manifested bodily in 'the pattern of fear and minimal movement'. For some, neglect and exclusion resulted in a 'withdrawal into stillness [...] into a "no-man's land" where there is little or no recognition of their existence' (Paxton and Kilcoyne 1993: 9). As will be seen, the four principles helped to re-assess and modify the process when a loss or lack, or an unmet need was witnessed. To meet the magnitude of this challenge would be contingent on an ability to mediate the impact of societal prejudice on individual need, or degree of impairment, or adaptability, or on access to services and opportunities to participate.

It became apparent that lack of mobility and of opportunity for self-discovery were not unique to visually impaired people, but were, to varying degrees, also apparent to the sighted participant group. Inevitably, the *first* basic principle of 'Learning From and Through the Body' defined the focus of the work (Paxton and Kilcoyne 1993: 11). The diversity of the participant group made it imperative for Paxton and Kilcoyne to work inclusively within the interdisciplinary CI framework. A concomitant shift in attitude towards inclusion took place through movement-based dialogue, both subjectively and interpersonally with a partner.

The *second* principle, 'Integrated Exchange' (Paxton and Kilcoyne 1993: 15), identifies the reciprocal relationship between the sighted and visually impaired participants in an equal partnership for mutual benefit and learning. This challenged the unequal partnerships habitually experienced in a non-touch, visual world. For these perceptions to shift, the *third* principle was necessary: 'An Attitude towards Learning' in which the 'student focus is brought to the present moment, to sensations that they are feeling now towards patterns in the body which later form the basis for technical communication while moving' (Paxton and Kilcoyne 1993: 16–17). These principles grew from an imperative for new, non-visual methods of facilitation and learning, making movement and touch essential. In a non-touch culture and environment, the fear and taboo of touching necessitated the *fourth* principle defined by Paxton and Kilcoyne as 'Communication through touch – the distinction between the sensuous and the sexual' (1993: 22). These principles were the foundations of the 'Workshop Process and Structure', they acknowledged the necessary modelling of methods that were based on dialogue and insightful reflection, rather than prescribed standards of achievement (Paxton and Kilcoyne 1993: 25).

As I go on to explore, the four principles identify a reciprocal relationship between innovative pedagogical methods and new practice-led research methods which, in more recent times, evolved into the membrane construct. Paxton and Kilcoyne contended with psycho-social polarities and the exclusivity of traditional systematic research methods and approaches to learning, in order to subvert them. Their published ethnographic accounts in *Contact Quarterly* (Paxton and Kilcoyne 1992) and the *New Beacon* (Paxton and Kilcoyne 1990a, 1990b) introduce the interdisciplinary discourses inherent to CI practice and indicate the issues of social injustice, social inclusion and cultural diversity that are congruent to participant-led studies in learning disability (Walmsley 2002, 2004; Williams et al. 2008) and visual impairment (Heller and Gentaz 2014).

As an interdisciplinary praxis, CI lends itself to ethnographic research[50] (Novack 1990); it provides a context for situated enquiry into human movement in which habits, cultural influences and personal histories arise, both implicitly and explicitly. The plurality of practice-led learning methods rendered CI accessible and inclusive to visually impaired and sighted research participants. However, inclusion is also contingent on the imprints of social and political discourses on these individuals. How these manifested in the process and were managed through the four principles is discussed respectively in the sections that follow.

Principle 1: Learning with and through the body: Towards an interdisciplinary research activity and inclusive pedagogy

The praxis of CI evolved in a culture already opening to the body through somatic practices and drew on both traditional and new disciplines. Touchdown Dance offered an adaptive and accommodative process using techniques from four main disciplines:

1. Embodiment practices taken from somatic movement educational approaches, such as Body-Mind Centering®.
2. Breathing techniques, body conditioning and balancing based on yoga postures and somatic approaches.
3. Principles of the centre of mass (or gravity) and awareness of pathways of energy flow, from Aikido (a Japanese Martial Art) and tai chi (Chinese Martial Art).
4. Inversion, balance, lifting and falling (response to gravity) from gymnastics.

All these disciplines interlink to bring both specificity and breadth to the experience. They all involve body-based learning methods and provide a palette of accessible, adaptable approaches of appeal to a diverse audience, taken into more detail in Chapters 4 and 5.[51]

When the workshops opened up to sighted non-dancers, a general demise was observed as arising from the contemporary sedentary culture; a general increase in physical and mental ill-health exacerbated in the case of visually impaired people by a lack of accessible provision. In confronting this lack, Paxton and Kilcoyne's approach appeared all the more radical, as a process of revaluing corporal well-being and inclusion, whilst revolutionising the learning environment. The three-dimensional environment of the moving body became the 'class room' and, once awakened, the body became the starting point for communication, displacing attitudes that privileged the authority of the mind in learning. Having access to their physicality, each person learnt to be proactive and discovered the potential of other inhibited selves and how these can change and transform over time. Symbolically, the diversity in these responses reflected wider social/political issues (Albright 1997), specifically the disadvantage faced by the visually impaired participants in a visually prejudiced culture. The non-visual methods evolved to confront this disunity and to provide an inclusive realm founded on trust.

Adjusting and adapting the form: Breaking new ground with kinaesthesia and touch

CI in its initial fast and reflexive form for athletic and physically 'literate' people like Paxton, was not an appropriate model in this context. Paxton and Kilcoyne drew on the other disciplines to serve those less literate and needing a slower pace in order to engage in an experience of the mechanics of movement. To address individual idiosyncrasies and bring them to the foreground in a unique sensory and perceptive process, it was necessary to slow down and break each movement idea into parts, in a time frame conducive to accommodating the learning differences and diverse bodies. Touch was used to identify the anatomical points of reference and movement to identify the level of mobility and range of movement each person had available to them. Then, gentle subtle movements give insight into the level of tension that inhibits flexibility and response to gravity or to a partner. Learning in this somatically informed way introduced the body schema, its form and mechanisms for movement – kinaesthesia.

> The centre of our work lies in the kinaesthetic senses; i.e., the body's own sense of its interior tissues, processes and movements. We also give primacy to the sense of touch. The main sensory organ that we work with is the skin. We move the focus of attention to the internal and the kinaesthetically sensed.
>
> (Paxton and Kilcoyne 1993: 13)

The process of attending to sensation was uncommon to both visually impaired and sighted people, it required introspective methods that became core to the practice and research. Paxton and Kilcoyne learned to move away from the external focus of sight and hearing, towards an inner listening state accompanied by the use of touch, reiterated here:

> We are working therefore with the 'inner ear' and for this reason we do not use music to accompany exercise or improvisations [it] would [...] detract from the results of inner listening towards which we are aiming. Teaching is often through touch and learning is largely through the body.
>
> (1993: 13)

In practice, 'inner listening' arises in all aspects of body-oriented methods; rather than listen to sounds, it refers to the act of attending to bodily (sensory) experience. The act of inner listening is a sensory receptive position of attending to the felt sense of an experience whilst it takes place, unlike copying the teacher. In this way, the experience may ring true for the mover who is engaged in their bodily experience and available to discover something new based on the teaching point, unless they default to a more habitual movement pattern, in which case the learning starts there.[52] The starting point for the praxis was the subjective and unitary sense of the body as a realm of self-discovery. Each exploration informed and widened the scope of this experience into new realms of corporeality, introducing new insights and revelations or altered perceptions that establish new self-identities or feed into other social or collective ones. The significance of these different existential layers became more apparent in the research, as individual dancers learned to interpret meaning from their bodily experience and took this dialogue a stage further in moving with others. A dialectical process is required for a democratic and liberating exchange in which embodiment methods interlink perception, emotion and consciousness. Using CI, Paxton and Kilcoyne offered a space to explore the sense of self, and, second, to integrate the bodily self, in an 'integrated exchange' (1993: 15).

Principle 2. CI – An integrated exchange

By focusing on the intelligence of the body in movement, the pedagogy of Touchdown Dance embraced an Ontology within a reflexive approach to learning that was breaking new ground as a phenomenology and shaping an epistemology on touch and movement. The collaborative exchange between Paxton and Kilcoyne led to the integration of physical and psychotherapeutic methods in the praxis. In the act of sharing the dance, they shared perspectives on the bodily sense, including the affective psycho-emotional impact of empathy and the wider psychosocial effects of inclusion (Novack 1990). In integrating their ideas, they found 'a common learning process' that was 'basic to [their] methodology' (Paxton and Kilcoyne 1993: 15), an ethnography in which they observed the gaps between visually impaired and sighted participants regarding the 'natural physics of movement' (Paxton and Kilcoyne 1992: 47) and the benefit of finding ease and greater bodily awareness. Paxton used his years of movement experience to understand that a lack of flexibility and response-ability relates to the loss of sight and that not being able to see inhibits the will to move into invisible space. This mental and physical block resides as tension and may become customary. Kilcoyne was aware of this trauma, so care was taken to ensure that the participants were not pushed beyond comfort but moved 'gently' in slow incremental steps. Needless to say, sighted people also experienced forms of trauma and also stood to benefit from the gentle pace. The discovery of such commonalities displaced the estrangement of difference with an opportunity for an integrated exchange.

CI necessitates partner work and brings to the fore different aspects that each person either struggles to do, or can't do as well alone, but also what they can do. Lisa Nelson facilitated this experientially for us in her workshop with Karen Nelson; when a sighted person walks forwards, the world moves visibly towards them and disappears behind. As a sighted person who closes their eyes, or as a visually impaired person walking in space, they do so without knowing what is there; the world is moving invisibly towards them and is encountered through proximity and physical contact. As both walk together arm on arm around the space, discovering its dimensions, where the doors are, etc., they are surprised by what they learn from this exchange, about the space, each other's perception, body and persona and the foundations of CI practice – the tacit and ineffable become explicit in movement.

In the research format explained above, participants take movement-based information and explore this in their own bodies, in real time with a partner. The workshop content evolves in the present moment, from the invitation and facilitation of the teacher who creates the container and provides the supportive matrix for ease and motility, for an inter-corporeal event (Deleuze 1990). At times, the sighted mover closes their eyes, and the visually impaired mover opens theirs to use what sight they may have; both focus on the experiential process, to witness the movement interplay and dialogue and to grasp 'how effective non-visual communication can be when movement is the subject being taught' (Paxton and Kilcoyne 1993: 16).

Vignette 1: An integrated exchange and inter-corporeal event – The three reciprocal membranes

The following vignette captures an exchange between myself and a visually impaired partner that works on three levels: first, it shows the continual examination, revelation and adjustment to the internally sensed movement that enables conscious, self-perception; the subjective reciprocal membrane. Second, the verbal exchange or 'test' between the partners enables an objective accommodation of each other's intrapersonal self-sensing process; the objective – reciprocal membrane. Third, working with the senses held in common, 'moves the senses of touch and kinetic perception away from the personal' creating a space in which a shared attitude arises for all the movers; 'it puts a new set of perceptions centre stage' that everyone buys into (Paxton and Kilcoyne 1993: 24); the third collective membrane. We see how '(t)he reciprocity of roles is crucial to the dynamic process of learning that the group is involved in' (Paxton and Kilcoyne 1993: 16).

I write this vignette from memory of a real dyad; I am the first person 'I' and 'you' is a visually impaired partner doing a back-to-back exercise. What is evident is the continual flow of consciousness, of feeling and emotional content, all inseparably inter-woven in the experience (Damasio 2000; see Figure 4.1).

> You sit behind me and I feel your back pressing into mine. I feel the pressure of some of your spine against my rear ribcage to one side. So, I move and adjust my position, not knowing if that is OK with you. I wonder if you feel comfortable with my back. We wriggle and rub our backs together as instructed, 'like a bear against a tree' you say. We share a bit of light humour. We settle back-to-back again and, as instructed, we attend to the small movements of our own breath and then each other's different rhythms very subtly at the points of contact between our backs. Then, very gradually, with each breath you lean forwards, a fraction of a centimetre at a time, slowly by degrees, over your outstretched legs. I follow, as instructed, keeping in contact with your back. I wonder 'is it too much weight?' and you say 'no', so I can rest back on your back. On the out breath we allow the stretch to lengthen further with my weight, the pull of gravity brings us closer towards the floor. The front of my body opens up and I face the ceiling now. On either-side of my body, my hands are now behind me touching the floor, with my arms hanging from my shoulders, drawing my shoulders backwards. After a few breaths I feel you start to come back up and I follow actively, bringing my body back up to lessen the weight. We sit up, pausing to feel the change in our bodies. I don't feel your back so much now, I feel open in my own body and feel your support behind me.
>
> Then I reciprocate, I lean forwards, slowly, but not as slowly as you did, my hands on the floor to support my back and your weight. It feels uneven so I come back up a bit to support you more evenly. I feel you sigh and rest back a bit more. I take a few breaths, and feel the weight. I feel my body lower each time I breathe out and your weight helps this. I feel a strong stretch at the back of my legs and breathe slowly.
>
> Prompted by the teacher I feel it's enough and come back up slowly to sitting. I feel lighter, I feel more open in the chest. My legs have stretched too, underneath. We speak briefly and agree that it was better after we had settled. I prefer going back on you, but felt the benefit of the stretch of my legs from the forward bend. You prefer the forward bend as you like the weight. But we both feel good, grateful and embodied. You mention I went 'a bit too fast to begin with' but then I came back up to support you, adjusting to your pace and feeling more at ease. I agree to your perception, clearly, I was faster and had broken the connection. This deep listening brings consensus and connection and a greater inter-corporeal understanding inheres in our conversation and gives value to the idea of co-operation.

A demonstration preceded this experience and implanted a structure that our experience filled and integrated. Now sight seemed more superficial, a surface percept that pre-empted and even distorted the relationship between us before it happened, as in 'seeing' my partner's initial inhibition I had deferred to an existing socially constructed perception, expecting us not to move very much. This didn't last long and liberation from those 'veils of perception' (Harre and Gillet 1994: 42) came spontaneously and reciprocally with 'you'. The reciprocal role-taking generated so much shared information that our relationship grew more perceptive, mutual and more coherent.

In hindsight, writing this, I have become aware of how my understanding of the key role of the senses informs this type of learning experience; how I relate to my sensory affinities, to my feeling-responses and how their expression in movement at the edge of my self-membrane enables me to relate to my partner in theirs. In this exercise, 'I' attuned to my inner felt sense which, when shared with 'you', led to a more integrated exchange. The experience of moving in contact with 'you' established an intimate touch-based form of relating, an empathic, deep listening state; a state of being that affirms the principle.

This vignette witnesses how the dialogue is iterative of the three 'membranes' of integrated exchange noted above and already in the case of Gerry (see Case study 1 in Chapter 1, Part 2); the first is self-directive and subjective and the other two are interpersonal, the second lies between the partners and the third around the group. To begin, I reflect on my experience of being in movement, at the two-way membrane of movement and consciousness. Second, at the objective reciprocal membrane, I open to our mutual presence, to our interplay and non-verbal dialogue, and become conscious of 'your' reciprocal movement. Third, as we discuss our shared experience, paying attention to the moments of transition, we note how touch, movement and a facilitative environment support changes in bodily perception. During the experience, change took place on either side of the three membranes, and at the surface of the shared outer membrane, peripheral to our partnership. If left unspoken, much of the understanding would be dismissed and left separate, but through dialogue, emergent sense is referred back and enables coherence.[53]

This vignette captures, in essence, how the inner felt sense discovers a new truth that renders questionable existing social constructs. It witnesses how, through touch, 'I' and 'you' as self-percepts moved towards unity, balance and self-regulation – which was not always smooth or protensive (extending in space and time). Our self-perceptive minds adjusted with regard to the conditions imposed – to the familiar and unfamiliar, to comfort and discomfort, unity and disunity, whilst the body opened up to direct action and dialogue, to new ways of being in the world that effect change.

At this early stage in the process, the initial discoveries and learning are clearly body oriented and unique, with CI methods that explore 'movement generated unpredictably' (Kilcoyne 1989: 18) and in a facilitated exchange. According to Barrett and Bolt, unpredictability is a recognised feature of practice as research, as creativity arises from methods that are 'experiential, action and problem based' ([2007] 2012: 3). In their view, innovation arises ineffably from 'subjective, emergent and interdisciplinary' pedagogic approaches as 'tacit knowledge is always present in human activity and learning' ([2007] 2012: 3). Such methods reveal 'the potential of studio-based research to demonstrate how knowledge is revealed and how we come to acquire knowledge' ([2007] 2012: 2). Evidently, in being improvisatory and tacit, CI operates effectively across practice led and practice as research paradigms, its creative potential transfers to improvised and devised performance making (see Chapter 6).

In the vignette, the social status of the dyad participants is not 'seen', but bodily states and idiosyncrasies are, indicating that the wider socio-political impact for this approach begins with the intimate integrated exchange between people from diverse backgrounds. In learning through the body, CI was the cause and liberation the effect, impacting on attitudes towards integration, and further reflection upon this experience admits the next principle.

Principle 3: An attitude towards learning

The shared experience of learning through the body-iterated inclusive attitudes towards learning and shifted perception as to who can participate in research. Participation in research activity was generally exclusive to visually impaired people due to assumptions that lack of vision precluded them (Heller and Gentaz 2014). In Touchdown workshops, it became apparent that such assumptions were not only 'short sighted' and an imposition of the medical model of disability but also discriminatory against those living with sight-loss, blindness from birth or partial sight. Exclusion may be justified statistically, but undermines the value that may arise experientially, which only inclusion can inform. The degree to which each person considered themselves disabled impacted on how they accessed and responded to the process, it was not for any sighted person to prejudge. It became important to recognise that some felt deprived of social and cultural activity on the grounds of their visual impairment and others felt more autonomous and active. Just as some sighted people lacked awareness of these issues and others not, and many, in their own way, experienced other forms of exclusion. Illuminating to observe, these variables informed the research and led to a revaluing of dialogical, experiential learning and practice-led research as a paradigm.

Paxton and Kilcoyne sought to provide an accessible model for learning movement and in so doing exposed attitudes underlying the under-acknowledged continuum of social and cultural inequality. The reality of teaching CI to a group of people, half of whom are usually excluded due to visual impairment, inevitably touched on sensitive and divisive territory and deeply held perceptions that perpetuate exclusion and inequality, putting any hope of 'integration' in the balance. They sought to realise an ideal attitude for inclusion by creating an environment in which all were at least 'nearly equal' (Paxton and Kilcoyne 1992: 49). The body proved to be the territory in which exclusion is incarnate, as a socially given identity, embodied singularly for each person (Foucault 1995, 1998; Cooper Albright 2019). The body singular (Principle 1) experiences self-liberation with other (Principle 2) and allows for wider societal change to take place (Principle 3). The body becomes the micro-cultural basis for equality, mutuality and mutability, the root of an implicit moral sense (Sheets-Johnstone 2008).

The preceding vignette shows an attitude of mutual interest in the exchange and how each person tracks the sensory information from the proprioceptive stimulation.[54] During the exchange, the participants are guided into the given 'picture', whilst focusing on the sensations and becoming conversant with them. This attitude stems from the role reversal and is seminal to the process of becoming more motile; the body's usual movement range is stimulated and focus is brought 'to patterns in the body which will later form the basis for further technical communication while moving. Conscious observation of the physical experience here and now is the basis for the study' (Paxton and Kilcoyne 1993: 16–17). Acceptance of the movement experience is contingent on accepting to move with another person, and this precludes being receptive to the experiencing self – so the body becomes mutable, and as it changes, so does the self-percept. During these intense condensed moments, the physicality of movement contains sensations and emotions. As the body finds ease and completes the movement, then fear of change, of failing or of falling releases into curiosity, and preparedness to learn more.

CI provides a dialectical model of learning in which to experience the mechanics of human movement, from the micro-movement of the fingers to the macro-movement of the interconnected, whole body. In so doing, attitudes to the self-body relationship are placed in focus. According to Paxton, any movement, gesture, in fact 'all muscular states – are intrinsically linked to feeling states – or moods' (Paxton and Kilcoyne 1993: 17) and represents the subjective self of the mover. In the direct experience of self-movement, 'other selves' (or attitudes and identities) unexpectedly arise in association with the feeling states and relational dynamics. Other identities arise with the discovery of alternative and new physical possibilities that become translated in what is spoken or perceived as new or changed. The research activity brought awareness to these multifarious aspects and worked with any underlying attitudes that impacted on the experience of CI and the learning process. In embracing these new attitudes towards learning, the work invited greater permeability between individuals and communities.

As documented in the historical overview (Chapter 1), the general participant experience indicated a willingness to 'do' but a lack of knowing 'how to be'. The lack of availability for movement was assigned to habitual inhibition and withdrawal and to a lack of familiarity with learning by experience instead of by rote. Paxton and Kilcoyne found ways to facilitate this double negative through a positive and different learning experience. Their pedagogic 'model' embraced the subjective experience of each person in their direct experience of movement making, with an acknowledgment that this was impacted upon by ingrained attitudes to learning by training the mind and by enforced passivity – the literal embodiment of a *lack* of social mobility and societal motility.

In the context of a Touchdown Dance workshop, visually impaired people of all ages have experienced a change in identity and transmitted to others a sense of how things are phenomenologically different in their world. In Touchdown Dance, the differences are part of the learning and they need to be the basis from which the work takes place: each task is moved differently by each person, so that both the visually impaired and the sighted participants discover more from their inherent diversity. By giving space for visually impaired people to learn to move in ways they have not accessed before, society itself starts to alter – as in the case of Peter (see Case study 2, Chapter 1, Part 2), who having discovered the body-oriented methods of movement and touch, trusted his new insights and became a massage therapist. For people like Peter, Paxton's ambition and CI proved to be compatible and shared a pragmatic purpose. Through an underlying ethos of social justice and of co-responsibility through empowerment, a sense of potential and transformation was realised, even if quite passively at times. From the enriching experience of the movement practice, greater awareness of body–mind unity led to a shift towards greater autonomy. This took place in the space that movement created for new or renewed thinking and self-realisation. As the vignette indicates, inter-subjective relationships are the key to the development of the subjective sense of self (Stern 2004) particularly through touch.

The pedagogical process was held within a group, in a collective context, to promote values of equality, trust, empathy and belonging through movement. The next section explores answers to the questions that arose as to how CI was able to provide for this kind of shared experience and learning. It meant investigating non-visual teaching methods, such as learning through touch, directly body-to-body. It meant staying open to the reality of the present moment, rather than learning something already old, tried and tested. It led to a process of self-study, through movement, so that when old habits 'die hard' moving was used as an effective way to shed or shift them, to move from a state of mis-use or dis-use to 'good use' (Paxton and Kilcoyne 1993). Radical movement and touch methods were seen as a form of *detraining* (Hulton 2018: n.pag.), which was alive in the environment in which the work was spawned[55] and this undoing required trust, which went hand in hand with touch communication.

Principle 4: 'Communication through Touch – The Distinction between the Sensuous and the Sexual' (Paxton & Kilcoyne 1993: 22)

The three previous principles underlie this fourth and are interconnected within the developmental domain of CI praxis. The four principles support the iterative nature of the research, and from within the shared membrane an amalgam of different continua becomes apparent; traversing time and space, connecting the past with the present.

When Paxton and Kilcoyne introduced touch communication methods, they proceeded with awareness of the conflicting discourses they inevitably encountered. CI confronted the dominant societal non-touch agenda and taboo-ridden preconceptions, often used as justification to *not touch* someone or as a device to exclude (Nelson 1996; Dymoke 2014). As the core process and structural device, CI embodied a permeable boundary, a membrane that breathed with possibilities and two-way channels of communication, to both receive and respond. This established the 'natural resource' aspect of touch in the work, a sense of equality rather than privilege of access, and neutrality to render mutable apparently rigid and exclusive belief systems, forces of control, with 'natural' human understanding.

From her background in psychology and psychotherapy, Kilcoyne's research brought touch-based practices out of the closet of societal taboo, thereby instating a careful and respectful process, to enable meaningful exchange.[56] When the question of sexualised touch appeared in the process, it was due to the ambiguity of subjective experience; of touch as sensory communication or as a sensually affective connotation. As touch can be either or both, clarification was given in the process of guiding the experience with clear intent:

> The instructors guide the students toward senses we have in common, and in the act of guidance there is a quality of objectivity, of neutrality [...] This commonality further moves the senses of touch and kinetic perception away from the personal. We go on to examine the sensing apparatus as a theatre of operations for one person, which is to be used in conjunction with another person who has a similar attitude.
>
> (Paxton and Kilcoyne 1993: 24)

The perception of touch shifts from being taboo and undefended, a troubled social construct that veils the interpersonal and personal levels of relating to it. In the partnering, touch is an integral aspect to experiencing weight, energy, pace and movement, it opens up sensory pathways that underpin more complex movement and extends the physical experience of each individual and the group. In role-exchange, sharing physical contact with another nurtures 'a similar attitude' of empathic understanding and bridges the gaps between participants on all three

FIGURE 2.1: Stuart Jackson lifts Katy: High Bridge.

levels of engagement and inquiry – to reiterate once again, self, self-other and group with 'a quality of objectivity and neutrality' (Paxton and Kilcoyne 1993: 24).

The following vignette is composite of having learnt this exercise with Paxton and Kilcoyne in 1989 and having undertaken it many times with different partners and re-experienced the neutrality of touch. In this instance, I am partnering a member of Touchdown Dance, Stuart Jackson in 2019.

Vignette 2: Touch – On the gap between physical and verbal language – The motile membrane between states of consciousness

> Standing, back-to-back, I move with my partner who is blind, shifting weight side-to-side, and forward-and-back. The movement is not large or spacious; it is relatively small, we do not move our feet, our heads are moving the most in space, but only a matter of centimetres. The smallness of the movement adds to our focus;

FIGURE 2.2: Stuart Jackson lifts Katy: High Bridge.

smaller is larger in our tactile perception of it. He leans into my back and I adjust my feet, moving them further away so I can lean back against his weight. We adjust and explore without verbal discussion. I feel the compression into my feet and the length of my back and his. I feel supported. I feel strong. I feel safe that we will not topple. I feel rooted, through my own feet and his. He then shifts his centre lower so that his knees bend and his pelvis is below mine. Naturally this means he has to widen the space between his feet and knees. He can then bend forward and I sink backwards remaining back-to-back, then back *on* back but I find I am slipping off onto my feet. We return to standing and try again, a few times, till I remain on his back, my centre aligned with his, my pelvis resting just above his as I am higher. I feel like I am floating on a raft, as if weightless, and my feet lift off the floor. He lifts his back up, my feet return to the floor and we stand back-to-back taking our own weight. I then do the same for him. I carry his weight and he vocalises as his feet come off the floor, momentarily reaching into space, reflexively responding to the sudden sense of being off the ground. Once stable, he can let go of the sense of falling and find the ground beneath his back, that is, in my support. My hands on my knees, my arms brace my back to carry the weight. I come back up to stand and he follows, coming back onto his feet without any wavering as they naturally find the floor on his descent.

We converse afterwards. I say: 'How was that?'

He replies: 'Yes, that felt good'.

Me: 'I was surprised I could take your weight.'

He: 'Yes and I got you up on my back.'

Me: 'We can try again and see where we go next?'

He: 'OK. It helps to be the same size.'

Me: 'Or to have strong legs!' (I tap my thighs hard so he hears.)[57]

In this experience, there was much to learn from each of the small moments that accrued into a reciprocal exchange. The 'shared partnership' (Paxton and Kilcoyne 1993: 15[58]) gave a sense of a situated and contained inquiry and our learning was directly accessed in the practice as it was experienced (Barrett 2007). The unpredictable nature of CI is apparent in this practice; we both attend to the present and seek ease. Certain limitations are revealed and grappled with as we extend our habitual movement range and encounter hesitancy and other forms of psychological restraint. We both have some pre-sensory awareness based on prior experience and postural patterns, particularly in moments of transition from standing to bending forwards or taking weight and standing up. We also have limited and different understanding of body parts and their inter-relationship – the legs, back and head. My recollection of this dyad was informed by physical memory and our feedback – as neither of us 'leads' we are both responsible for finding balance and mutually interested in the movement that achieves this. Cooperation is made possible by the trusting touch-based relationship and admits the fact that complete balance is elusive, if momentary and transitional.

A touch-based learning path

This recent vignette exemplifies the ways in which CI is located in natural movement and how, through touch, the perceptual gap between 'verbal and physical languages' is explored (Gallagher 2008: 7). I notice that I fill in verbally to affirm some core methods, such as giving weight and finding support, to ensure against the loss or absence of corporeal content to our shared learning. The sense of agency and trust is evident in the exchange; we explore transitions through deep listening to the corporeal dialogue we have going on and discover the interdependency of movement and touch events. We share the challenges and then feel empowered to try again, we learn more and fill in the gaps. Having relived the experience in writing the vignette, I wrote the following interpretation:

> I perceive the familiar expansion of kinaesthetic awareness, and increasing motility in our movement. The two of us form a combined 'body' at times. Our perceptual membranes move between physical and verbal languages, whether teacher or learner, visually impaired or sighted, we aren't anymore of separate yet collective identities, or segregated societally into two estranged groups that we may no longer wish to subscribe to. There is a sense of a perceptual realm in which we both move and become more of ourselves, with more in common, without losing a sense of singularity. I sense the motile membrane that brings light and energy to the space between subjective and inter-subjective experience – the potential for something new to unfold is available in both the movement and our verbal exchange.

To conclude, the 'almost equal' sense of the partnership is validated in an experience that unifies, for each of us, our sense of a shared perceptual realm that we co-create through touch. We experience the first and second principles as a continuum of shared bodily contact, connecting interior and exterior realms, filling the void of pre-sensory perception with bodily-sense, to find the possibilities available to us (Jones 2009; Gallagher 2008; Bainbridge Cohen 2012). Our shared partnership awakens perceptive awareness and brings feelings and emotional resonance to conscious coherent expression – the third principle and continuum of new attitudes to learning. There have been times when this exercise does not complete in a sustained lift, where the shared 'bodily-sense' is not compatible. This does not deter the attempt to find a way, to undertake direct action to make adjustments, and it reminds us all that perfection is not the 'goal of this journey' but responding 'appropriately' to the sensed experience is (Paxton and Kilcoyne 1993: 25) such as giving space for it to be related, discursively.

The exercise involves physical contact – the fourth principle, a 'touch-based learning path' (Paxton and Kilcoyne 1993: 47), which takes place in a tactile-kinaesthetic state of consciousness (Sheets-Johnstone 1990). This is the emergent epistemic field, the ethos of humility and 'human communication' that is innate to CI (Paxton and Kilcoyne 1992: 47) and essential to the mutual appraisal that established key attributes of empathy and trust. Paxton and Kilcoyne placed visually impaired and sighted participants together in a literal partnership in a space shut away from the social body where they are estranged and subject to preconceived perceptions that impede the possibility of communicating. The CI domain is an environment conducive to healing, where 'emotional, personal and subjective concerns' are exposed and acknowledged, appraised, transformed or shed. The philosophical membrane enables the cycle of perceptual change to open up a space for interpretation, and for the development of new expectations, influenced by the positive experience and its conscious reinforcement (Bainbridge Cohen 2012).

Over the past 50 years, the practice of CI has operated in the context of socially and culturally diverse groups, and more generally, with non-vocationally trained dancers. In such inclusive contexts, CI has been described as providing a 'coherent physical language for all participants' (Ptashek 1992: 41). Inclusivity and coherence relate to the fourth basic pedagogical principle; which clarifies that when teaching CI, communication through touch is safe. This principle negates the possibility of adverse, incoherent and deviant touch experience and validates the pro-touch agenda in which movement and touch are fundamental languages of communication (Deleuze 1990; Bainbridge Cohen 2012), providing a natural resource that is generally accessible and available to 'all'. The Touchdown Dance programme revealed to Paxton and Kilcoyne, as well as to the participants, how touch-based communication brings perceptual and conceptual insights, taps into the imagination and indicates the significant interrelationship of perception and consciousness in the process of learning (Gallagher 2008). However, this can be absent from participant accounts as physical language can be hard to verbalise coherently, particularly when touch is commandeered by society as taboo. The purpose of the second vignette is to attend to these issues, to show how embodiment has a positive influence on the learning experience, and how touch remains present and essential to the methodology. It is intended to illustrate how '(t)he sensory material – sensing the body's state in the present moment – ultimately allows the body to move more freely, more appropriately and more safely' and this was the type of learning that Paxton and Kilcoyne intended (Paxton and Kilcoyne 1993: 25).

Pedagogy of inclusion and the facilitator role: Shaping a process and structure

> Consciousness and emotion are not separable [...] it is usually the case that when consciousness is impaired so is emotion.
>
> (Damasio 2000: 16, original emphasis)

The next layer in this overview of methods is the teaching approach; how the pedagogy of inclusion actually worked to establish the practice-led research program of workshops and the foundational process and structures that we continue to work with over 30 years later. CI has always been experimental and pluralistic – it does not have a fixed curriculum or any vocational centres or ambitions, as Paxton defines it in a 'backward' he writes (with self-irony), in a book on CI published by his career-long dancing partner, Nancy Stark-Smith, called *Caught Falling*:

> (CI) is not predicated on the structure of art dancing, which retains from its marital heritage the pyramidal form of rigorous hierarchy. It's more on the order of the

> ancestral motifs of groups, picnics, potlatches, potlucks and holidays. Temporary but renewable structures rather than grand institutional edifices.
>
> (cited in Paxton 2008: 88)

CI is a practice that evolves horizontally, opening within a community rather than depending on a hierarchy and is defined by experiential methods of facilitation. Naturally, some facilitators carry more information than others, so this is the only hierarchy in terms of who facilitates the work in the Touchdown Dance context. Facilitation methods evolved from the accumulation of ideas that arose inherently from doing the work with conscious awareness of, and attention to, the diversity of responses. Each individual's way of moving was significant in terms of the subjective nature of the material it generated and for the specific attention it required. The facilitators supported both singularity and variability in response. In each dyad, having witnessed a demonstration, the selected facilitator may be a partner who has clarity, or a workshop leader who co-ordinates a response with the use of hands-on guidance and verbal invitation. When collated, each considered response offers further insight and adds depth and richness to the overall epistemology.

The ability to be spontaneous and specific demands knowledge of the body, of movement initiation and support processes. It requires both skill and an empathic ability to be able to communicate on the physical plane and effectively engage each individual case. Like a witness with an open and informed mind, the facilitator has the skills to observe and discover what it is that is intended in each present moment. It takes courage to suspend what one knows, to 'bracket' the first response that feels appropriate (Depraz 1991) and to invite the mover who appears to 'not know' what to do, to pursue the present sense and feel the pathway get clearer. The facilitator has, in many instances, to shelve a precious a priori idea, the idea of a completed and fulfilled action that is the topic of exploration and allows for other solutions, for the unexpected, as more often than not these arise. Variant perceptions of freedom or inhibition determine the combined movement experience. Associated feelings of ease or withdrawal may intervene in the dance, making the work dependent upon finding compatible states of consciousness which will arise when there is a shared sense of purpose, of 'unified mental patterns that bring together the object as the self' (Damasio 2000: 11). This may not arise if there is too much reliance on ideas, or on information already in the body memory that redirects the movement into another already learned pattern. Or alternatively, due to lack of information and experience, more time is needed to pattern proprioceptively first, for the movement to become familiar and then integrated. The least important is to clone an idea, the most important is for the body to experience new movements that bring in a sense of connection and inter-connection, a sense of commonality with diversity respected.

A playful mind-set in the realm of improvisation permits the journey that each person undertakes enabling access to the material as they excavate it.[59] The facilitator can be there with them discovering the bodily shape as it forms, to share discoveries they believe they have made. This is not to evidence any specific truth, so much as to evidence the nature and value of subjectivity within a humanistic and phenomenological pedagogy and research methodology. In this way, the experience shared by any dancer, whether visually impaired or sighted, is likely to contribute new knowledge and this all the more so from a visually impaired dancer who has previously not been able to contribute or been considered as a potential source of knowledge. Accepting diversity and difference in learning proved complementary to the uniformly goal-oriented 'attitude' towards learning inculcated in educational systems, with a potential shift towards more inclusive pedagogies, cultures and society.

The initial purpose of the research was Paxton's ambition for visually impaired and sighted people to co-operate in a movement-based dialogue. This was held within a wider membrane of a group in which all may learn from each other.

PART 3

The role of discursive, ethnographic methods

New iterative trends in research (Smith and Dean 2009)

I now explore the value of participant accounts within the new interdisciplinary and non-systematic trend in practice-led research that Touchdown Dance undertook and the contribution that touch and movement make through mindful consideration of complex bodily experience. First, I refer to the book *The Discursive Mind*, in which Harre and Gillet note the advent of a synthesis of new discursive methods in psychology research that embraces the complexity of human behaviour in ways that the 'lawlike predictions' (Harre and Gillet 1994: 4) of quantitative research could not achieve. The resonance is evident:

> Once one sees the task of understanding human behaviour as involving interpretation and empathy rather than prediction or control, the self-reports of the people one is studying become very important in any psychological research project. And these should not be taken as (falsifiable) reports or states of mind but as expression of how things are to the subject. Thus, the experimenter or observer has to enter into a discourse with the people being studied and try to appreciate the shape of the subject's cognitive world. But at this point it no longer makes sense to talk of

> observers and subjects at all. There are only co-participants in the project of making sense of the world and our experience of it.
>
> (Harre and Gillet 1994: 21)

Dan Fishman (1991) makes a 'case for a pragmatic psychology' and speaks in defence of 'a pluralistic approach to inquiry that seeks to generate initial reflexive knowledge about what works in practice' (Fishman 1991: 28). The complexity of an interpretative and reflexive process involved consideration of the empathic inter-relationship with the partner; and in the case of Touchdown Dance, during the experience of moving whilst touching another person. By 'experience', I refer to the membranous realm of sensing, feeling and moving, the pausing to self-witness, interpret and ideate, and share with others who may share a very different perception of the same incident, based on its affective impact. In interdisciplinary contexts such as these, multiple subjectivities and an interest in the value of in-depth practice-led research has grown, based within a mixed methodological paradigm (Hansen et al. 2005; Cooper and McLeod 2011). Phenomenological insights arising in data collection and discursive analysis serve to add new understanding and to modify existing theories or beliefs. In short, reflexivity has become an established method for examining the phenomena of human experience in both practice-*as* and practice-*led* research (Barrett and Bolt 2007; Smith and Dean 2009).

With respect to Touchdown Dance, the exclusion of visually impaired people from participating in dance was experienced as the outcome of the sighted person's perception of what is possible or not. In their inquiry into research involving visually impaired people, Heller and Gentaz (2014) document how the research world is controlled and constrained by sighted people and remains exclusive to them.[60] Through involving visually impaired people in an inquiry, Touchdown Dance 'enters into a discourse' with them to discover the 'shape' of these different 'cognitive worlds'. However, the discourse is not only cognitive; it materialises through a corporeal experience within a body-oriented medium, in which new self-perceptions arise, namely, the radical type of activity that CI entails. This is achieved through participation in a carefully held process involving a wide range of new movement experiences, reflective dialogue and discussion, or 'self-reports'. Due to the subjective nature of the movement experience, each participant will bring their current sense of self, their beliefs and self-expectations to bear, as Gerry and Peter did (see Case studies 1 and 2, Chapter 1, Part 2). When considering the various reports made of the work by any participant, it is necessary to be aware of any hidden or apparent antecedent ideas or theories, which emerge and constitute another form of 'prediction or control' (Harre and Gillet 1994: 21) and predicate the research outcomes.

In reading other forms of documentation, archives, interviews and published accounts, the ethnography reveals how the methods are seen to work. It was through careful negotiation and attention to even subtle differences in learning input that Paxton and Kilcoyne discovered what form of inclusion applied to CI practice; how to make the space and the work accessible; how to manage the diversity of the real world and move away from the old trends of domination and oppression. It is not surprising to find that the old trends of stereotypical prejudice are at times perceptible, seeping into the room in the moments of meeting, of choice, of touching and of moving with others. Within a practice-led research paradigm, such 'particularities' (Barrett and Bolt 2007: 4) arise and can be challenging, particularly in research involving marginalised and disenfranchised groups (Walmsley 2001, 2004). This is true in terms of bringing visually impaired and sighted people together to move, who may in fact consider themselves polar opposites in regards to this idea. Once differences are disclosed, this information adds breadth and depth to the findings, as a case study does, and if not disclosed, a sense of deception or manipulation of one participant by another may undermine the experience.

Any breach of trust can be circumvented, by ensuring an open, safe and transparent environment, so no one feels used to serve a prescribed purpose, reducing the complexity of their experience to an unrepresentative and disembodied reality – i.e. unlike the one experienced in the vignette. Had either person in the dyad lacked empathy with their partner's need for clarification or been over cautious in their regard, the sincerity of the mutual exchange would have been lacking. Such clashes in relationship could undermine the purpose of the work and yet reveal important information about the type of challenges that need to be confronted and managed. These kinds of perceptual disconnections cause conflict, isolation and the perpetuation of hierarchy, unless expected and attended to within the process.

This is also an example of when implicit pre-understandings or pre-expectations surface from within the mind to the 'membrane' and become conscious and perceptible. Self-observation (which means objectively looking at self) or introspective reflection (which means contemplation and internal self-sensing) during this experience serves to inform, re-pattern or reform a previous percept or belief (Depraz 1999; Bainbridge Cohen 2012). It is satisfying to witness how these pre-expectations generally seep out and disappear into the past, leaving a change in the person as a new bodily-sense transmits at the self-membrane. For example, the cases of Gerry and Peter in Chapter 1 indicate the role of embodiment to the process of altered self-perception and regaining self-movement. Part of their learning experience was to manage these new body-focused events with ease, rather than subjected to intransigent 'veils of perception' (Harre and Gillet 1994: 42) that imposes self-exclusion, such as loss of sight implying loss of mobility. However, through interactive movement experiences, these 'veils' can become responsive, membranes – a construct that has evolved through reflective analysis of the type of permeability that emerges in the minds of educators and participants over time. The level of engagement between visually impaired and sighted participants was generally open and responsive, as Paxton and Kilcoyne modelled with humility, and any contention was pursued for information.

As an illustration, I cite a pragmatic review of a weekend residential workshop that I taught in 1995, written by a visually impaired participant, Eleanor Tew, in 1996. It was written in response to a request by the *New Beacon Journal of Blind Welfare* (a journal published by the Royal National Institute for the Blind) to hear the voice of a visually impaired participant. Tew had attended a workshop in 1989 in Devon so this was her second weekend, held this time at the University College Bretton Hall in Wakefield. Tew succeeds in giving context and emotional content to provide a more immediate sense of the experience to visually impaired and sighted readers. The title *It Feels like Flying* is like a launch pad into the experience and this account captures a sense of the corporeality and new insights that she gained in words, which I highlight in italics.

> Touchdown Dance specialises in the form known as 'contact improvisation', where the dancers work mostly in pairs, maintaining almost constant physical contact. The *whole body* is used, and although contact may be made *peripherally* through a hand or foot, much of the time it is more *central* – through, for example, forehead or shoulder, back or hip. Usually, the movement is slow and flowing, with the dancers *slithering down* one another's bodies and *rolling over* one another, though *fast running and spinning* also have their place. *Degrees of weight* in the contact are explored, and the dancers sometimes lift one another right *off the ground*. The form was not devised for visually impaired people, but fairly early in its development it was noticed that sighted dancers quite often close their eyes and it was surmised that this would be something that visually impaired people *could do well*.
>
> My first encounter with Touchdown Dance was a weekend six years ago, and I feel the movement has matured since then. There is a greater recognition of what is of real value and what just a passing fad. Teaching techniques have also become more broadly based, drawing from a *number of disciplines* including yoga and aikido to explore natural movement and relaxation.
>
> (Tew 1996: 39)

Tew gives witness to the liberating aspect of the work and the potential for self-making. She had immersed herself again after six years following her first attempt; she recalls the process and speaks eloquently, as well as with caution, introducing the phenomena of the experience. Her account relates to her encounter with CI through a 'tutor' (myself), it is subjective and personal, indicating her physical experience in descriptors such as 'slithering down one another's bodies and rolling over one another' which, out of context, could read like the world of childhood or 'out of the box' for many adults whether visually impaired or sighted. Her embodied memory is expressed with energy, a sense of satisfaction and self-appraisal that indicates a positive learning experience. I recognise the breakthrough for many visually impaired people to be lifted 'right off the ground' – as they literally no longer know where the ground is. This 'whole body' experience may well test every ounce of courage in herself and trust in the partner but it is expressed as if habitual to this context and provides insights into physicality of the experience to be had in a workshop.

This 'semantic work' (Gallagher 2008: 7) gives witness to the liberating aspect of CI and the potential for self-making and for knowledge production. There is no perceptual gap between Tew's physical experience and her rendition of it in words; she describes parts of the body, as if recalling a moment in the dance, of a hand touching a foot, or a back to another's back, the changes in dynamics as well as spatial relationships. To bring it to a close, she writes somewhat ruefully:

> We did some mad things but we did them in a fairly sensible way. We were encouraged to explore what our bodies could do but never forced to go on when we were too tired or to do what we were genuinely not ready for. Also though our tutor said little about safety, I am sure she was always mindful of it.
>
> I think she enjoyed teaching and dancing with us. She said it was a pleasure to teach people who threw themselves into everything so wholeheartedly and who were so obviously enjoying themselves. As for me, well, there were moments when I found myself underneath one of the heavier people and did feel a bit squashed. But there were moments which really did feel like flying!
>
> (Tew 1996: 40)

Tew's concern for safety touches on fear, as if 'thrown in the deep end', which she expresses in her interpretation of the 'mad things' that were done, even if 'in a fairly sensible way'. She is empathic with the general need for both visually impaired and sighted people to trust others in the face of trying something new, particularly something that has originated and operated in the sighted world. Such feelings run deep and form the antecedent control mechanism that may inhibit a participant or enable them to jump in, depending on whether the sense is negative or positive. Tew reveals the empathy and other affective states that Shaun Gallagher (2008) refers to as *prenoetic* – the state in which a decision to participate in an activity or not is influenced by a bodily felt sense or affect. She had a pre-expectation based on antecedent positive experience, which may have pre-determined her decision to participate and then to write this report. She has learned the language from her physical experience, her body remembers the totality of it, and she writes from this place of bodily sensation and its recall in the context of a culturally inclusive environment. Tew writes as one of many 'co-participants in the project of making sense of the world and our experience of it' i.e. discursively and meaningfully for others (Harre and Gillet 1994: 21).

When given an opportunity to explore new modes of expression involving movement, verbal interaction and group physical activity, as in this case, it is no wonder that the version of events given by visually impaired participants is of great interest to themselves and others. Tew's language expresses her conviction and appropriation of new movement terms, she also alludes to the disabling non-touch discourse when it becomes the rationale for 'not-touching' at all (Principle 4). Her self-report provides an example of an account that presents a view or the physical experience from both within and from without, (Principle 1), and a reflective, critical view that is constructively in support of the work going forward, in naming aspects that need further attention to enable it to grow further (Principle 2 enfolded within 3). Her account reveals three membranes; of the self, the influence of social diktat and the availability of being able to recall and reflect on the breadth of the experience for others to learn from. These aspects constitute the type of contribution that CI makes to discourses as well as to changing lives.

In this review, Tew provides insight into another type of world in which visually impaired and sighted people can co-exist; it is meaningful and enticing for some, and not for others. Offering up an opportunity to participate in movement does have knock on effects on the lives of those who try it – either it entices them to travel and seek out further opportunities to be part of the CI community, or it provides a language for the body to use to inform other lived experiences which are not open so readily to visually impaired people. The research undertaken in Touchdown Dance achieves a great deal in terms of the impact it has on the quality of life of the participants. Initially inspired by the changes in bodily sense, participants become more open to explore the outreaches of their body from within and without. This may then translate to feeling actively involved in breaking down social division, by bringing this body awareness into other contexts to inspire change or transformation. Some who take the practice further, may understand better how 'intentional action' such as making a performance, incorporates both implicit and explicit aspects; the private 'thoughts' that do not manifest yet underlie the event that takes place and are made explicit through other modalities. We will discover more in other chapter sections about how the practice continues and opens up to other contexts. Touchdown Dance practice serves the same overall purpose – to show the potential for self-growth and healing through commitment to becoming aware of the body and its role in our experience of being in the world. From here, we find our own discoveries and pathway, and I return to mine.

New aspects arising from the practice-led research that shapes the membrane concept

In my understanding, CI is about spatial orientation and relationship, about the motility and mutability of relationships in space. In CI, the quality of the relational dialogue is relative to the energy, the physicality and the ability to make sense or deduce meaning, which can be non-verbal, as with the movement itself, or verbal exchange (Fulkerson 1996). Within the relational sphere, I experience layers of permeable membranes that are the pre-set perceptions and pre-expectations that situate me in either of two positions. The first is a protensive readiness to bond and willingness to relate before the experience takes place, the second, an attentive sense of the unknown, potentially rejecting the unfamiliar as I remain unsure, may be protective, careful about participating, introspective and retentive. In either case, I am reflexively responsive towards or away from the experience according to the state I am in. These preconscious or reflexive states of being are primary to the sensory information received and rendered conscious through reflection, identified by Gallgher (2008) as

> structures of consciousness that are more hidden, those that may be more difficult to get at because they happen *before we know it*. They do not normally enter into the phenomenal content of experience in an explicit way and are often unavailable to reflective consciousness.
>
> (Gallagher 2008: 2, original emphasis)

Later Gallagher explains how '(s)uch prenoetic constraints necessarily shape my experience' which I translate to veils or membranes of perception, plural, permeative and available. If I am bodily aware I have a felt sense of the way I am in the room relative to the CI experience. I may track the 'phenomenal content', as it moves and changes when I connect with others (Gallagher 2008: 2). My membranes move in communication with others as we connect – and my felt sense, or my self-percept, changes tangibly. I may feel more at ease and open to more input, or less. In Body-Mind Centering®, the prenoetic defers to an embodied approach to cellular consciousness, a more fluid concept than constraint, the membrane is the place of containment and of potential for change either towards myself or outwardly towards others – a place of transition, as Bainbridge Cohen puts it: 'transformation happens at the membrane, is just that thin membrane away' (Bainbridge Cohen 2012: 75).

This embodied approach and accompanying theory are foundational to my experience and the use of CI as a practice-based research method. CI provides the context for an experience, providing the sensory input and facilitating the perceptual process: the conscious registration, interpretation and reflection upon the experience, its positive or negative evaluation, which influences how it is considered in future or shared with others (Bainbridge Cohen 2012).

CI is, in my view, a perceptual experience, a process in which bodily movement is made conscious through touch, in ways that may not be customary to the participant. The movement experience is a type of 'semantic' account of how it is to have a body, to have awareness of the body and for this to become a more 'habitual' experience (Gallagher 2008: 7). The CI experience becomes further enriched by the context in which it takes place; again Gallagher concurs with Bainbridge Cohen but from a phenomenological perspective:

> Perception is not a simple reception of information: rather, it involves an interpretation, which frequently changes according to context [...] the empiricists were correct to suggest that perception is education by experience [...] One should think of this as perception enriched by experience and by habitual, as well as customary, ways of experiencing things [...] The phenomenologist would say that perceptual experience is embedded in contexts that are pragmatic, social and cultural.
>
> (2008: 7)

The appearance or not of phenomenal content is where human interaction, reflection and engaging in debate can serve to make such content more available rather than stuck within the membrane. This point of view assists the endeavour to carry out experiential research with people who may, through choice or neglect, struggle to identify the constraints that impede or influence the choices they make. Awareness of such subjective inhibitors/excitatory aspects also enables the researcher to negotiate and facilitate as a guide or witness when there is a lack of ease, confidence or self-assurance in body awareness. With a strong intersubjective relationship, it is possible to identify and understand underlying factors that emerge within the accounts, to move them out beyond the membrane, for growth or healing to occur.

Now it feels timely to consider touch more fully – having concurred with Paxton and Kilcoyne's findings, I explore the experience of touch and touch-based relationships. It is possible to grasp the significance of touch communication and so then mediate deprivation.

3

Touch Communication: The Reciprocal Membrane of Inclusion

PART 1

Touching the skin is touching the membrane of the inner body

The first part of this extended chapter explores more specifically the aspects of touch that arose in the practice-led research of Paxton and Kilcoyne and yet remained initially in shadow – as if residing behind a screen of culturally imposed non-touch agendas, the 'prenoetic constraints' that become available to change (Gallagher 2008: 2). Touch-based research methods were taken into more depth as participant experience exposed a range of themes, insights and principles which I have alluded to in previous chapters and reiterated later. These new waves were concurrent with an increased interest in the sensuous body (in phenomenology and philosophy), which implicated touch, and manifested in pro-touch discourses against its undervalued status. Many practitioners from this 'clan' have demanded further research into the nature of touch experience, its ontology and approaches to healing (Caldwell 1996; Juhan 2003; Sheets-Johnstone 1990; Classen 2005, 2012; Johnson 2006; Levine 2010; Westland 2011).

Having taken part in the pro-touch discourse in psychotherapy and allied arts contexts for many years (Dymoke 2000, 2014, 2017, 2019), I concur with the perception of touch as an intricate and variant experience (Montagu 1986; Autton 1987; Caldwell 1996; Classen 2005; Johnson 2006) that is primary to development and self-regulation (Schore 2003; Stern 1985, 2010; Levine 2010; Bainbridge Cohen 2012). From this experiential standpoint, I revisit the touch territory established by Paxton and Kilcoyne and explore the space between the polarities that emerged and were confronted in the praxis. Again I refer to the archives, to a moment of touch interaction that captures the experience of Tony Rahim to illustrate how the research moved from an idea to a question and an interest in exploring what 'touch of the blind is like' (Paxton quoted in Halliley 1991: n.pag.). This background contextualises the later research that I write about in the following sections.

In Part 2, I include another Contact Improvisation (CI) interaction that I shared with a deafblind woman renamed Ida. I describe the phenomena that came to mind in our delicate touch and movement conversation and the underlying small dance in our fleeting moments of contact. This is to illustrate the layers of preverbal and literal communication that took place and founded my interest in arguing for touch as a neutral and carefully negotiated realm. This pro-touch ontology serves the needs and rights of those with untapped potential or troubled interpersonal relationships, where verbal and visual methods aren't enough (Levine 2010; Totton 2015).

From this standpoint, I discuss the context of the wider cultural and societal discourses on *pro*- and *no*-touch, informed and influenced by many years of Body-Mind Centering® (BMC®) touch practice and research, in touch-averse clinical, educational, and health and social care settings. In my experience, one of the main challenges of embodied approaches and practices like CI is to communicate the experience of touch events. In the moment of touching we may identify a touch experience and recognise the familiar aspects (temperature, texture, motility, etc.), but touch is variable, often transitory, and an experience of unfamiliar touch can be disarming (Patterson 2007). Certain touch sensations can be brief and slip from consciousness and may leave a preconscious felt sense that evades reflective thinking. Against the backdrop of the no-touch agenda I argue that if touch is not discarded, it is possible to recapture the moments, even those that seem to settle back into the depths of consciousness, to appreciate and revalue touch as essential to life itself, to growth and maturation. In the pro-touch discourse and BMC® approach (Bainbridge Cohen 2012), we share our experiences of touch, of what it means to touch and admit its many manifestations – in trust, love, care and empathy. It meant accepting that the ontology of touch is primary to human development, communication and relationships (Bainbridge Cohen 2012; Sheets-Johnstone 2010). The significance of touch in the Touchdown Dance context helps to reinstate its value as a safe and integral aspect of human interaction when contained within an ethical, dialogical and permeable relational field (Dymoke 2000, 2014).

The common ground of embodied knowing

Much has already been said in Chapter 2 about the multi-sensory modality of CI, and the integral place of touch methods in relating to others, particularly when sight is not the dominant sense. The current aim is to revisit the four basic principles and their illustration in personal experience, relative to the two participant case studies of Tony in Part 1 and Ida in Part 2, which reiterate how CI established a common ground of embodied self-knowing that is primarily transmitted through touch (Paxton and Kilcoyne 1993). In documenting the memorable aspects of their experience, these moments share the subjective bodily sense each person brings to the work. Analysis of the available descriptive accounts elucidates the aspects that are integrated inwardly and documents a more reflexive process of interpersonal assessment in which cultural conditioning, such as the absence of touch, is revealed. To put this process in a research context, Nelson recommends both the subjective and interpersonal forms of documentation in the arts, as a process of 'research inquiry based in practice' (Nelson 2013: 6). Nelson indirectly validates the research perspective that Paxton and Kilcoyne used: by purposefully combining these forms of documentation they avoid losing sight of touch. Paradoxically, by emphasising touch experience they document its absence and so the absence of dance for visually impaired people is explained. Whereas a dry depersonalised reconstruction would render such experiential praxis 'elusive' (Nelson 2013: 5).

Herein lies the motivation for this chapter; the elusive nature of touch accompanies the lack of practice-led research on it, which has, in turn, undermined its value. Touch taboo has taken precedence and the dominant no-touch discourse has imposed its conviction to regulate touch. Inevitably, attention to such ideas has displaced the bodily sense of touch from consciousness (Caldwell 1997a/b; Classen 2005, 2010; Piper and Stronach 2007; Totton 2015). Touch praxis (as Nelson indicates) requires its own space for consideration, particularly to rectify the dearth of research into its intricacies (Johnson 2006) and to support the pro-touch discourse.

Without seeking to convince, I hope the stories will provide information to indicate the sense or priority that is required to 'prove' a point rather than create rules, to be pragmatic about the experience and what 'truth' this speaks, as an alternative 'assessment' to the risk-averse trend. I recommend to the reader to practice touching, be that yourself, or other object, as you read. It may be that even now you are surprised by what comes to mind when you consider what a touch event may entail. You may already be in dialogue with yourself as to what intention such touch may hold and dismissive of the 'pro'- or 'no'-touch interpretations that you already have in mind. Self/other touch may then 'document' your awareness of the unity of movement and touch, rather than their functional polarity, and you may bring this reflexive process of examining a direct experience to your reading of this chapter.

In this reflexive process, a continuum between 'the interior void of the soul and the exterior void of the absolute possibility' (Jones 2009: 25) is encountered. Another continuum inherent to CI praxis-led research arises between the interior subjectivity of the bodily sense, and the diversity of external objectivity encountered in touching and moving with the other. This continuum could stand in the way of finding a 'common ground or sense of knowing' – yet bodily 'knowing' was evident from each touch-based encounter, and the 'common ground' was inclusion. To assist in finding the substance of this 'knowing', I refer to the documentation I made of my personal experience and practice-led research that determined another 'common ground' of touch practice. This is aside from those already mentioned, of which Maxine Sheets-Johnstone and Bonnie Bainbridge Cohen predominate. I borrow a descriptor from Irene Dowd's chapter on the metaphors of touch, in which she captures the type of knowing about touch that is elusive; a state she names 'lucid neutrality' (Dowd 2007: 78). In my experience of touch, like Dowd's, an open, neutral presence avoids ambiguity and establishes a common ground for both people involved, for the experience to be 'lucid' and open to dialogue. I mention this now to set the tone of this chapter, to position the topic of touch in a neutral place, knowing how much the experience waivers between the extremes of the continuum and between polarities that are rarely explored elsewhere.

It is pertinent to note how different polarities arise within the three-dimensional continuum of CI praxis, such as the ideas of benign and malign intent (or interpretation) and the constructs of 'erotic' and 'non-erotic' touch. The intricacies of these are discussed in Part 2 with reference to experience and reflexive critical analysis. This process of practice-led inquiry is the means by which the content of touch experience in CI is presented, and new 'knowing' revealed.

To illustrate, I return to Paxton's epiphany and how, with the incisive use of principles that evolved, he and Kilcoyne handled the polarity of touch and vision.

Towards an ontology of touch and the other senses in CI

I briefly return to our online conversation when Paxton recalled the trip to Boston in the early 1980s, to an institution where he demonstrated CI with a blind boy.[61] They engaged in moving together and at some point became disconnected. Paxton continued to move and intuitively waited, watched by a member of staff. Later he recalls:

> I was commended by their instructor because when we got out of touch I didn't take measures to reconnect. She saw that I let him find me, and we rejoined that way. I remember the dance fondly, and it was probably an impulse toward the thoughts which became Touchdown.
>
> (Paxton 2019: n.pag.)

Like an epiphany, the experience left a sense of potential, encapsulating the possibility, that others lacking vision could respond in the same self-directed way, and through CI, move independently, free of mobility 'aids'. The boy had grasped the nature of the movement dialogue without instruction, to the surprise of the 'teacher'. He moved naturally in physical contact with Paxton – there was no refusal or resistance to touching another unknown person. In this regard, the concept of CI was acceptable, available and conducive to a young person who is otherwise restricted from such access – like the 'hatching'[62] metaphor, the blind boy was ready to follow an inner impulse to crack through the protective 'shell' imposed by the sighted world, and be out in his world, moving freely with and around another. Regarding the neutrality of touch, the nature of the boy's experience is indicated in his decision to reconnect with Paxton, to continue to move with the connection, such that touch was integral to his way of being in the world, connecting his inner and outer 'void'. Paxton's experience revealed the potential value of this approach to an underserved group in a touch taboo society. It indicated how the practice of CI could provide a common ground, where touch, a common language, established a relationship that would not otherwise have taken such a direct, interactive path. This 'knowing' was transmitted through the process of CI, a way previously unavailable to 'blind' people, as far as Paxton was aware. This may not have been the only time this had happened for the boy; however, it was a remarkable event – the ripples of a stone hitting the water started here for Paxton. Even in this brief moment the basic principles of CI changed the expected course of events and led to a life-changing decision to launch an inquiry into how, and why, CI would be of use to visually impaired people.

Touch – What the sense of touch of 'the blind' was like?

The initial idea to make CI available to visually impaired people assumed that touch was a significant aspect of their day-to-day life, in lieu of sight, and that their sense of touch was different to someone sighted (Paxton and Kilcoyne 1993). In conversation with Mike Halliley of *The Guardian* newspaper (Halliley 1991: n.pag.), Paxton puts the premise thus:

> In contact improvisation you are looking for a development of the movement in such a way that the movement itself is telling you what to do. It started as an experiment to see if there could be a non-hierarchical, improvisational form of dance based on touch rather than sight. Touchdown came out of wanting to explore that with blind people, we were very curious about what the sense of touch of the blind was like.

To reduce visually impaired people's experience of touch to a generalised 'touch of the blind' may appear a presumption, removed as it is from the sphere of experience. Paxton's dance with the blind boy indicated that there was much to learn about the nature and role of touch. The context of this curiosity was how the experience of CI as a 'non-hierarchical form' translated into agency in the touch-related world of a blind person, as indicated by the epiphany. There would be no single answer as to what this would mean in practice; to want to explore touch 'with blind people' confronted the polarities of the visually dominant world; of touch suppression and eroticisation; of the province of sight over touch, of the sighted over the blind and the polarity of touch itself – the no-touch taboo against the pro-touch view. It meant that all kinds of adjustments had to be made to address the complexity of subjective experiences, and required personal, as well as researcher-led interest, to elucidate this elusive and troubled realm.

Touchdown Dance workshops dealt with touch and movement, the language of movement in time and space based also on the *idea* that people without sight see with their hands. With Braille, each dot reveals an aspect of a wider picture, a letter and a word, from which meaning is communicated through surface-to-surface contact. Paxton and Kilcoyne translate this surface-to-surface contact to touching at the two-way permeable membranes of the body, such as the skin, which is tangible, and the kinesphere, which is intangible.[63] These 'membranes' locate perceptive spaces where mutual processes of communication take place, through the receptive sense of touch and its reciprocal movement responses. At any moment the body's state of receptivity or of readiness to respond is revealed in movement, in the quality and direction of the movement itself. Paxton and Kilcoyne studied first how touch brought awareness to the body's physiological tone, to the responsiveness of the muscular system, and secondly to its affective impact on the underlying reflexive responses upon which ease of movement depends (Bainbridge Cohen 2012). This 'bodily-sense' (Gendlin 1999: 234) introduced the intrinsic role of touch in determining the role of other perceptive senses in the dialogue between the inner self and the external world, which underlies perception, thinking and learning. The primary basis of touch and movement was foregrounded and in so doing a change in tone, motility and permeability changed the self-percept from a state of restraint and withdrawal to a sense of liberation and engagement.

Paxton and Kilcoyne placed their curiosity as to what the 'touch of the blind was like' in a learning environment in which touch of the sighted was prevalent and typically exclusive of blind people. The sighted could see the blind and experience themselves and the touch of the blind as they moved; the blind experienced touch as a way of seeing themselves and the sighted and experienced the touch of the sighted as they moved. With this complexity in mind, Paxton and Kilcoyne knew that in the visually dominant, societal realm, touch had implications beyond the individual's sense of self, and that of their partner. However, Paxton's epiphany had indicated that in the practice of CI, touch had a primary purpose, a neutrality in the reciprocal realm of relating, when removed of prescribed perceptual constructs such as intimacy, deviance and taboo. Touch in this case involved body-on-body contact in movement, as the means to create and inform the interrelationship.

The other senses of sight and hearing are less so (Merleau Ponty 1952; Montagu 1986; Schore 2003; Bainbridge Cohen 2012, 2018). CI enabled each participant to experience the primacy of touch and the right to access an effective and purposeful experience that the non-touch discourse deprived.

As the participants came from a context of social exclusion, the touch of a blind person and the touch of a sighted were implicated differently – the sighted could represent the oppressor, and the visually impaired the dependent or passive, 'oppressed', their relationship an embodiment of societal injustice. Paxton and Kilcoyne took the opportunity to understand better the implications of how touch may be useful in this situation, to level this imbalance and in particular to empower those disempowered by sight loss to move or to level down the dominance of sight for sighted people as they move and impose it on the dance (Kilcoyne 1991; Heller and Gentaz 2014). The participants presented the complexities of touch, from all its perceptual contexts and manifestations; fraught with taboo, paradox and polarities, whilst heralding a new prospect. Touch was on the one hand a 'touchy subject' (Dymoke 2000) and on the other, a place of 'neutrality', where the usual societal inequalities are questioned and mediated through the dialogue in action and later in ensuing debate (Paxton and Kilcoyne 1993; Novack 1990; Nelson 1996; Albright 2019).

The neutrality of touch, the mask of complexity

As a place of complexity and polarity, the body is also a place of neutrality, comfort and ease, where touch is primary to consciousness (Montagu 1986; Bainbridge Cohen 2012).[64] Paxton and Kilcoyne focused on the body, and on its particular intelligence; the duality of the sense of touch in the perceptual process. Touch is both sensory receptive (at a tactile receptor) and responsive – an act. Tactile information is transmitted from the receptor in the skin to a synapse in the dorsal root of the spinal nerves, then enters the spinal cord and either passes through to exit ventrally or travels up to the brain and returns, to be transmitted ventrally to targeted muscles and to trigger a movement response (Bainbridge Cohen 2012). When touching another person however, this intelligence resides in the two-way sensory-motor pathway between them; each feels their own singularity whilst simultaneously feeling the other – as an object, a partner or an extension of the self; together they create an interpersonal/inter-subjective realm that is motile and mutable, unquantifiable.

To manage the inherent complexity of touch, this intelligence of the body was harnessed in CI and the distinction between the two ways of transmission became part of the solution. Paxton and Kilcoyne worked with the sensory-motor aspects and created principles that embraced their significance; the different responses that arose from the partners became part of the 'vehicle' for learning and documented how it worked. The tactile sensory pathway conduits a touch intention which can be clarified, and the response options are also subject to factors which determine whether or not the CI experience is conducive, such as familiarity with 'non-erotic' touch, comfort in the relationship and environment, and so trust. By engaging the participants in their corporeality CI was purposeful and the process was imbued with a focus on neutrality and to demystify assumptions of difference. Touch bridged their reciprocal 'voids' with trust and mutual learning.

In our online conversation, Paxton (2019) stated his sense of trust in the human-to-human relationship of the partnered participants. This was not complacent; it indicated his trust in CI, in body-on-body contact, a process that evidently engaged 'hearts and minds' (Paxton and Kilcoyne 1993: 29) in knowing how the primacy of touch underlies trust, how trusting touch underlies empathy and empathy requires strong relational bonds. To adjusting oneself relative to this experience essentially required assuming a neutral state, a state based on mutual respect, equal status or symbiosis (Stern 1985) – a reciprocal relational realm. CI engaged people variously in a re-experience of their corporeality, in the phenomena of their multi-sensory being and in its innate ability to respond, to find its way, to inform the mind of what else is present, of how else to live (Novack 1990; Fulkerson 1996; Keogh 2018; Koteen and Smith 2008). Paxton and Kilcoyne (1993) identified the role of touch in establishing the preverbal sphere. It is useful to reiterate these touch communication methods before coming to the case of Tony, to illustrate the trajectory from the Boston boy to the discovery of 'what the sense of touch of the blind was like' (Halliley 1991: n.pag.).

Touch literacy and the four basic principles: From polarity to commonality – Applying an emergent understanding of touch-based communication

Paxton and Kilcoyne documented their inquiry into 'touch literacy' (Kilcoyne 1991: 47) and sought to elucidate their findings by identifying significant diverse aspects or commonalities. First, touch was introduced by embodying '(t)he main sensory organ [...] the skin' and with 'the focus of attention towards the internal and the kinaesthetically sensed' (Paxton and Kilcoyne 1992: 13). Through exploring tactility and sensitivity they defined the first basic principle of working with another commonality – the body.

The many attributes of CI outlined in preceding chapters were applied early on to show how increased self-awareness and an enhanced bodily sense were attained by using touch and movement as a means to 'listen' to the body, and through the skin connection, to the partner. In this case 'listen' is a metaphor for a 'multi sensorial mode of perception', implying seeing through touch and movement, and listening to tone change (Copper Albright 2019: 182). From this perceptual realm, touch reclaimed its primary role, its sense reminded and expressed through movement before being imbued with other meaning, such as thought provokes (Sheets-Johnstone 1990).

In paying attention to the senses common to all, the idea of the tactile and proprioceptive 'inner ear' replaced the perceptive 'inner eye'. The sense of 'listening' to changes in body tone was adopted as the focus, and replaced the language of 'seeing', in the context of perceiving, believing and understanding. Listening took on a multi-sensory perceptive sense that was available to both visually impaired and sighted participants to relate to in learning CI. This proved to generate vitality and empowerment in the CI experience, as in the case of Peter for example, that lack of sight had numbed.

Second, in line with the second basic principle of integrated human exchange, touch instils a need for cooperation, for negotiation and accommodation of the other. In discussing the use of touch Paxton and Kilcoyne used descriptors like 'non-erotic' and 'natural', to indicate the intended stripped back, neutral dimension of the touch relational space. In this way, the social implications of reciprocity and role change were experienced in the workshop in different 'democratic' relational configurations, to provide 'a structure [...] that can work within the parameters of concentrated, non-erotic touching' (Paxton and Kilcoyne 1993: 29) and which constituted the touch revolution documented by collaborator Karen Nelson, in the Contact Quarterly journal (Nelson 1996).

Third, touch as a tool, in line with the third basic principle of creating new attitudes towards learning, gave equality of access to an educational experience of learning the mechanics of movement in new ways. Participants discovered a previously unknown language and territory together, within an inclusive sensory realm. Hierarchical preconceptions of sight over the other senses and the polarity of sight/no sight were inevitably challenged. Given the opportunity to work unsighted, with a blindfold, the sighted had an opportunity to learn from the visually impaired partner, how non-visual forms of communication work effectively as means to 'see'. The equality of status apparent in these compensatory roles gave space to validate each person's subjective experience. This was reinforced by sharing through talking about how the bodily experience had gone for each of them. Interestingly, as I come to below, touch was not discussed so much in accounts of the benefits they experienced.

Fourth, as each partnering exercise involved a role reversal, each experienced both facilitator and mover roles. A clash in sensitivity could still arise, so all participants were encouraged to change partners for different exercises during the workshop, to experience individual differences and to learn from the diversity of touch and movement qualities. In depersonalising the CI experience touch remains a sensual and non-sexualised form of communication, and in essence, another valued commonality whilst, paradoxically, affirming the value of our different responses.

Paxton and Kilcoyne brought together visually impaired and sighted people, who are usually estranged by assumed differences, to meet in a trusting relationship, using touch and its inherent intimacy as a tool. The four principles above instilled the sense of safe non-erotic touch that served to nurture a space in which a new understanding of difference appeared. The ethos of support sustained the sense of a 'safety net', and with fear removed bit by bit, both visually impaired and sighted could participate in a collective experience. Touch became a conduit for shared experiences, for socialising (Paxton and Kilcoyne 1993), for breaking down 'veils of perception' (Harre and Gillet 1994: 42[65]) that may have otherwise undermined this research experiment. The tired and defiant no-touch discourse and touch taboo caused by touch being considered, narrowly, as a conduit for sexual deviance, and more broadly as a conduit for psychological oppression, or physical abuse (Piper and Stronach 2008; Dymoke 2014, 2019).[66]

Inevitably, issues of deprivation and neglect appeared, in the absence of touch, in the void of no-touch, as another counter-current to the no-touch agenda, as the cases of Fred and Ida will illustrate later. I come to the case of Tony, written originally by Paxton and Kilcoyne (1992), to illustrate the principles and learning experience in the process.

Case study 4: Tony

Facilitating inner listening, reciprocal learning and adjustment: The space–time orientation to self and other, the group and the social body

There is a photo in the article 'A common sense' by Paxton and Kilcoyne, printed in the *Contact Quarterly* journal in 1992. It shows one man lying on his right side, reaching his left arm out in front of his body at shoulder height. Tony kneels behind his partner's head, his arm extended over this partner's head, having traced along his arm, and his hand rests on the upper surface of his partner's reaching hand. The description of the photo is by Paxton, as follows:

> Tony, the visually impaired partner, traces the same pathway while sighted partner does spiral roll [...] No person in Touchdown goes untaught, untouched, even Ann and I. This reciprocity is informative and regularly reverses the teacher – student hierarchy. Integration, reciprocity, intimacy, physical care, are the basis for Touchdown Dance workshops.
>
> (Paxton and Kilcoyne 1992: 49)

Again pragmatic and resolved, in the last sentence Paxton preserves a set of values that their research ambition into reciprocal, intimate and physical touch embodies. Paxton reduces the value of the experience of the two movers to a metaphor for the collective context. As already mentioned, he admitted in our conversation that it was not possible to know what exactly each person experienced at any moment, but captured on camera, the photo transmits the attentive relationship that the narrative makes clear. It is then possible to accept this moment as an illustration of the values Paxton asserts with this endeavour – Tony is learning from direct experience, to be full 'of' that experience, with touch as the conduit for reciprocal, intimate and careful physical learning with a sighted partner who also learns. The sighted participant assumes a position for Tony to trace and then embody, in this case the 'spiral roll', the picture is of one step in the process. However, the sighted partner is not only serving Tony's learning need, as in receiving Tony's tracing touch he fills in a proprioceptive picture of his body schema. By slowing the movement down to be traced, the sighted partner experiences his body's position more fully. In slowing the movement down both participants explore the pattern more deeply, through inner listening – it appears pragmatic and functional, with a sense of lucid neutrality. This relational sphere is the space for a form of transmission, and of transition; it is the place for re-patterning habitual and possibly limited self-perceptions, and a sense of self. The sense of comfort and shared attention is intended to support a sense of 'ease' and clarity, to minimise the possibility of confusion (Paxton and Kilcoyne 1993). This latter is rather inevitable with sight not being available to both for verification – hence the need for focus and time, for the mind to grasp what is happening in the here and now through adjustment of both to touch as the primary perceptual tool.

In relating through touch and the skin (Principle 1) the visual perception of the world peripheral to the CI partnership became secondary (Paxton and Kilcoyne 1992, 1993). The focus away from sight lessened the amount of movement in space and placed the attention upon the detail of the sensation of movement, and upon the body itself, its inner spaces and the space close to the skin. These are the spaces and gaps which were discovered by listening and negotiating in the sphere of the moving partnership, inhabited by the partners and shared (Principle 2). As the dimensions of touch perception and communication are slow then time was taken with the transmission of information and in the approach to movement instruction. This enabled both the visually impaired and the sighted participants to find gaps in movement awareness and to discover subtlety and new qualities to fill them (Principle 3). By moving slowly, new patterns became attainable. It was discovered, from moments such as this with Tony, that the blind used their hands to trace the movement on the body of the person demonstrating the movement, and then explored it themselves – the four basic principles were covered.

In the article 'A common sense', Paxton and Kilcoyne generate insight into the experience of touch had by the visually impaired participants, to show how the active partaking in movement gave a 'direct experience in the body of the student', and illustrated the 'hands-on or body-on approach' (Paxton and Kilcoyne 1993: 14). Featured in the *Contact Quarterly* journal, the form assumed a sighted readership, so Paxton and Kilcoyne gave a glimpse into the discrete world of a workshop. I return to inquire of you, the reader if it may serve as a revelation to assume the roles of Tony and his partner. It appeared to me, as a sighted person assuming both roles, as real as Paxton's epiphany, a realisation of what the experience can bring to my attention that my conditioned mind prevents. Evidently, the root of this process is bodily initiated and had I not experienced this myself, I may not value it in the same way. As the small dance awakens the 'internal and the kinaesthetically sensed', so too, sensed experience revealed in this associative way, is relative to your, the reader's, lived experience of touch and movement in your world.

The case of Tony and his partner has given us an account that 'includes an awareness of or sense of the body' with careful, indicative use of touch (Gallagher 2008: 2). Witnessing the moment captured in a photo and describing it from personal experience of doing the same exercise is a secondary way of re-experiencing and exploring the content for perceptual insights. However, in its original form, this type of documentation is only readily accessible to sighted readers who can see the image objectively, and follow or grasp the new knowledge arising reflexivity, from a sighted narrator. For descriptive language to be accessible to visually impaired participants, Paxton and Kilcoyne discovered that the process worked better when comments were directed individually, on a 'personal' basis. The process had to be adjusted with 'methodical steps' and 'careful verbal introductions to any exercise' as 'overly general directions result in a number of different interpretations which, if not corrected, can lead to confusion and loss of confidence' (Paxton and Kilcoyne 1992: 50). The privilege of sight gives access to language and experience, and herein lies the paradoxical lack of opportunity for visually impaired people to access the same or to participate in research about the role of touch in their lives (Dymoke 2000; Heller and Gentaz 2014). This would account, at least in part, for the lack of descriptive accounts from visually impaired people about their experience of CI and of touch in particular, and the long learning process from being a participant to becoming a teacher, which Jamus and Holly share in Chapter 8.

From their reports, it becomes clear that Paxton and Kilcoyne provided an experience that focused on what the body can tell us and, in this context, focused mindfully on what touch could tell us – they experienced how this was limited particularly in the workshop when a movement is demonstrated. This led to a search for verbal descriptors to give 'clear pictures in the language' they used, to find out in practice 'how quickly the language itself flounders' (Paxton and Kilcoyne 1993: 49–50). In answer to this, they discovered the importance of 'methodical steps, careful verbal introductions to any exercise, coupled with an attempt at a casual conversational air' and by 1993, they have evolved strategies, to be 'clear and soft', to undertake 'analysis of exercises to try to determine what exactly is suitable', and to consider the 'proportions and durations of things' (Paxton and Kilcoyne 1993). In this way, 'the teaching and learning move further towards a practice-based, touch-based language of communication' (Paxton and Kilcoyne 1993: 13). This may well indicate the absence of direct discussion about touch in isolation from the other senses, as it was integral to the process as it is in day-to-day life – a neutral topic, until it touches on more complex interpersonal issues and perceptual differences – such as a sighted person focusing on 'what touch of the blind was like'.

Questions … on documenting inclusive practice

Two questions arise from this ethnography to take forward: first, is the task of describing touch visual so difficult to language? Paradoxically the intricacy of touch experience outlined earlier is barely visible externally, yet with so much information arising from the skin to be internally sensed and mediated all at once, touch information is hard to decipher and articulate consciously. It is no wonder that from infancy we learn to experience a touch more subconsciously as the other senses overlay it, placing touch secondary, however vital, and integrated in our awareness of a felt bodily sense (Montagu 1986; Autton 1987; Schore 2003; Stern 2010). With sight last, and fast, vision entices connection to the external environment and becomes predominant to our relational self-awareness in the world (Bainbridge Cohen 2012). As touch responses arrive more globally within the body, at an intuitive, preverbal or 'pre-literate' level, as Kilcoyne described it (1991: 47), would it therefore not be inevitable that touch becomes suppressed in a *literate* sight dominant culture, uncomfortable to talk about and judged as aberrant? Would this partly explain the difficulty in talking about touch, or the tendency to patronise those without sight – treating them with disregard or like a pet to be let out on a lead? The curiosity about 'what the sense of touch of the blind was like' (Paxton in Halliley 1991: n.pag.) continued like a baseline or lens through which CI was applied and touch theorised, and some answers provided disturbing insights into their touch-deprived world, as I come to after question 2.

The second question arises after a CI experience: How to capture the moments of touch in words when the sensation is fleeting and easily discarded or, to the other extreme, overwhelmingly affective? This may well be why Paxton and Kilcoyne took it upon themselves to write about touch so lucidly and didn't press it upon participants to discuss the topic – all the more so when the experience for some, touched on 'deeply felt, difficult emotions and painful social configurations which accompany the loss of a person's sight' (Paxton and Kilcoyne 1992: 50). Such psychological factors, as with Gerry and Peter, were met carefully, with direct reference to the participant, as the influence of deeply felt responses were complex and personal with implications to do with social injustice or the institutionalised prejudice of sighted people. I pause to consider accounts gathered from visually impaired people on touch, found documented in the archives, with insight into the lack of touch for the blind that was disclosed to bring these two questions together.

Mike Halliley of *The Guardian* newspaper reported on the response of Fred Raffle, who, born blind, worked successfully to become the principal of the National Mobility Centre of the Blind in Birmingham. After a workshop, he commented to Halliley that 'Many blind people are deprived in terms of touch [...] Sighted people who work with them see to it that they don't touch each other. It's seen as somehow grotesque' (1991: n.pag.). He then talks about the irony of the so-called 'caring' touch of well-intentioned strangers: 'We get handled by so many people whom we wouldn't choose to be handled by, that we have to anaesthetise ourselves to it. If you attributed a loving or caring meaning to it, you'd be completely wrong in most cases' (Halliley 1991: n.pag.).

Separately, Paxton and Kilcoyne also report on Raffle's feedback and refer to his disclosure of the situation at work, where sighted colleagues show 'paternalistic attitudes which infantilise, patronise and stigmatise'. However in the CI workshops, 'he found the politics more nearly equal [...] he was met as himself, rather than as a person with a disability' (Paxton and Kilcoyne 1992: 49). Raffle speaks openly on behalf of visually impaired people in society more generally, and gives insight again into why touch is troublesome, and why Paxton and Kilcoyne were focused on how to facilitate touch in the CI experience.

In the wisdom of what sight perceives, some sighted people appear to feel aversion for the touch of blind people – the touch of the blind would naturally be different as they touch to see, not as sighted do, who see to touch and so may touch the blind person differently. The uncomfortable truth is in the difference, in the occurrence of a level of control, or presumption, which indicates an aberrant position – and an abusive one so familiar to the disability rights and social injustice agenda.[67] The no-touch discourse misconstrued touch experience and deprived the blind of access to it, so touch of the blind was banished. Paxton and Kilcoyne wanted to reclaim touch, first demystified as a form of interpersonal relating and second, to recalibrate the social status of the blind and sighted – in a democratic and equal partnership – and put touch at the heart of that endeavour.

Other answers ... telling the tale

The value of these forms of documentation lies in their revelation, in telling of new ways of knowing self and the body; as this changes inwardly (touch perception) so does the scope for change in relation to the external world (e.g. touch as a political issue). Tony was present at this meeting and said, ‘Here you can find other people and interact with them – you cast away all your shyness’ (Halliley 1991). This indicates the success of inclusion in his interaction, and even though touch is not directly referenced, its positive affective impact is reclaimed in ‘feelings of self-empowerment, joy and satisfaction’ (Paxton and Kilcoyne 1992: 50) which affirms Fred’s political statement.

Like the vignettes in Chapter 2, the story that unfolds in the moment of the direct experience resides in the body memory of both the visually impaired and sighted participants. It can be documented by revisiting and showing the learning in movement, or by giving first-person accounts of the moments unfolding. Both are usable forms of documentation of the research that has taken place (Butler 2005; Sheets-Johnstone 2000, 2009; Nelson 2014). However, without the four principles to identify touch as a common sense, a means to cooperate, a tool for learning movement and requiring reciprocal exchange, the significance of touch within the interaction is easily lost, or discarded, due to self-censorship, fear or deprivation. Paxton and Kilcoyne noted in the description of the photo that ‘no-one goes untaught, untouched’ to reiterate the unity of touch and learning for all (Paxton and Kilcoyne 1992: 49) whilst reducing the previous polarities to a continuum.

This was not glibly done. The CI experience, in its new role as a pedagogic approach, required a process of adjustment between using touch and setting vision aside, for the interaction between visually impaired and sighted people to work. The accounts provide a space to return to, in which this potential can grow further. It is hoped that the reader obtains insight and inspiration to practice CI, using the above principles, to start practising the skills of inner listening and pre-attunement to the world through touch. Then to attend to what comes to mind that is new, or affirming of the already familiar, to allow the possibility of transmission into the world beyond the page. Whether visually impaired and hearing this read, or sighted reading yourself, you may rethink the next time you are in physical contact with another and how you may ensure the touch relationship is mediated by consideration of the conditions in which to undertake to do so – i.e. far removed from the polarities imposed by society and in tune with listening to the neutrality of touch and to a neutral social status.

The reciprocal process of adjustment away from sight and towards touch continues as the topic to revisit, and again, paradoxically, the significance of touch still tends to be invisible and disregarded, indicating its low status in regard to vision.

Touch – The paradoxes and polarities

This is a commentary on how CI kept touch safe and neutral, by displacing 'erotic' and 'non-erotic' touch and finding a new language.

To counter the invisibility of touch neutrality, Paxton and Kilcoyne presented touch as 'natural' and supportive of wellbeing, as another way of perceiving and so 'seeing' self. This has been discussed from the point of view of the experience, and clarified with reference to societal taboo, in the hope to encourage further reflection and 'careful' consideration of the value of touch in the world of those without sight and of sighted society more generally (Kilcoyne 1991). For example, the magnitude of the polarised perceptions of safe and unsafe touch, of 'erotic' and 'non-erotic touch', was made explicit in the documentation (Paxton and Kilcoyne 1992, 1993). These labels placed touch firmly on the map in the research, as topics they were keen to open up on paper, or in reflective dialogue, interlinked with the polarity of touch and vision. Eroticism is inherent to touch experience but complex and problematic once it appears, alarming and destabilising for some, and so not a topic to address directly in the workshops when ease was the focus and touch the tool to achieve this bodily sense. The act of touching to support ease and relaxation was introduced pragmatically in the face massage technique, and brushing to awaken the skin, but the topic of erotic touch or non-touch, was rarely a conversation had with participants and rarely presented a dichotomy in terms of the work moving forwards safely; it was addressed discretely when needed as the case of Bonnie illustrated in Chapter 2 (Kilcoyne 1991). The shadow of 'erotic' touch is present in the term 'non-erotic touch' therefore not using either term with touch disables this polarity from getting in the way of the focus at hand. However later, in the context of working with vulnerable adults, young people and children, this issue is magnified by safeguarding concerns, making explicit the focus on neutrality and careful oversight of all partnering, to abide by risk assessments, as indicated in Chapter 5.

Touch and vision

Touch in all its layered complexity is apparent and further coverage of touch here serves to bear witness to the continuum of touch, and its role experientially in personal growth, development and transformation, for all. Paxton's intention to create an experience in which sight was secondary, succeeded and the polarity of touch and vision merged into a duality, as was inherent to the deliberate pre-tuning of the sighted dancers towards touch and away from vision. In this process the benefit of touch would continue to validate the work and have wide social significance, as has been apparent from Paxton's epiphany, through to the early participant accounts of Peter, Gerry, Bonnie and Eleanor. In essence, the accounts express the rediscovery of the 'lost' movement, a reclamation of a sense of self-worth and belonging, a sense of active self-nurturance and a rebuilding of self-confidence. It is possible to say that CI supported sighted participants in similar ways and the added benefit felt by all came from the sense of commonality – through the touch and movement experience. Whether visually impaired or sighted, the CI experience was conducive to change, was more or less therapeutic, and with the attention to bodily states, the impact was felt as a whole, by all. It appears that participants felt more at ease, more engaged in self-movement, with more agility and agency in the world around them.

And so I ask, as I did at the time of doing, what about the role of touch, specifically, in this outcome? The modes of touch (like face massage and limb manipulation) and the exercises involving touch have provided points of reference for further application and discussion of the guiding principles and of how to touch, to remain safe, with lucid neutrality. Alongside the embodied accounts of 'touch' experiences that shape our sense of knowing about touch, more detailed accounts of touch will enable its presence in the world, so we can really grapple with it carefully and respectfully. Neutrality is one thing but ambiguity is another, which lack of 'knowing' how to distinguish between the sensory and motor aspects led to introducing greater clarity in the touch methods – as I explore in Part 2.

How motion meets emotion – The effect of affective touch on the state of lucid neutrality

In the practice of CI, touch is consciously adopted as a means to communicate when sight is absent, even partially. In the Touchdown Dance context, this is non-negotiable; touch is a mechanism key to CI praxis and pedagogy. CI validates the fact that touch and movement are inseparably realised, as human development indicates (Sheets-Johnstone 1990; Bainbridge Cohen 2012) and the complexity of touching manifests in the movement that enables touching to happen; if hesitant and unclear this mind state manifests in the lack of clarity, willingness or availability of movement, and so in the touch quality it accompanies. In CI, the physical tasks of doing the movement are experienced in the relationship with another person, and so relate, essentially, to the emotional aspects wrapped up in that challenge. When two people move in CI, their experience of touch and movement may concur at the membrane in which they move, or not (as I encounter with Ida below). CI is the method by which the experience of two people moving is inclusive. When not, such questions arise and require negotiation to provide answers that will then permeate the dialogue and seek attention for learning and understanding to take place and with affect on the sense of neutrality in some cases.

How touch neutrality impacts the duality of touch and vision

Experientially the inherent duality of touch and vision arose in teaching CI when touch was experienced as benign, safe and inclusive. This allowed other positives to be effectively perceived and shaped new understanding that wards off malevolence, fear and assumed 'deviance'. The duality of vision and touch serves greater potential and meaning than their polarity, since what we see or think remains deceptive until we touch and experience to perceive and understand that world (Sheets-Johnstone 1990, 2010). The touch of the blind is variant, a way of seeing the world proximally, with movement as a means to calibrate the landscape and 'see' it (Heller and Gentaz 2014). Likewise, the touch of sighted people is variant, but sighted people discover the duality of touch and vision, and can share this in their dialogue with the blind. The societal polarity of touch and vision becomes suspect, as the dominance of vision that subjugates touch to taboo status and regulation is reversed when the outcome of no-touch (once lauded as a means to protect all from abuse) is witnessed as a cause of deprivation for many, imposing a continuum of an unmediated void.

In the Touchdown Dance research, the polarity of touch and vision has been mediated by the specifics of touch neutrality that became pertinent to the CI experience; specifically regarding the haptic perception of weight and density from the lines of force that penetrate the body at the point of contact with another. Heller and Gentaz (2014) concur that the haptic sense relies on the duality of vision and touch, or, for blind people, on the optic aspect of touch, that operate together in the calibration of shape, time and space. Without vision, the density of the physical touch connection can substitute loss or lack of sight, and give a physical purpose to touch that is functional and mechanical and serves the CI experience. Touch then has a neutral intention; to make something happen and to create movement from receiving movement – and displaces the taboo. How, is the next.

Touch density– From the polarity to the duality of vision and touch

I begin by referring back to the experience Paxton had with the blind boy in Boston. This vivid memory in which touch became the common denominator for the dyad equates touch with the usual rank of sight. This also lent a physical purpose to touch that is functional and mechanical, serving the dance purposefully. Touch then had an intention with the dyad relationship – to make something happen, to realise a desire, to create movement from receiving movement, as sight would. In displacing sight, touch is normalised; denuded of its alarming associations it slips into a natural undifferentiated position in the experiential realm and displaces its taboo. Paxton wrote that when he and the boy lost contact he let the boy find him rather than actively dominate and reconnect through sight. This had been a seminal moment in which the blind boy was empowered to follow his desire and was able to self-direct his experience. Paxton adjusted down the visual sense to open up the possibility for other sensory modalities and pathways, for preconscious responses that later are realised consciously to have occurred, as in the conversation that followed with staff, and our discussions on this topic. Touch eventually beckoned attention, there was literally more to it than met the eye: a superficial sense of connection and a deeper sense of communication, affect and 'recalibration' had to be realised first.

Paxton wrote in an e-mail discussion in 2010 how he perceived the dyad in CI as a mutually affective way of being, as fundamental as being 'at one' with an aspect of the environment. Rather than in a dependent relationship, both were 'part of the same system', one in which 'desire' causes the interaction and determines the focus of attention. This 'system'

> recalibrates subject and object, noting the dyad mechanism and collapsing the dyad into an interdependent union in a process happening pre-consciously.
> A student may be dancing in the space. One may wish to put oneself into the space as well. Both will be aware of this change, although no exchange or communication is required.
>
> (Paxton 2010: n.pag.)

On this basis, Paxton and Kilcoyne established a 'system' within which to nurture conscious learning. They had to stand back and take a look, to define building blocks that evolved over the first years of workshops, and the practice of CI became enhanced with new modalities. Further discovery of a wider range of responses and levels of interaction, followed by the flow of movement in the dyad, later described as the 'juice' of CI practice, potentially including both readings of 'fuel' and 'fun' (Paxton and Kilcoyne 1992: 49). It involved playing with the 'optic' sense of sight and touch, with separating the senses in movement modalities, working with milliseconds of the preconscious moments, the sensing and feeling undercurrents in the moment of responding in movement.

In seeing the mutual 'union' of the participants, Paxton and Kilcoyne's extensive experience brought awareness to touch as part of this mutual relationship, and responded with touch modalities, which I come to here, the realisation that practising CI provided 'an education in touch "density"' (Paxton and Kilcoyne 1992: 49). The variant intensity of the physical connection between the dyad indicated a choice of response, determined by negotiation or desire, both contingent on the relational sphere/the environment and also on the teaching. The degree of density refers to the tone of the touch connection, the amount of force being passed through from one to the other, the amount of weight and the intended direction of the movement flow. As an 'education' it required being attentive to the immediate physical relationship at the point of contact, to the dynamic of the touch-based movement dialogue, rather than to what is happening in the wider field. This was also referred to as the third centre (in addition to the two 'centres of gravity', one within each partner) around which or through which movement densities would pass. From this point the choices were to converge with the density and roll around the surface of each other's body (as with the forehead to forehead rolling contact exercise),[68] or for one to move under the density for the other to flow over the top, or vice versa. The other choice is for one to gather the other partner into a lift and use the momentum of the movement to travel upwards and into space. This is an example of when 'Contact Improvisation for sighted people involves "pre-tuning" the senses from the normal sighted dominance to the sense of touch' and when in a class with visually impaired people, 'this occurs very consciously because of empathy with a blind partner' (Paxton and Kilcoyne 1992: 49). Paxton and Kilcoyne deferred to the sense of touch and to the literacy of tactility so that the sighted people learned to be less dependent on seeing their partner's body. In learning about touch density the sighted movers had to adjust to moving in an unfamiliar way, without vision, and the visually impaired movers adjusted to an unfamiliar world of touch density and movement by simply stepping into the CI experience. Touch density was part of the language of adjustment, and 'tuning', which re-established touch in the experiential realm in which it started in childhood – that is, in its role in establishing the 'primary sense modality for mother/baby bonding' before we learned to speak (Kilcoyne 1991: 47). The density of touch was a point of focus for both dancers' learning how to move together, which reduced the polarity of having and not having sight – but with implications.

With the blind boy Paxton had adjusted down his use of sight, but over longer periods this conscious adjustment process to suppress visual referencing was not always easy. Paxton and Kilcoyne highlight this plight for those with sight:

> It can be an intense experience to assume a state of mind in which the visual world plays no part at all in the reference systems. This shock is a most salutary one, giving a high contrast to the differences between one sense, vision, and the other, touch. (1992: 49)

In the photo of Tony and his partner, eyesight is not evident at this moment of perceptive exchange. Both movers use touch to access the inner realm of practising CI together – to explore a proximal and less spacious relationship. As the dance expands externally Paxton noticed that for the sighted 'concentration [...] will waver between touch and vision'. At times wavering towards sight was useful to the experience, as vision informed the sighted mover of their shared spatial orientation; the availability of vision also supported times of disorientation that could then be rectified. Sight provided reassurance first to the sighted as their usual means to make adjustments relative to space, and then to both as it enabled them, reciprocally, to become more daring. The mutual benefit of sight was acknowledged by both partners as different to the familiar-sighted guide role; in this context, they both become freer in the space with the assurance that it provides.[69] With this mutually supportive agreement, the visually impaired dancer could focus on the 'density' of touch in the moments of contact, which in turn became grounding and supportive for the sighted. The sighted partner had to accommodate touch and the visually impaired partner had to accommodate vision, to be true to the '50/50 balance' ground rule (Kilcoyne 1989: 18).

The 'shock' for the sighted, of not being able to use vision in the usual way was, in my recollection, relative to how easily the sighted person perceived the necessary barter that was required to establish the 'democratic human exchange' in the dance – indeed the equal relationship was an essential aspect of the experience but took time to establish (Paxton and Kilcoyne 1993: 29). Human developmental research has shown that sight is instrumental from an early age to establish relationships and resonance with others through face-to-face expression and recognition (Schore 1994, 2003a). The absence of sight requires a significant adjustment, as in this case where a sighted person tries to relate to a visually impaired person who can't respond in this way. For some sighted people, empathic resonance through touch and hearing, movement and voice, was not readily available and was not as primary as seeing, at least initially. If so, it took time for the partners to find how to establish the terms of their '50/50' relationship, before any gains became evident.

If the visually impaired person was unfamiliar with the space, and being blindfolded was also temporarily an unknown aspect for the sighted, then each felt reassured or doubly concerned by the fact that sensory-motor challenges were to be had by all. With adapting to the 'shock' of not seeing the space, the sighted attuned to non-visual forms of communication (i.e. touch) and were therefore more informed about how else to be in the unsighted realm, that is *more* in the sensory and *less* in the 'doing' and concerned with aesthetics, so more balanced and considered in their partnering role. If the visually impaired person complied with touch to support their movement and spatial orientation, then the CI experience worked well, the aspects of touch relative to the principles all contributing to a 'rounded' experience.

Regardless of these exceptions, touch was instrumental in establishing equality; whilst the visually impaired participants were unable to perceive the whole body of their partner beyond the point of contact, they became more competent in reading this through the density of contact and positioning. They became more adept at allowing for the unexpected, to take decisions and to relate more confidently through movement and touch. They became less fearful of connecting to sighted people and were more trustful in themselves of their own ability to engage and disengage with touch-based contact. For both, this adjustment process enabled them to discover more about how to engage physically and mechanically in movement with the different energy states that are presented, how to respond to different qualities and spatial orientations, how to transition in space through different levels and how to move in ways that would not be possible without each other, as the next case study illustrates.

PART 2

In search of a natural attitude towards touch

Introduction – The polarised attitude to touch conceals its absence – The paradox

Over many years of practice, I have explored the problematized polarity of 'non-erotic' and 'erotic' touch, in which the potential scope of a 'natural attitude'[70] towards touch experience (Sheets-Johnstone 2011: 41) is overshadowed by attitudes of deviant, sexualised or eroticised touch (Dymoke 2000, 2014, 2017, 2019). If, as Constance Classen discusses from a historico-cultural perspective (Classen 2005, 2012), touch is further perceived as malign in intent, then malevolence remains alarmingly 'visible', rendered obdurate in the dominant societal diktat on touch taboo and enshrined in the dogma of non-touch policies (Manning 2007; Piper and Stronach 2008). In a risk-averse culture and 'no-touch' alert environments, like a school or health care setting, both in which I have worked, interpersonal touch is cautioned, perceived as potentially deviant, associated with primal, feral or 'untamed' behaviour, negatively sexualised, or interpreted as a sign of an inappropriate attachment (Dymoke 2017, 2019). Evidently, in these settings, touch is regulated with the view to safeguard service users and protect staff, but at a price (Piper and Stronach 2008; Totton 2015; Dymoke 2014, 2019). As I witnessed, in practice, teachers and care staff may either (ab)use the no-touch policy to justify not touching when it would be beneficial, or, conversely, use touch with an awareness that touching is a necessary form of human interaction and ethical (Dymoke 2017, 2019). Where both of these positions are defensible and justified, both can also be abused, and from this polarity, conflict, suspicion and confusion arise, preventing open reflection and discovery of effective, benevolent and 'neutral' touch.[71] Touchdown has operated with this disparity, using CI as a field in which these polarities are 'suspended' to allow for 'free variation' (Sheets-Johnstone 2011: 41).

Evocative language matters and instils attitudes of resistance and distance in human relational situations (Dymoke 2017, 2019). For example, I cite the word erotic and its assumed opposite in the term 'non-erotic', in which the reductive and invariant perception of sexualised touch is perpetuated over other variations. The possibility of a primary, preconscious and dialogical attitude towards touch experience is to be defended here, to highlight its primacy to interpersonal relationships and to admit the variation and diversity of subjective experience. By its very nature, touch requires mindful, cautious conditions – but not the kind that removes its essential human purpose, or reduces it to a metaphor, like a ball batted about between competing discourses that discount reference to practice-based, experientially informed frames of reference.

The human purpose of touch is evident when safety, security, survival and pleasure are considered (Tuan 2005; Patterson 2007; Classen 2012; Bainbridge Cohen 2019) or the significance of touch to self-regulation, and so wellbeing (Schore 2003a; Gallace 2012; Totton 2015). To deny touch is also to deny how dependent the other senses are on touch (Patterson 2007; Heller and Gentaz 2014) to enable communication, interpersonal relationships and inevitably, within the continuum of human existence, social cohesion (Manning 2007; Patterson 2007). These aspects can inform the perceived realities of benevolent or malevolent behaviour; indeed case examples indicate how informed use of touch can provide answers to the problems it appears to create (Dymoke 2000b, 2021). For example, when touch is implicated in physical abuse, it is the abusive intent that wounds and maligns touch. When used with healing intent, touch is instrumental to recovery, not only of healing the scars but of an ability to relate, to trust 'other' – at the individual and societal level. Evidently, benevolent touch practice needs profiling, to validate concepts such as 'lucid neutrality' (Dowd 2007: 78) and others I come to later, for greater awareness of these methods of implementation and to create a more convincing frame of reference.

The Touchdown Dance story provides examples of this in action. The next chapter section presents insights that informed further research into the phenomenology of touch and informed the application of relational touch to CI. The aim is to stimulate further debate on the intricacies of touch, to find traits that can substantiate the pro-touch discourse and eradicate the need for 'non-erotic' as a descriptor.

Touch, reflexivity and the recalibration of the subject–object divide

To begin, I return to Paxton's e-mail comment, that the dyad in CI 'recalibrates subject and object' into an 'interdependent union in a process happening preconsciously' (Paxton 2010: n.pag.). Paxton contributes to the current philosophical and more metaphysical discourses on touch as inherent to CI, and more broadly to the experiential process of preconscious inter-relating. Touch underlies the relational bond that is established in this recalibration process, as has been already discussed to flesh out the fundamentals of Touchdown Dance's pedagogic intention and praxis. The 'interdependent union' not only recalibrates the sense of a unitary self but also the subject–object societal 'split' separating visually impaired and sighted people. As a benevolent interactive art-sport, CI provides a space for touch-based experiences, and a source of language and meaning, to permeate the membranes between the subjective, internalised preconscious states of awareness and the objective, externalised, conscious states of self-other awareness.

Paxton took his observation into further discussion with professionals in other fields where this recalibration was also happening. As already referenced[72] he discussed the primacy of touch in human development with Daniel Stern (a psychoanalyst who specialised in infant development). This affirmed the comparable nature of touch experienced in CI to the symbiotic relationship of mother and infant in which the infants' bodily sense evolves from their preconscious, touch-based 'interdependent union' (Paxton 2010: n.pag.).

Equally informative, Bainbridge Cohen positions this primary and preconscious relationship as foundational to learning and establishing a sense of self: 'How we are touched and moved as infants as well as our "random movements" underlie our perception of self, our behaviour and how we focus on what we will perceive in the future' (Bainbridge Cohen 1994: 8).

This primary relationship is therefore lucrative. It is an investment and the possibility of emulating this experience later on in life, a gift since this is not the only time such a primary relational event may source a renewal of self. If preconscious experience enables self-discovery it may also indicate a change, a new path ahead, since '(a)ctivity needs to occur before it can be perceived. What we perceive is what we then reproduce or control' (Bainbridge Cohen 1994: 8). Perceptual change is an ongoing possibility, if the environment allows for movement and touch, for the mutable membrane to be reciprocal and operative.

In the case of CI, the inter-subjectivity of interpersonal touch lends itself to practice-led inquiry and experiential learning, in which meaning manifests in 'essence', recalibrates previous perceptions and establishes new ones. CI practice cannot, however, be totally unblemished by a priori theoretical constructs that an alternative discourse would bestow, nor can subjectivity evade bias or preference.

The purpose of this ongoing inquiry is to revive memories and speak to them, to open up the meanings that lay hidden or were hard to perceive at the time (in the mid 1990s) as touch neutrality was still new territory. In this process, other pragmatic positions arise and bring familiar conditions with them that may be worthy of further reflection, or challenge the existing ontology to change.

This discursive process and its incumbent subject/object perceptual challenges have also been addressed in the field of consciousness studies, using methods from phenomenology to evolve self-reflective techniques to use in the moment of experiencing. Reflexivity involves introspection and the suspension of interpretation during an experience so that by distancing our awareness from the already known, we become aware of new meaning, of new ways of being – a form of reducing perceived reality to allow for the possibility of others – for the possibility of change. If suspension of judgement begins the *hermeneutic turn* (Depraz 1999), a cycle of embodied perception and cognitive awareness follows, in which the first understanding assigned to an experience is bracketed, to give space for new aspects to be perceived. This continuous cycle may repeat itself many times when contemplating an object in the here and now, and so new layers of insight may arise over time in which the process, also referred to as the 'phenomenological reduction', displaces logic with new transcendental readings, or 'knowing':[73] Reflexivity is useful in situations of continued, open, reflection; in this continuum the perceiving subject is freed from old perceptions as new dimensions emerge – reflexivity is a process of re-tuning attention, a new form of objectification (Depraz 1999: 97).

Bainbridge Cohen (1994: 8)[74] takes this further in the BMC® approach to experiential anatomy and human development. Consciousness is separated into two interdependent realms: 'nervous system' consciousness and 'cellular' consciousness. Like the hermeneutic turn, the former is suspended for other possibilities to arise in moments of 'cellular' listening:

> (The nervous system) records and stores patterns of movement and behavior, and once stored, it can retrieve them and thereby control experience through habit, memory and projection. The nervous system records present experience, integrates it with other previously stored experience, interprets it within the context of its history, holds it in memory and projects the coordination of the whole into the future as expectation.
>
> For new experience to be obtained, release of the nervous system control is necessary so that old habits are no longer directing the outcome and new cellular experience can come into being. By cellular experience, I mean initiation and awareness directed locally by the cells. Cells learn new experience and tap potential patterns of behavior stored in themselves and in the nervous system. This dialogue between present cellular and past nervous system experience is, for me, learning.
>
> (Bainbridge Cohen 2012: 9)

In this process, the imagination creates bodily sense 'directly' from the 'cellular' level of the experience, giving value to the subjective over the control of an objective truth, such as a given or learned belief. In the realm of the preconscious–conscious continuum, intercellular communication enables again their 'interdependent union' (Paxton 2010: n.pag.).

The convergence of these three approaches is evident and subjectively experienced when I met the deafblind drama group in Denmark. In the face of this 'new' experience, my visual and auditory systems of communication were unavailable and set aside. A new, essentially intuitive way of experiencing CI emerged – reduced in essence to touch and movement languages. This 'natural attitude' became stored in my memory to shape the next 30 years of practice-led research into the 'density' of touch and its role in dance teaching and psychotherapy. The 'cellular' nature of this touch-based experience was the foundation for a subsequent, unexpected exchange with a deafblind individual I call Ida, and contributes valuable material to this epistemology. I return reflexively to the practice of CI, to an embodied experience of touch, where new insights into touch remained hidden from view, in the cells. Deeply moved at the time, it needed time, and distance, for the inherent attributes and profound complexities of her story to reduce, for neglect to emerge in essence and be brought to mind through repeated reflection. The membrane emerged in this relational field to enable communication across the socially imposed divide, the distance caused by disunity, to enable emergent felt sense to permeate, be identified and admitted, to unify the self-experience and its emergent meaning (its essence) through reflection.

The case of Ida – The lucid neutrality of touch

This case, involving myself and a deafblind young person I name Ida, took me into the realm of the lucid neutrality of touch. Our open interaction is relocated to the present, to track it moment by moment. This experience was seminal, furthering but different to the experience of working with deafblind adults the year previously in Denmark. Like Paxton, I had an opportunity to explore CI in a one-to-one session, which became a touchstone for the later direction I took into BMC® and Dance Movement Psychotherapy. I came to understand how Paxton and Kilcoyne invested eight years in touch research and CI. Ida teaches me patience and resolve and how to show love and care through primary empathy, without infantilising her. I became aware of how elusive touch becomes and how much more there is to learn from the unique relational conditions that diverse subjectivities bring to the experience. Ida brings hidden complexities that affect how the subject/object unity is conditioned. We had no shared verbal language, only each other's presence, unknown to each other, subject to the limitations of our prior experience, unaware of subjectivities that may not be realised due to 'blindness' caused by lack of time and access to these differences.

In our encounter, Ida's responses illustrate the significance of the four basic principles outlined in Chapter 2, and of the role of touch density in the experience of CI as Paxton taught it. In this recollection, I consider that what I bring to the dyad pre-conditions the experience for Ida. Only by reflecting hard, using the introspective analytical tools of the hermeneutic turn (Depraz 1999), and consciousness theory from BMC®, may I discover, in her absence, some possible sense to what Ida experienced bodily, preconsciously, or cellularly. In order to acknowledge the 'conditions' that Ida brought to me, I allow for the possibility that for her, our 'dance' was a meeting with an unknown 'teacher', a totally new experience. I also admit the pragmatic purpose of this case is to examine the elusive aspects of touch communication rather than our cultural differences. These are the 'conditions' referred to above which I attempt to explore by setting other beliefs to the side to give space to the unique and extraordinary aspects of this experience that have changed me and become fundamental to the work.

Ida

My meeting with Ida takes place in Iceland, at a workshop that was organised around the tour of the production *Poetry in Motion* (1990–91). The tour from Denmark to Nordic countries included community workshops delivered to the disabled community in each country. Ida is one of the adult participants; she has an assistant, who is also her interpreter and guide. Their sign language is very minimal with mostly finger spelling and some words, and the carer speaks English: 'Ida is just starting sign language.' They come to join in the workshop but Ida wants to 'watch'. After the workshop ends I ask if Ida wants to try moving with me, and Ida consents without much apparent enthusiasm; I offer a taste, just a few minutes. I stand by her left side and her carer stands back. I believe that Ida knows I am there; she waits quietly still facing forward. I place the top of my right hand gently into the palm of her left hand, hanging by her side. Ida takes her hand off quite quickly but stays where she is. I do the same gentle initiation again and again, and she removes her hand, after a moment, without any sign of distress. It takes several times before we remain connected long enough for me to give a subtle impulse to lift her hand up and flex the arm at the elbow, to move in a forward direction. She takes her hand off again. This continues a few times until she remains in contact and we take one pace forward and then take her hand off, and stand still, quite 'blank' and silent.

Each contact attempt is light in density and consistent in intention so as to not confuse or meddle. Ida eventually understands, takes one step and stops, as if back at 'base' before the next impulse leads to the next step, successfully taken and landed. By dint of taking time and listening through our connection, to the cells local to the point of contact, I attune to her gait, and she decides to move forward together.

We continued and eventually walked further into the space a few steps. Ida remained engaged and did not seek her guide. I changed sides and after a while came to a halt. This was clearly a new experience and I felt this was enough. I recall the intensity of this experience, my anticipation, feeling the sense of the 'small dance' the pull of gravity, the rising away forwards into a step and returning to 'base' by bringing both feet together.

I wrote the following introspective account of these moments:

> The bodily-sense is the starting point, the territory I return to, to recall our encounter in which the pre-set conditions of the CI dyad arise and shape our meeting. I have glimpses of touching as I have often experienced it in my practice. I do so to reiterate the in-flow – out-flow pathways and to capture aspects that pass quickly but none-the-less vibrantly when recollected in text. It is a surprise how much is enfolded within a few moments, and how this places a pre-set condition upon my perception of the moments of touch density shared by Ida and myself, that follow.
>
> Our skin is a continuum, the surface of the body that is touched, the place of transmission, where initiation into movement flow arrives and passes on its pathway into the body or out into space.
>
> It is at the place of contact that the perception of flow arrives and passes through, erasing the sense of touch intention so that the trace of the touch may linger but is washed away with the on-going kinaesthetic sensation.
>
> Once moving with maintained contact, the point of contact becomes the point of transmission back and forth, like a centre through which movement flows or around which our two bodies move.
>
> The experience of motility, the sense of constant modulation or change, is monitored at the membrane we share through the tactile sense, so the two senses feel like concurrent streams, or, possibly and less comfortably, like streams in counter-flow.
>
> I have no visual memory of this moment, of either Ida or myself, but an imprint in my hand, my memory, in the cells that sensed our movement, in the conscious sense of the connection, of the usually subconscious realm of being that vision overrides with the sensation of seeing what is happening,
>
> Touch appears to be significant in the landscape of movement, in the direction of energy 'traffic', in the generation of affective responses, and in the dynamic shaping of the inter-relationship between self/self and self/other. In this case I re-appropriate the connection of the inner landscape with the outer; in the meeting with the unknown, the unfamiliar. I note that CI is a place of potential that is

> nurtured through relationship, a means to explore and learn about the deepest internal aspects of self, and to discover the infinite possibilities of touch hues, qualities, encounters and relational responses in practice. These aspects have potential, as intricacies of touch, and through self-reflection at the surface, at the point of touch, their assessment can be revealing, if inevitably subjective. There are times, as in the case of Ida, where I am aware of bringing too much of myself to the case discussion – and have to step back to return to the memory – to allow the experience to reappear as the substance to explore, so that the touch aspect, the issue of touch density, does not get lost or mystify what in essence took place.

During our moments of connecting, I looked at Ida's carer from time to time and notice her intently watching and tearful. When we eventually finish she joins us and confirms with Ida that the time is up and if she liked it. Ida signs yes, still very little emotion comes. Hearing my feedback – I relate that I didn't want to push it further, unprompted she tells me a small piece of Ida's story. Ida stands linked to her carer's arm but is not part of the discussion. I hear that this is the start of a new life for Ida, coming out into social situations. She has spent most of her life enclosed in a cupboard under the stairs, her parents assuming her 'demented'. I reflect, had she found refuge there in some way? But I feel the sense of restriction, imprisonment and control. In this new life, Ida was discovering a way into the wider world, of her feelings.

I thanked the carer and having had no idea of this when we started, it made sense of Ida's hesitancy to accompany me, each step seemed a singular event, far enough at first. I am happy not to have persisted and not to have given up. In my perception, we had gone from being two people connected at the hand to two people moving together in an interconnected unit, guided by my past experience and present intent to allow Ida to respond, for movement to find a pathway, for the flow to find form. There was an emergent sense of our preconscious attentive state, and of an awakening to a shift in cellular consciousness, with the in-flow of sensory stimulus from the meeting of our external 'membranes' whilst creating a 'field of operation' in which our inner selves also met. Tentatively sensing each other at the point of contact, this stimulus was initially spent in the pause of non-response, but over time the potential path of communication opened from cell to cell, registering a positive relational experience in 'nervous system consciousness', the one we then repeated (Bainbridge Cohen 2012: 9).

Hindsight – The affective–reflective process

My initial interpretation of this meeting was affective as I felt deeply touched by Ida's story. I have since taken much from the memory by revisiting it using the three approaches mentioned earlier. I have also allowed myself to suspend all of these thoughts and to dwell in the memory of the bodily experience, the place of interconnection, noticing what arises when I recall being sensed and felt, mutually met, by Ida. I then reviewed that felt sense with reference to the four basic principles, the concepts of touch density, to find more of Ida in the analysis. In the absence of Ida in this process, her presence in this epistemology is a means for me to process and integrate this memory into words.

Having connected hand to hand, the gentle initiation I gave into movement was a suggestion, implementing the first principle listed earlier in this chapter, of meeting through touch and movement and then being receptive to other sensory occurrences. Little by little, through quiet negotiation, and by remaining open to 'yes' and 'no' signals, I work with Principle 2 – to find a reciprocal relationship. Ida found a way to trust that we could move together, and as we did it became difficult to notice who was then leading whom; I allowed myself to follow her moving hand intermittently whilst sensing that she followed mine. As in Principle 3, I honoured the sense of a shared sphere of the unknown, an attitude that gives space for difference. Our movement really opened up with the contact of our hands, and then I gently tried to connect a little further from her arm to her whole being. The space and the time were not a dark empty void, but one of light and awakening to potential. We did not undertake a role reversal, as in Principle 4, at this point, but I felt a mutual sense of acceptance, even if momentarily, and once established the other three principles continued to interweave. I explore these in more detail with more descriptive reflection and content, in revisiting there is inevitable repetition but further variation in essence emerges and value given to the membrane construct as the mechanism I add to the mix. In this encounter the initial point of contact of the hands transmitted both energy and engagement of the mind – and we both received aspects of each other – as if traces transmitted both ways at the two-way membrane and generated a sense of relationship that remained in the body. For the first time Ida accepted moving with a stranger, into space, trusting and responsive.

Touch Principles 1 and 2 – Acceptance, negotiation and communication

I felt open to explore, but when our hands met so briefly I had to suspend any pre-expectation of dancing, to meet Ida openly and authentically. At the moment, I reconnect and wait to 'see' how Ida responds. I do not know if Ida responds differently to any other person she has not met before. In this instance, gently resuming hand-to-hand contact works; it communicates 'yes' from me to Ida who responds with a brief yes, unaccustomed as she is to further dialogue. I do not feel the disengagement as a 'no', it was rather like looking away or attending to something else.

Due to the lack of sound and sight, the tone of our meeting is established through the quality or 'density' of our touch connection and the movement that Ida brings to it. Initially very light and brief, I take each moment at a time, to establish a sense of 'unconditional' expectation, I stay with her awaiting her response to initiate mine. The gentle signal I give is clear in intention, 'I am here, with you', and Ida responds 'OK' so the connection is made. I am aware that touch is an aspect of deaf sign language spoken into the hands, but one she is learning. After a few attempts to shape my hand and remain under hers, the slow touch impulse I use indicates to move forward, intending that we may take a step. We don't move forward and she takes her hand away. We resume contact; I move a little further and take a step; she follows attentive and follows the contact of the hands. Her touch quality is tentative and instils a sense of careful attention, which I meet gently and leave space for her response. I feel that in time, with ongoing non-insistence, she will find herself moving further into a relationship.

We attune to each other as we move and the transition from passivity to motility indicates how the senses of touch and movement are affective at any age, and this affect can transport the body into expression and into relationship, from the micro to the macro, from the motility of the cells in the skin to the whole body moving in space. At each contact Ida experiences in-flow and out-flow into movement, each occurrence motivates a voluntary response, and she meets herself potential, her own agency in this reciprocal cycle.

This way of meeting, through her available senses, defines a positive experience and positive affirmation, which her care worker confirms. I assume as we progress that Ida feels met, even if partially, in each accruing moment of our movement relationship. Her demeanour is different at the end, softer and more attentive. Considering the sensory deprivation that Ida experienced growing up, this meeting felt like a moment of acceptance and self-discovery.[75]

Touch Principle 3 – Touch communication – A mutual learning experience

The essence of this relational experience is in the nature of touch communication; the reciprocal sensory-motor response cycle establishes the interrelationship of cellular and nervous system consciousness – from receiving the touch stimulus to the movement response that we both sense, feel and register. I may be working with the benefit of past experience and so 'nervous system consciousness', but this knowing allows space for Ida to enter the experience and I can then set this knowing aside (Bainbridge Cohen 2012: 9). Our dialogue is a testament to the nature of cellular communication and consciousness – by staying with the quality of providing space for her to begin to 'initiate' and respond from the cells, Ida found her way into a reciprocal relationship.

In the moment of hearing Ida's story I became aware of how useful this experience could be for her, even preconsciously, and then more outwardly opening to the environment she was starting to explore. If neglect had been her life, through being shut away, denied of autonomy or relationship with the world, then that reality may be all she knows and would be self-perpetuated by an expectation of being left alone. When Ida broke contact and stood alone, this attitude appeared. I adjusted my expectation, I did not want Ida to feel an object of my attention, nor did I want to move her as one. I persevered repeating the initiation that started to be receptive to something new occurring – at each touch stimulus, she felt able to respond a little more; each time a minor reciprocation and choice to open and move in space ensued. I witnessed her partial autonomy increase as I relinquished more of mine, sensing the possibility of a change in response from passive dependency towards active interdependency. This membranous process exemplifies the 'dialogue between present cellular and past nervous system experience' that is 'learning' (Bainbridge Cohen 2012: 9).

I felt happy that we kept to the taster we had agreed to, we had ventured into a new experience that we co-created and Ida's latent autonomy naturally evolved. I had already experienced this delicate process turn into a mutual and reciprocal one, by remaining consistent and open, as I recorded about an earlier experience with a deafblind partner that informs this one:

> Dance/movement sessions become a place of natural ease and involvement, offering a chance to relate through touch beyond the restraints of functional social relationships (such as when sign language is spoken into their hands). Dance/movement offers an opportunity for the whole person to interact with their immediate environment.
> (Dymoke 2000: 85)

In the present constraints of her 'immediate environment', Ida showed accruing acceptance and ease. I was aware of the possibility of coercion and the need for affirmation from Ida which came at the point of contact and was expressed in her body as a whole. Lack of ease had been a sign of sensory deprivation and trauma with other participants, indicated in facial expression, withdrawal and tense body tone. I perceived Ida's initial detachment and reticence to be a sign of accustomed relational passivity, the withdrawal of her hand is a sign of unfamiliarity with staying interconnected, due to dysfunctional social relationships. Paxton and Kilcoyne (1993) also observed familiarity with disconnection and immobility, as an expression of the exclusion of visually impaired people from participation in dance; a societal restraint which CI reverses.

I remained considered and intuitive. I knew that touch could be used effectively as a means to relate self to self and self to other – as a means to 'see' and 'listen' to the inner being, to sense and feel movement responses and give space for expression. I didn't expect to go far in external space as I focused on the internal space to find the reciprocal cycle and motility shift from our hands into our bodies, the back and forth from a yield state towards Ida into a gentle initiation of a direction, a subtle impulse to move into space, without going beyond her present field of awareness.[76] Our combined sphere was still nascent the sense of space beyond our relationship felt absent, we would need far longer to pursue any desire for or sense of seeking beyond our meeting.

Touch invites communication at a primary level and is integral to the founding of a trusting relationship. If this is reciprocal and allows for accommodation of difference, as with Ida, the possibility for a relationship would manifest over time. Ida may eventually bring other aspects of herself into a relationship, to nurture herself in a mutual self/other context. She may also experience desire, curiosity and agency, become more self-directive and find more bodily sense in relationships with others. I felt positive about her future possibilities and her assistant agreed, as indeed her relationship with Ida was on similar terms – touch-based, calm and spacious, indicating compassionate care.

Touch Principle 4 – Reciprocal embodied learning – Touch density

It felt questionable to me, from Ida's hesitancy to relate, that 'benign' touch had been a feature of her relationships with others. As her primary source of sensory information, touch would be expected to feature more in her ways of reciprocal relating, other than the usual hand-on-hand talking with her sighted guide. In our meeting, she had no expectation of remaining in contact; I was there, then I was not. Touch was on or off at first, as if a functional way to listen and respond, an underexplored potential language. Ida's initial response communicated that she didn't know she had a choice to engage or to redirect the stimulus; she initially kept returning to self and didn't move away from me. Ida had an opportunity to explore the basic relational process in an unforced and unhindered environment. We didn't exchange roles and I was the 'lead', the facilitator and educator; however, each step held the sense of an interconnected unit, of a reciprocal, mutual process conditioned by 'touch density', the process underlying the intricate signalling outlined above.

Touch density – The preconscious condition that informs my analysis of this experience

Touch density was apparent in the quality of our touch-sense connection, at a very subtle and localised level. This I recognise in retrospect, as the deep listening process; I attuned to her touch and felt the impulse disappear in her withdrawal, until she felt clearer, from the space–time of the neutral contact, that if she reciprocated something else happened – a change and dialogue followed. It felt like her body took time to register what could happen next. The touch connection was like a place of meeting, and then once that was established the next step could literally take place. My initial intention was not to move but to sense her response in my hand, either a movement or a shift in tone from passive presence to a state of activation, a shift in density. In response I met this tone and followed any slight movement of hers, implying that she had started the dialogue, and then we moved forward – the usual direction of movement from standing side by side. These small steps were familiar to the leading/following exercises and prior experience of how some blind and deafblind people orient forward, not backwards, for fear of falling.

I realised later how intuitive the process was; we learned from each moment for ourselves. I learned from holding the space–time frame of the experience, attentive to what happened but also kept an open mind as to what could happen in the event that Ida didn't want to move. I explored the nature of making a connection of communicating a signal and allowing space for Ida to take an initiative, for her passivity to shift and for us to move as a unit. This was the limit of our reciprocity; each time we sustained a connection the quality of touch changed only very slightly, but enough; just as tonal density shifts in a conversation with someone new, the relationship will not shift instantly, it takes time to unfold.

Reservations

Touch is an effective form of communication when words are not available and when little is known about the person. In the moment of meeting Ida, I assumed her to have some capacity as she communicated with her guide/interpreter. We didn't speak the same language so her support worker was our translator. I accepted Ida's consent as an a priori as she continued to respond; however, Ida may have had an underlying feeling that I had not been party to, that didn't surface bodily, or that she suppressed preconsciously. As the interpreter didn't show concern I didn't consider Ida's compliance a covert act, or a veil to hide her vulnerable self, to avoid conflict or to please me. Ida went off happily, however, I am aware that there is much left untapped in such situations, that there can be far more complexity hidden deep into an inner membrane and that this complexity unravels over time. Ida may have well gained more confidence in experiencing her autonomy in dialogue with me, and she may also have been totally confident and aware of the situation we were in. The inherent issue here is the ambiguity of touch – as the potential conduit for 'unconscious' as well as more conscious control taking. The facilitator has to be skilled, to avoid manipulation, to give time and space for the vulnerable person to self-direct – they may have been abused and not learned to say 'no', or lacked opportunities to experience autonomy, to choose, or to say 'yes'.

By offering this kind of positive experience, we have a responsibility to ensure that where possible we seek verification of what we perceive, in verbal exchange. This can be attained via a significant person in their life, or integral to the way we continue to work. By offering the possibility to say no, or the possibility to disengage and to stop, space is created for autonomous responses and self-direction. This has been the case with other individuals and groups which feature in the workshop in Chapter 5.

Rite de passage

This experience has stayed with me, as an epiphany, as another illustration of the 'hatching chick' metaphor. I feel that touch equates to that human connection called *love*, in essence; where there is touch love manifests at some mediated level, where there is no-touch love manifests in absence or neglect, and both are veiled by convention and regulation that touch can permeate (Dymoke 2000, 2014, 2018). In my mind, this meeting with Ida retains a glow of light and a sense of continuum, a light that illuminated the pathway that we took together. It may just be mine, and as such, it remains a rite de passage for myself as a practitioner, as I have no idea how it was really for her, I can only wish Ida well from this fleeting meeting. This is the first time I have brought this experience to light in detail, to describe the significance of touch that otherwise remains veiled and unknown, as touch was essentially all we did, affectively and deeply, to communicate in movement.

In the case of Ida, I experienced how touch creates unity from difference and duality from polarity; the sense of body within a shared membrane. This body found a sense of aliveness in being able to access *self-knowing* from deep inner listening to the *unknown* or, as yet, not *given* to consciousness (Depraz 1999). Ida and I found the lucid neutrality of touch language; as her ears and eyes, I provided a safety net around a place of play and replay, where we related through the combined realm of the senses to get beneath the superficial states of consciousness. Finding an ability to receive and respond through touch enabled our movement, where the permeable membrane and touch density concepts are encountered. The experience of touch was then captured to provide insights and forms of understanding that are needed to permeate the culture of restraint, a culture that intends, ethically, to predetermine safe human relationships, but also prevents people like Ida from experiencing what more there is to life.

By documenting and reflecting on this epiphany the socio-cultural context which enabled our meeting had conversely caused Ida's deprivation. The culture of control was in polarity with a culture of social justice and an apparent remedy was our meeting through touch. From a triple negative, a positive can arise.

Questioning the questions – Touch resolves the counter-flow of confusion

Ida's case shows how benevolent touch is naturally shared between humans who share a mutual and reciprocal relational bond, whilst giving space for the difference it enables inclusion at the social level as well as the bodily felt sense (Autton 1989; Montagu 1996; Dymoke 2000, 2014, 2018; Bainbridge Cohen 2012; Tanner 2019). In such relationships, touch has an intention to reassure a vulnerable person, or arises intuitively and mutually as an expression of that bond, expressing trust and safety. Conversely, as Ida may have experienced, touch between those who have not experienced secure and safe relational bonds, or those who do not know each other, may become thwarted by self-conscious feelings, doubts and conflictive feelings so that touching as an act becomes uncomfortable, hesitant, ambiguous and confused. The intricate nature of touch is revealed in these acts of touching, and naturally, the sensitivity of the topic causes discomfort and brings into question the potential implications of what acts of touch intend to do. In this heuristic, touch appears to be more than a physical connection that provides reassurance and presence of a mutual and empathic other, and begins to carry other, more suspect intentions, such as control, or manipulation, deceit or fear of perpetration of abuse, a threat to survival. Touch is considered as a conduit for malign intent rather than for natural human interaction in which a person may flourish. The preconscious and neutral lucidity of touch-based practices are readily sidelined as complimentary, rather than a key to feeling at one with others, recovery and maintaining overall health (Patterson 2007; Tanner 2019). In this regard, the non-touch protocols contradict the explicit meaning of a duty of care to the vulnerable and can detract this right (Johnson 2000; Piper and Stronach 2008; Tanner 2019), subsuming the cellular level of communication where self is nurtured, with ingrained intransigence – nervous system control.

In CI, the nature of the touch interaction is not of a healer to a patient, or of a sighted guide leading a blind person. CI as a method brings touch to the foreground in such a way that it isn't generally problematized in the way that society dictates – it begins with instigating physical contact as the means to connect and communicate where vision is not available; it then continues with this reality so that in the first instance physical contact is facilitative of a dialogue that continues in movement; touch provides the cue where a gesture or eye contact would signal for a sighted person. However touch density, as it is taught, is not commanding – it is inviting and showing the way, supportive and engaging of a progressive cyclic process that is essentially developmental. As with Ida, touch establishes a mutual place to 'yield', into sensing and feeling the presence of self and other, to 'listen' to the bodily sense, to the motility of the inner streams of energy and vitality. The response to sensing is in further yielding and dwelling in listening, to yield into the support, or to expand into space – the primal 'push'. The expansion into movement is a natural primitive reflex, to move away from, or with, the support of the yield (Bainbridge Cohen 2012). Self-awareness expands and the whole bodily response is to follow this impulse into expression. Whilst touching and moving with a partner, a simultaneous sense of 'reaching' into space becomes implicit, as it did for the blind boy with Paxton so that either person, at any time, can move more peripherally and still maintain a sense of continued listening whilst being disengaged from the physical connection. This defines the lucid neutrality, the open mind that CI brings to the dance experience and within a humanistic psychotherapy approach, it pre-ordains the participants with the potential and possibility for change.

Why look at the 'touch of the blind'? (Paxton quoted in Halliley 1991: n.pag.)

Regarding research involving visually impaired people and touch, Paxton and Kilcoyne are unique, as is the ongoing work of Touchdown Dance. This includes the collaboration undertaken with the Royal National Institute for the Blind on the tactile sculptural exhibition of Willem Boshoffs' *Blind Alfabet C* in 1998, and ongoing workshop projects with visually impaired, adults, children and young people. Noted is the absence of any other research on touch of the blind, other than Heller and Gentaz (2014). Heller and Gentaz (2014) write about the psychology of touch for blind people and note the lack of research in general due in part to the lack of respondents to make a systematic study. They note that the focus of research has been more about what blind people can't do, rather than what they can, so touch they can do, and Heller and Gentaz research its significance and insights that arise from the haptic sense which they explored with blind participants. This excludes research undertaken with deafblind people by Dr Paul Hart (Hart 2010) of Sense Scotland, whose thesis explores the dialogical framework of touch communication and the gaps that occur when objects that are referred to in conversation are not 'present' to the deafblind person to otherwise perceive.

The purpose of bringing touch to the foreground is two-fold; to demystify the taboo of touch in society and to address the non-touch discourse that promotes the incongruence of touch in day-to-day life. The purpose is to assert, instead, the 'lucid neutrality' of touch (Dowd 2007: 78) which determines the value and conducive role that touch has in learning, in defining a balanced non-hierarchical relationship, and as a language of communication and self-regulation. My pro-touch discourse, as already mentioned, embraces the idea of lucid neutrality in the permeable membrane concept so that the insights that arise from the experience can admit both sides of the argument, for a careful embodied exposition of the issues to be located where the value of touch can also be addressed.

The purpose is also to advocate for the pro-touch discourse, in which touch is a tool for inclusivity, for 'social awareness' and 'self identity' (Bannon and Holt 2012: 9) in both integrated dance contexts such as Touchdown Dance and exclusive contexts. If this is the case then should visually impaired people be able to access higher education vocational dance training? When used in the teaching and learning process, it may be assumed to be functional and invariant, as touch is touch and movement is movement. However, touch admits plurality in participant and so variance in the embodied experience; it allows for the plausibility of both diversity and subjectivity; it connects those who are different, who would not otherwise touch, and brings insight to each. I don't 'see' someone if I don't look at them attentively and I may not be seen in return, likewise I can't not 'touch' someone when I touch them, but then 'we' touch – when I become conscious of their presence and I am changed so are they (Dowd [1981] 2007; Caldwell 1996). Touch is innately reciprocal, unless this is denied along with choice, in its neutrality it is a metaphor for connection, affect and relationship – it brings us back to Autton (1989) and the continuum of Jones (2009: 25), to the nature of human touch as human need, and CI an endeavour to meet it, through shaping, transmitting and moving.

Addressing the absence of touch ... what has been learned?

I made note earlier in Part 1 of the lack of mentioning the affect of touch and touching in the participant accounts, and that this exposed some key factors relative to the practice. One in particular was the *lack* of any direct reference or description of the use, or of the role of touch, or of how or if any direct benefit was gained. Another factor was the apparently unrelated disclosure of positive feelings about changes in self-percept and letting go of inhibition relative to the external world – in other words, of finding a positive connection to the world through touch. Despite the plurality in response, touch-sense tends to be absent from the apparent and explicit benefit of 'doing' CI, hidden behind the sense of agency and social engagement. The apparent answers are that touch is complex to language, silenced or left in shadow, veiled by other more conscious concerns such as trust and ethical interest (Kilcoyne 1991; Bannon and Holt 2012).

Paxton and Kilcoyne had responded to the need to deliver training in 'human tactile contact'. In her article 'A common sense' (1991) she writes this up, she refers to the benefits assigned to touch generally: 'The sense of positive self-worth is enhanced through touch, as are trust and communication with another. These psychological gains are paralleled by physical benefits' (Kilcoyne 1991: 49). She writes about the technical aids that had been produced for visually impaired people to connect with the world, which 'capitalise on touch' and yet the 'primacy of touch', that is human touch has not been considered in their education or rehabilitation process (Kilcoyne 1991: 49). Evidently touch was erased and forbidden, it was against a culture of denial and deprivation, in which touch was made unwarranted or invisible to blind people, that touch was put out of mind. Kilcoyne points to the psychotherapeutic value of 'human touch' evidenced in the Touchdown Dance pilot workshops, which serve, in her view as more than an antidote to the non-touch agenda, as a conducive environment for learning and healing (Kilcoyne 1991: 49).

I would add that touch taboo silenced the possibility for open discussion of the topic, as if once brought into the open, to consciousness, the dominant non-touch agenda came to the top of the list, erasing the significance of the fleeting traces of touch and any value assigned to them. CI is still a touch–movement–touch form and the role of touch remains inseparable and integral. However, the troubled topic of touch continues to resurface, demanding further excavation of the non-touch – pro-touch polarity. Perhaps then, the art of lucid neutrality will be accepted as the basis for a pro-touch cultural shift (Patterson 2007; Classen 2012; Dymoke 2014, 2019).

Touch in psychotherapy

Like Paxton and Kilcoyne, I had a pro-touch discourse at the heart of my research. Back in 1995, I wrote quotes in a notebook from a book I really enjoyed reading many times, called *Touch an Exploration*, by Norman Autton (1989) and I write:

> Norman Autton. Touch – the way we touch.
> 'Being touched by another human being satisfies the need to be desired as a physical presence'.
>
> Physical touch links us with our humanness: it helps us share ourselves and our feelings with another.
>
> (Dymoke 1995: n.pag.)

These words express the self-reflexive curiosity I experienced as I started leading Touchdown Dance workshops, and sought inspiration from others. Significantly,

Autton affirms the value of exploration of the 'how' we touch; it is the way we touch that has positive benefit if it 'links us with our humanness'. This attitude recognizes that the desire to be accepted is human need that touch satisfies, and simply through 'physical presence'. However, in psychotherapy contexts I encountered a dichotomy; the idea that the therapist might satisfy a clients desire for touch is questionable (Bloom 2006) as the 'physical presence' indicated by Autton may be tainted by the abuse of status and outweigh the possibility of sharing 'ourselves and our feelings with another' (Hunter and Struve 1997; Caldwell 1997a/b; Eiden 1998; Westland 2012; Gallace 2012; Totton 2015). As I write myself, 'touch links us with our humanness', and the unconditional regard for the humanity of the client is key to a person-centred, humanistic or integrative, body-oriented psychotherapy model as I went on to practice and continue to research, with touch at the heart of all I do.

The culture of CI has provided a space for desired physical presence, for being touched whilst moving, for perceiving form and shape, and so each other. CI facilitates a culture for the movement of ideas; by being 'open' it has provided a radically different 'body' of knowledge, contained by the theoretical basis of 'democratic human exchange' where curiosity was woven into the fabric and narrative as it unfolded (Paxton and Kilcoyne 1993). To avoid losing sight of the how touch experiences were the same for visually impaired and sighted people, the case has to be made for the fact that touch is a common, relational sense. It is the difference in attitude and idea of how to touch that creates non-neutrality, and ultimately a discriminatory non-touch discourse prevents debate and change.[77]

Common to the accounts given so far, there are concurrent experiences; as we touch we appear to ourselves and to others, as we move we transmit these traces, as an exchange or affirmation of knowing, a means to displace old perceptions and replace old membranes. In CI practice we can give and take space and time in order to create a forum for a new emergent empirical and ethical movement practice. As we shall see, CI challenges the existing etc. to the existing discourses that make claims without exploring the intricacies of touch density, even though such detail is there, materially, felt and sensed.

4

The Pedagogic Process in Practice

PART 1

Introduction

This chapter is subdivided into parts to identify key aspects of the pedagogic practice that have evolved and introduce new influences that have enabled a more bespoke service in more specialised contexts, such as working with children and young people (see Chapter 5). These continue the original ambition and integrate new aspects that I have introduced, such as developmental principles from Body-Mind Centering® (BMC®) (Bainbridge Cohen 2012, 2019) to respond to the participant need.

Importantly, I acknowledge that these developments were in response to the creative input from the many different artists and collaborators who have been involved. More directly, the visually impaired and sighted company members have travelled nationally and internationally to teach, curate and perform to a wide range of audiences in dance, theatre, education and health settings. This means that information from a wide range of contexts has been experienced, further developed and applied to make the work accessible and available to people of all ages and abilities. Touchdown Dance workshops provide a realm for exchange, where differences are bridged; we experience both diversity and uniqueness and where we teach, we also learn to deliver a more rounded experience, and some go on to perform.

The nature of the pedagogic approach is explored more in-depth with reference to examples in descriptive accounts. Whilst attending to the detail and intricacy of the methods, I have sought to affirm how movement manifests the mutability of self-awareness and perceptive states and how the mind can focus on the felt sense of this process, to both unify the sense of self and acknowledge this learning in the presence of other. There is inevitably some reiteration of themes, principles and aspects already covered, as I explore the pedagogy in more depth and discursively revisit the practice – I start with accessibility and move on to revisit the adjustment that is conditional to its success.

Managing difference and meeting uniqueness – Finding commonalities and new social values

In earlier chapters, it was established that Paxton and Kilcoyne's basic principles supported each person, in each dyad, to work through discomfort and dissimilarity, towards finding ease and some commonality. It is as a result of the movement experience and the reflective dialogue at each step of the way that difference was not perceived as a limitation to the research, nor as a predefined attitude imposed by rote. This philosophical framework ensured the workshops provided a safe space in which 'difference' was accommodated by the availability of processes, and principles that nurtured new perceptions and possibilities. For example, the limitation of not being able to see imposed an existing perception of difference, but this was challenged by reverting to movement, to corporeal commonalities – such as having a back, having an arm, being able to touch and move together with prescribed and unprescribed elements. These commonalities, grounded in touch and movement and in the sensory-motor aspects of perception, underlie our ability to relate, communicate and ultimately to create identity and community. Bodily knowing is foreground as a source of potential for the nurturance of new relational and societal encounters, and for new, or shifting, emergent societal values – a paradigm shift.

To this day, Touchdown Dance provides a space for change, for the most unlikely type of experience to occur – a space where blind and sighted can move together and an intimacy that seldom materialises elsewhere. On returning to the 'status quo', following a workshop, the process continues in day-to-day life or is put on hold as society lags behind. The gaps in the availability of cultural provision for visually impaired people, in general, are still sustained by the lack of belief in the possibility of creating new and inclusive cultural forms (Heller and Gentaz 2014) like Contact Improvisation (CI), in which the value and role of movement, of body-oriented activity, and touch, enable this gap to be bridged (Albright 1997). CI started to shift the cultural landscape in the 1970s, it permeated the once immovable membrane that sustained and protected the cultural ecology from 'other' contemporary dance forms (Albright 1997; Kuppers 2011, 2014) and embodied the idea of cultural integration and social inclusion. Initially perceived as a touch revolution, a counter-culture pitching against the wider cultural ecology (Nelson 1996) in reality CI serves the needs of those who are deprived by, or excluded from, this ecology.

It is evident that, until recently, the mainstream dance arena has been deploying non-touch policies (Paxton and Kilcoyne 1993; Siddal and Rist 2002; Bannon and Holt 2012), thereby propagating the idea of touch taboo, resulting in segregation and exclusion of other identities, of which both visual impairment and CI are examples. The self-preserving membrane of the 'status quo' has admitted and adopted CI and the democratic proposition that it embodies. However, Touchdown continues to serve this cause, seeking to eradicate exclusion, to enrich and enliven the cultural ecology, and show we are 'more than' what we otherwise believe or consider to be possible – in the face of an ongoing failure of the 'mainstream dance' culture to provide a citadel for it.

Working with movement – A path towards change

Touchdown Dance's mission to provide opportunities to visually impaired people to participate in dance arose from witnessing how inhibition shifts as they learn to inhabit their moving bodies. Something quite extraordinary takes place, which I have referred to as a form of 'hatching'[78] to communicate the literal sense of breaking into a new realm of existence. For some the sense of injustice surfaces and a desire to reclaim a lost realm of experience, then we discover together how movement is perceived differently by each person, and how this is influenced by the lack of availability of resources and access to bodily experience more generally. Further, an initial place of discomfort is rooted in this lack, in the attitudes that cause exclusion, in a fear of the 'unknown' territory that movement represents prior to venturing into it. Once identified and acknowledged the 'problem' of reticence is addressed through movement – its natural antidote and a commonality. However, voluntary participation is rooted in accessibility, as in the case of CI, if not, the divisive use of visual frames of reference and its complicity with exclusion and social injustice is perpetuated. In all, moving voluntarily in space is a challenge, and fear of collision is a default response that is learned and re-enforced by the hypermobility and fast-changing landscapes current in a predominantly visual society.

If movement is a way to discharge energy and to regulate the body–mind, a visually impaired person has a right of access to movement provision and requires a safe environment in which to feel able to move out into the world. With the benefit of vision, sighted people are able to move voluntarily in space, to calibrate distance, time, and the size or dimensions of the spaces – including the sense of the horizon and outer space. Being without vision, movement in space is inhibited, even more so if objects are moving, and mapping the space through touch is less effective than sight (Heller and Gentaz 2014). Moving freely in a world of unseen objects that are constantly moving or changing is hampered by disorientation, and losing one's way.

Key phenomena revisited that established the pedagogic principles of facilitation

In Touchdown workshops, orientation to the space and teacher is essential, for those who lack exteroception (sight) and body awareness (which is not contingent on sight). This involves proximity to the teacher, for accurate transmission of the teaching point, and a reorientation away from mirroring towards facilitative and individual attention. By giving each person the starting point and then allowing them to explore the possible pathways into space, they find the ability to be autonomous in their exploration and to orient to space. The balance between facilitation and instruction is therefore key and variable for different participant needs. As teachers or leaders, we sense and feel the person in relationship to their own limits and possibilities, and are aware of even the smallest transition from static and immotile states. We learn to level down the sense of authority assigned to teaching, stand with them, within our own membrane, and accompany them in theirs, till we accomplish moving together with some of what each brings to the relationship. We discover the senses as systems of communication and language and ways to probe more deeply into the experience. As becomes evident, the process is a continual balancing act between giving space and time for the exploration – for the assessment of where each person is relative to the teaching point – and giving adequate input and support, whether to an individual, partner or group-based task. The subtlety and quality of attention required for this process will be illustrated later.

CI – A sphere for cultural motility and mutability

In the practice of CI, differences in participant need are readily accommodated if the facilitators follow a participant-led protocol; encouraging the dancers to negotiate and make constant adjustments to realise the possibilities that each present. The aim, collectively, is to establish a culture that allows the group to value diversity as a positive experience – for any ambiguity to be expressed and the experience to be affirmed. In the endeavour to support individual participants to explore and be open to insights that serve them on their journey, it serves to bring these aspects to consciousness, to shift any prior perception of what is not apparently possible or conceivable and give space to the ones that are. This may occur when the group is engaged in the same movement task, a wide variety of sensory sensitivity means that each person identifies a different insight from their response to the task and much learning is shared. For example, a sighted dancer learns to use vision to support a visually impaired dancer to orientate, who then, in turn, supports the sighted dancer's awareness of how touch supports clarity, alignment and self-perception. In this way the dancers share ways of 'seeing' that precede effectively articulated movement and support its execution.

It is the combination of differences between the two movers that enables their movement experience and defines their movement relationship. When the two movers are able to meet in a receptive and open dialogue, they orient to a shared 'centre' (Paxton 2018) and create a philosophical 'sphere'. Within this membrane, their relationship remains mutable and reciprocal, to accommodate new shifts and/or subtle changes as a new normal. At times, this dialogue can be a delicate negotiation and re-negotiation, involving careful navigation at the 'edges' of the experience. This may be at the moment of starting to move in a way that is unfamiliar to the body when it is important to take time, to move back and forth till familiar with the shift towards a new place of being or to connect with a new percept. An incidence of this can be when a visually impaired person is unsure of moving backwards. Uncertainty stimulates a reflexive reaction which inhibits this movement response. Then by stepping back and forward, repeatedly moving between the unfamiliar and the known, they establish more ease and open up a new perceptive space. When unsure or fearful, the emergent movement response will tend to be withdrawal back into self, as it feels too soon to move away from the support or to follow the lead backwards. This personal or subjective response impacts upon the combined experience of CI, but once identified, both movers learn to embrace this moment as another difference to accommodate so that where possible they are fully present for themselves and for the other. The sighted dancer, who feels comfortable moving backwards, accompanies the visually impaired person into that experience, to learn this bodily, by moving slowly, shifting their weight onto one foot, and taking one step back at a time, as they feel appropriate. The sighted partner learns how to be part of that positive outcome with encouragement and shared insight.

In all these relational dances, the dynamic that arises in the present moment is the ability to accept the experience 'at hand'. Being able to be present from one moment to the next is a challenge. To attend to the possibility of a new positive outcome is one potential 'truth' that our principle philosophy embodies. In his recent book about his relationship with gravity, Paxton talks of his own journey with movement self-observation (Paxton 2018). He observed how he slowed his movement down and how a 'normal thought/action relationship gives a mind time to step out of its habitual and practical relations to events and experience what were, before, transitional instants' (Paxton 2018: 28). When there are two people doing this together, slowly and with care, there is a sense of mutual interest in the non-habitual and in concretising new understandings of the discoveries taking place.

The transitional state – New ways of seeing, moving and being

In revisiting the focus of the early work, it became evident that the four founding principles – beginning with the body, finding an equal relationship, and addressing attitudes to learning, movement and touch – identified commonalities and new movement principles. In light of this, specific teaching modalities evolved in response to the need for new pathways and transitions to support the variation in participant response. The evolution of the work was contingent to the research process, and over time these ideas have nurtured new ways of seeing, moving and being that are now 'out there', integrated into pockets of society.

The practice of CI proved to be adaptable to different learning environments and needs because it was founded on the foundational and primary senses of touch and movement (Bainbridge Cohen 2012). Learning could therefore be transmitted through those same primary senses and in synesthetic relationship with others – creating the perceptual sense of the inner 'mind's eye' and 'inner ear' (Paxton and Kilcoyne 1993: 13). The process of attending to sensory information takes time but informs the learning experience with new inter-sensory and multi-modal sense data to reflect upon and integrate. This process was transmitted between participants and facilitators who could relate in the same inter-sensory ways and as expected of 'equal' human relationships. This in turn supported ease and adaptability, as by moving with attention to another mover, new pathways and transitions came by implication and with support. Sometimes the transition led to an unexpected place, and the movers returned to the starting point to try again. Through trial and error, they explored the limits and boundaries of their habitual movement patterns and gained awareness of new variations or possibilities as they became available to them. In time the basic foundational principles evolved from further interdisciplinary input to incorporate new aspects, and the inherent adaptability of CI enabled its broader and more specific application.

CI – The new and enhanced pedagogic aspects

Availability, accommodation and adjustment

The following subsections identify new aspects that emerged over many years of practice. These enhanced the pedagogic and therapeutic approaches with greater awareness of how to be inclusive. Firstly, in order to balance facilitation and direction, the facilitator had be be open minded and available to work with different expectations and to adjust the material accordingly. Secondly, in order to enable each person to be as immersed as possible in their experience, the facilitator had to accommodate the different degrees of receptivity and thirdly, able to adjust the pace and content according to the progress made. For example, some sighted people explored wearing a blindfold non-stop for a few days, with the aim to experience moving in the world with senses other than vision. This was sensitively facilitated as it could have appeared unethical to 'play' at being blind. The sighted movers sought to experience how it was to move without sight with the visually impaired partner if they agreed. This meant that both participants and facilitators accommodated this adjustment and were engaged in the learning taking place.

When contained in this context, CI deals directly with discrimination as perceptions shift in the doing of the work. When a sighted person moves with a visually impaired person, they become aware that they move differently and have a different perception of the body. A visually impaired person may never have seen dance, and yet they find their way to do so. Touching each other brings into question the societal touch taboo and also explores the right of any person to determine for themselves who touches them and who doesn't – which doesn't happen for a visually impaired person when a stranger grabs them to assist them across the street. In CI, illusory societal stereotypes and their associated assumptions are unveiled.

Motility

The kinaesthetic sense is inherent to communication and requires the availability of each person to engage with their moving body, with their own experience and also then with others. Motility is now explained in more detail to clarify the relationship of each participant to their learning process, to their ability to relate and communicate with others and reveal the undervalued significance of movement to our ability to 'be' in the world.

Motility of the whole being is experienced from the surface contact with a partner or the floor, and sensory feedback comes proprioceptively from sensory receptors in the joints, tissues and skin whilst moving. Through physical contact with a surface, the body is able to move in ways that may not have been experienced before. Both partners continue to move in negotiation with each other at the point of contact and in direct response to the shape and form of their bodies' surface. The sense of the 'rolling point of contact' does not require sight so much as an orientation to this proximal relationship and attunement to the proprioceptive process to register the movement experience. Movement and touch may underlie motility; however, a visually impaired person will not experience their movement in the same way as a sighted person who uses vision to support spatial orientation and balance. A sighted person can mimic the movement of someone they see, and close their eyes to focus and orient towards the internal bodily sense; a visually impaired person is able to orient outwards to the wider spatial realm and share the internal bodily sense in moving with others.

Motility is impacted by different modes of perception, relative to differences in the perceptual field. For example, the space and time aspects differ according to whether the movement is seen or felt. Seeing and moving occur simultaneously and provide the ability to perceive the movement phrase, to feel and grasp the more specific qualities through practice and through verification of its execution in the mirror, or from observing another. Depending on the availability of sight, a visually impaired person cannot perceive another person's movement phrase in this responsive self-directive way. Without vision, the kinaesthetic sense takes longer to register the flow, direction and position in space – but once the body has learned each movement and the transitions between each position then the phrase is more fully and effectively embodied. Once this process is experienced by all, it becomes the space–time modality for learning CI.

A developmental framework and the relational membranes[79]

The inherent developmental framework of CI invites movement through different states of consciousness and realms of being – between the corporal and cognate; the sensory or perceptual and the interpretative or imaginative; and between the introspective and expressive – and these vary, relative to subjective experience. Whilst working with this framework, the process is very much alive in the present moment so that the experience is the place of discovery, the story to describe and its content is rich and variable. In time the process becomes familiar as movement flows from moment to moment, with awareness of the transitions and points of arrival, of moving with and through self-membranes, which define the boundaries of self-perception and shifts in consciousness. Body sensations become identifiable within the context of the movement, relative to space and time, and relative to the supporting surface. In this relational realm the duality of motility and mutability enables the ebb and flow that allows for idiosyncrasies to be expressed. Examples of these foci are given in the context specific to CI methods in which commonalities arise that are also age/ability-specific.[80] This has been the case for members of the company as they develop artistically, as well as for workshop participants (see Chapter 8).

In CI, the three dimensionality of movement is experienced and gaps are filled in, providing the opportunity to move through them. Moving from lying to standing involves moving through the developmental stages of rolling, to crawling, kneeling and standing, but if I am 'parked' on the side whilst others are doing this, I will miss the opportunity and have gaps that I have never had the chance to fill. If I roll I do not need to see, but when I transition to crawling, vision brings the space towards me as I move. On the other hand, not having a vision means it appears to me in other ways as I move or not.

The human developmental movement patterns provide a time line for movement observation and exploration; each pattern has a relative aspect of 'mind', a relative form of mental agility that manifests in realms of awareness, consciousness and ultimately in thinking. In this way, the phenomenon of movement, the being in movement, initiates the sensory relationship with the world, involving other senses relative to the developmental stage explored. These 'spheres' of being are what I refer to as 'fields of operation' that underlie the process of perception and so the relationship of the moving being with the phenomenal world to which it is responding (Bainbridge Cohen 2012). Once more, movement moves the 'self-membrane' and opens the mind to new self-perceptions, awakening an inner agency that nurtures a more fully embodied, moving being.

The connection with the movement of mind

The implications for the participants in this facilitated environment are that the movement experience and relational context may bring unexpected insights and recollections of past times, which may be conducive to a positive outcome and learning, as well as potentially affective. A balanced and carefully timed process that navigates these aspects, allowing time to pause and for self-care, provides containment and a sense of safety; the 'door is open' and support is available for participants to meet their needs. Overall, the communal experience is generally self-regulating, under the guidance of the company members' personal experience and specialist training. Common for most workshop participants is the experience of how movement languages the body and may instigate a realisation of a need for change, or equally, of a need not to. Movement can lead back either to a voluntarily inhibited state, or forwards to a voluntarily responsive one, and the experience of moving between these places within a developmental framework supports the possibility to weave between them and to find new points of reference for further instigation and discovery.

This process is transformative and requires experienced facilitators to oversee and ensure that the developmental progression is expansive and productive. Even if regression back to lying on the earth's surface is an inevitable aspect of the process, it should enable positive maturational change for the mover (Dymoke in Daniel and Trevarthen 2017). The movement back and forth through developmental sequences can be a foundational maturational experiential process (Bainbridge Cohen 2012, 2019).

PART 2

The different modality-specific methods

The process of learning to move in space requires an initial return to the safest place, i.e. the ground. In learning to move with and against gravity, and on and away from the ground, we are re-experiencing the human developmental process and returning to the pathways that are efficient, effective and enabling us to move further into space. The modalities I have defined, start from lying to standing, continue with moving off the ground in lifts and provide the rationale for the modality-specific pedagogic process (Heller and Gentaz 2014: 57). The modalities facilitate and focus on inclusion and deal with the social stigma. This comes back again and again – the idea that, in general, visually impaired adults didn't have the same opportunities as a sighted child and developed inhibited response to being and moving in the world. This is also the case for people who experience sight loss and face the exclusive socially imposed restrictions as well as fear. CI also benefits self-perception and self-worth that comes from being able to move freely in the world and being able to affirm when it is safe to do so. All the above aspects appear through these modalities relative to each individual's experience, sensory affinity or impairment, and availability to engage. The key modalities are now explained with variations, and at times with cautions and examples, referring back to these new influences that I and others have noted and brought to the praxis.

Arriving in the space

Traditionally, the workshops start after each visually impaired person has mapped out the space by moving around it and then across it. As Paxton recalls, this may involve a sighted person accompanying them (see Chapter 1) and the opportunity to reciprocate as Holly Thomas gives an account from her point of view (see Chapter 8). We ask that the sighted guide does not give directive verbal instructions unless the visually impaired person requests it. Sighted participants may be shy and reticent, and talking is a challenge, or conversely, they may think visual impairment means their partner can't perceive what they can't see, and not talking can be a challenge. When the roles change, the sighted partner is given the choice to close their eyes, to 'open up' to the underlying support of their other senses. With the visual memory of the space, they experience a different way to orient within it and map it. Some peep and others inhibit looking, to trust in their visually impaired partner to lead with their memory of the space mapped in touch, movement and sound. The sighted guide facilitates the visually impaired person to perceive what is there, placing their hand on surfaces, and if they witness well, they realise other sensory information that the other senses provide.[81] Facilitation of where to touch, when to move, whether to move closer or to reach up higher, these indications help the visually impaired person to explore the space more fully, to initiate moving further than would usually be the case – and to encounter these initial transitional states together opens up a sense of potential.

Evidently, the availability of the sighted and visually impaired dancers for each other opens up new perceptions that transfer into daily life for both. The visually impaired dancer experiences freedom to navigate space safely yet without the customary devices that avoid collision with the environment and provide independence. Sweeping a cane over the ground in front as they walk is a way of perceiving changes in the terrain, a tactile signal transmits the rise and fall of the ground to the hand, to the arm, giving at least two metres, including the length of the arm, between the walker and any obstacle or level change. This enables the visually impaired person to 'meet' any uneven ground or obstacle before their feet and so body encounters it. As a sighted guide walks close by the visually impaired person, this can be convenient, time-saving and less stressful to navigate new environments, particularly if the relationship is negotiated. Alternatively, the visually impaired person may feel over-dependent and disabled by the environment. In CI the shared role of the guide is to indicate rather than control, as both partners lead and follow, and level down the prescribed unequal status, offering insights to all as to how this otherwise dependent relationship can be negotiated.

When not using vision to navigate the space, sighted movers slow down and engage inwardly to open to 'new' perceptual realms of being in movement and inner body awareness. Following this opening, they share their reflections upon their experience in either role, and this is often the first moment of clashing and meeting perceptual differences. Variations in this exercise occur depending on the group so that if the participants are shy and reticent then conversation during the exercise helps the movement dialogue. These examples of feedback from each partner are quite typical of the responses we experience.

Examples of feedback from the guide and guided roles

The sighted guide:

> I was afraid that we would collide with others or with the edges of the space but we found our way from the door, all around the edges of the space. We naturally went more slowly and it all felt much bigger. I must have gone too fast for her at first.

The visually impaired partner:

> I felt safe with my partner and needed much longer time with the exercise as there was so much to take in. It felt good to know about the space so I can find the door if I need to go out!

The visually impaired guide:

> I really enjoyed being the guide, it was easy once we found the wall and went around it. I lost track of where we started so we had to check together when we had finished. Going out the door and navigating the step was hard, but we managed by feeling with our feet and finding the rail.

The sighted partner:

> I found there was much more to touch than I thought. I also needed more time to perceive the space – I wanted to know what is on the walls, the size of the windows, what they are made of etc.

Finding support – The floor as the first partner

By referencing modality-specific processes, it is possible to combine perceptual information with a range of movement that is relevant to a specific spatial plane, and subsequently creates a relational field of operation. In CI, the process of starting on the floor (also referred to as 'the lower 6 inches') encourages movement exploration in a particular spatial orientation and with a specific field of awareness. Within each modality, the common sensory affinities are touch and movement, so these senses support different sensory abilities and sensitivities. Other potential intentions, such as relational support, arise with the invitation to explore the modality with a partner so that one moves and the other accompanies them with body-to-body contact. This enables both to experience the effect of an embodied exploration in the context of a relational version of the modality and to nurture interpersonal relationships with others. We have experienced how working with touch in this way provides a valuable learning experience, as everyone can optimise their senses.

By categorising the modalities according to different levels and spatial orientations, it is possible to identify the benefits for sighted and visually impaired people as well as the challenges. In each modality, we experience a range of responses, but the general trend is that the clarity and specificity of the content provide the structure and content to optimise the sensory-perceptual process for each person (Gibson 1966).[82]

Following meet-and-greet exercises and mapping of the space, we commence at the ground level. Returning to the floor, the body can dwell in relationship to gravity with a more internal focus on sensation. Lying on the ground the body can spread out, opening and closing with whole-body movement patterns. Equally, lying on one side can provide underlying support and enable individuated movement of a limb or a sequential flow from one place to another. By coming back to the floor there is space to regress to primal, fluid states, to let go of postural tension where possible, or to be simply lying and moving. We experience that some people have difficulty with lying on the floor for a long time, this may be due to a traumatic experience. Someone may consider rolling on the floor a childish thing to do, and someone may feel too uncomfortable due to injury or body size. In these cases, support can be provided to enable them to move towards and away from the floor, or to assume a position that feels safe and then to feel the natural movement of the body with the breath with the support of the floor. This is just an example, and it is important to note that even at this foundational place no one has to comply with the modality as written.

Moving in partnership with the floor introduces the idea of touch and movement as inseparable, and as the foundational senses for establishing a relationship with the outside world. As a mover, I become aware of my body movement autonomously and discover that in relationship with the floor, I can transition fluidly between movement and rest, if I feel at ease. If I feel uneasy, or my body's movement feels alien and unfamiliar, I may resist and choose not to participate. Either way, the response is always accepted, as the point is not to coerce or direct people into one way of being in the world or to serve a purpose or *a priori* outcome. The purpose is to share the remarkable aspects of our moving experience, each of us picking up tips and insights from each other.

Moving autonomously and effectively provides a sense of self-trust and self-support, and further potential arises from a sense of feeling stronger and more able. This involves growing into my own body awareness, and filling in my understanding of what my body can do, from its more delicate and subtlest shifts to more full-bodied responses.

Some examples follow, of practices that encourage each participant to experience their own limits in contact with the floor and to explore pathways to other levels. The floor is the surface for rest, to map connections and integrate the body, breath and consciousness. Paxton introduced these postural practices to build a sense of self-support that could be continued after the workshop at home or in day-to-day life. I describe each aspect as an activity that can be accomplished if there is an inclination to explore – or left for a hands-on workshop experience.

The breath and body awareness – The benefits of yoga

Practising yoga builds strength and alignment. Sustaining a pose with controlled breathing, specifically long exhalations, initiates the release of carbon dioxide from the muscles to the blood, to the exhale. This experience brings awareness to the breath and nurtures self-reliance and mental focus on the assimilation of a movement. The exhalation and influence of gravity support the muscles to lengthen or shorten, for the joint position to go deeper into flexion or extension, and for the muscles on the weight-bearing side to strengthen. The benefit of these practices integrates into a unified sense of bodily knowing that is usefully applied in CI.

Knee to cheek – lying on the back, one leg extended, the other flexed at the knee. Draw the knee towards the chest and hold with both hands without pulling, lift up the head and bring the cheek towards the knee on the same side, and use long exhalations, to support the front neck, abdominal and pectoral muscles to support the head. The other leg remains extended in counter balance. Hold this position for as many breath cycles as you can, lower the head when needed and focus on the front body support for the head, not just the neck muscles. Before feeling discomfort, pause, rest for some breath cycles and then do the other side.

The following are done standing. Based on martial arts principles such as tai chi, these modalities work with the sense of letting go of holding and restraint in the arms, whilst establishing a strong sense of stability and modulation of muscle tone.

Spinal twist – the twist of the whole spine around to one side is initiated from the turning of the ribs over the pelvis with the weight of the arms hanging by the sides. The arms are flung across the body and around to the side, following the swing to either side of the body. The hand of the arm that passes behind the body taps on the lower ribs in the back. The twist can be initiated from the shoulders/ribs or from the arms and supports spinal rotation on the horizontal plane. Commonly, letting go of the weight of the arms takes time, many hold their arms up to the side and need to be encouraged to let them hang by the sides of the body so that by turning the shoulders and ribs they are moved in space. However, this may take time and is not in itself a necessary objective.

Weight shift over the base of support – standing balanced on both feet, moving from the centre of gravity over the feet/base of support or swaying from side to side. Then widen the space between the feet, bend the knees over the toes and shift the pelvis forwards back, side to side and then in a circle tracing those four points. Feel the weight move over the feet in the same pattern and feel the top of the head circling in the same direction.

These fundamental modalities support the individual to feel at ease in the body, receptive to a dialogue and transition to working with a partner.

Working in partners – Giving and receiving support

We learn how to assume a stable support for our partner to move against so that they do not fall. We not only learn to lean against the support of our partner's body but also learn to support them in return. When we start to move with another, we use the body as a support surface. This enables us to discover movements that we cannot do on our own without support. It takes time to find the confidence to move as it requires trust in the support. We learn to move around the surface of the supporting dancer and then change roles to support them.

Two examples follow that initiate the dialogue, create a sense of confidence and improve awareness and coordination. This is similar to the exercise with Ida in Chapter 3, Part 2.

Meeting my partner and finding my feet

Standing side by side with my inside arm flexed at the elbow and my palm facing downwards, I place it on top of my partner's adjacent arm, which is in the same position. We stand for a moment and breathe, feeling the small movements taking place in our bodies, like a small dance. I am invited to allow my weight to come up and into my arm as it rests on my partner. Their arm remains firm, lifting up under mine and we feel this bridge between us. I sense my body shift position as my partner starts to move her arm forward and backwards, just a few centimetres. I feel my arm move and then my torso and spine, my head and then my weight shifts over my feet. We do not go as far as taking a step at first. As we start to move more, we start to take a step to catch our weight from falling. I feel clumsy at first. I find it hard to give my weight and to trust I will not fall. I am hesitant to move quickly and feel myself tighten inside. I find myself making lots of small steps to keep my feet under my body's shifting weight. Little by little I begin to feel that I can rely on my partner to rest against and we start to move together slowly into space. I feel the invitation to take a forward direction at the touch of our arms, so I follow – then we slow down, stop and breathe and she may choose to go backwards, up or down. After a while we stop, take our hands away and say a few words: 'I think I get it ... let's change roles.' We start from the same starting position, me leading.

I initiate moving in another direction and this exchange continues for a while, each time in a different direction, each time at a moderately different pace, and I start to feel more relaxed and steady on my feet. I start to feel lighter and enjoy the flow from one place to another, the slight turning, the changing of level. We pause and rest for a moment, breathing, and then discuss: 'That was OK, I am getting the hang of it now!'

I find it hard to feel where my partner's body is in relation to mine. I feel their surface, but I do not know their position beyond that. How to know if they are comfortable with my weight? I feel a need to better understand how to manage my weight, to experience it, as I fear I am too heavy for my partner. I may also not be strong enough to support my partner's weight – so how might I come to know?

Managing weight

After a while, we learn how to manage weight, to make a strong support by placing the feet wide apart and by moving the centre of gravity towards the forthcoming weight and to move with it or against it. To avoid too much weight, it is important to go slowly, make adjustments and ultimately to recognise when to move away, and to say 'no, it's too heavy' if needed. The risk is when weight is given too quickly or the base of support is not ready to take it, so taking weight is not to be rushed – as the body will collapse, fall and be injured. Too much weight also causes a reflexive response to move away or down to the floor, so it is important to learn how to manage the weight, to take responsibility to adjust and feel the pathway back to balance and the base of support. Through trial and error, we learn together what we are both able to achieve and what our limits are so that we stay within our comfort zone and can focus on learning to move freely rather than getting stuck with weight bearing. The key principle is to be aware of how much weight to give to the support, and to recall that weight creates momentum and the body gets lighter as we move. The same applies to a moving partner with whom we exchange the supporting role so that the dance continues and flows with the rise and fall of each person, the under- and over-dancing that generates gentle lifts.

Pace and time – Managing momentum

Pace is another aspect of motility and requires modulation. This is experienced at both extremes of being very slow in the transitions and of being able to use the momentum of the body moving on the floor or partner to carry the weight into space. When working with a partner, the momentum of our body moving around the support means we can pick up the pace or we can stay slow and easy going, feeling the slightest shift between us. The exactitude of the movement is felt by each person independently, and it is important when moving together with another person to keep track of each other and to accommodate the fact that at *all* times the visually impaired partner cannot 'see' more than the contact surface. After a while of getting to know one's partner, the exchange happens more by trust and feel, with a sense of co-responsibility for each other, for one's own and one's partner's weight, and to only go as far from the floor as can be returned to safely again.

At this point, the sighted person can be more at ease in moments where they too cannot see their partner.

With practice, the body-on-body lifts feel really exciting. It is extreme at first to be lifted, as I no longer know where the floor is and I can get scared; in these instances, my sighted partner is so important to my experience. We have to get to know each other's limits and work with those in the dance so that we both have a good experience.

Sometimes the physical contact and the kinaesthetic stimulation are saturating and I need to pause, step away and breathe a little on my own. This is also important – to learn to say 'pause', to take a break and be with myself again, just moving with my own body, or sitting on the edge and letting others continue. It always feels pleasant to 'watch' or observe others moving. Even if I can't see I sense and feel the nature of the relationship, when it gets edgy and one of us is not comfortable, or when it is bumpy and the dancers are really working to smooth it over, to transition more gently from one place to another.

When we dance, it is easy to feel over-stimulation from the contact as we pick up energy from our partner and this can be overwhelming. There are many ways to deal with this and you learn as you go. You can pause and sit back to back, rest and breathe into the support; or you can lie on the floor on the back or front and assume a constructive rest position. On your back, this is done by folding your arms over the front of your body and lifting the knees up so your feet come close to your hips. To rest on your belly, come to kneel sitting and then fold your body forward over your knees so that your belly rests on your thighs and forehead on the floor with both arms outstretched.

Using the imagination and proprioception

When vision is not available to demonstrate a movement, verbal description of familiar objects or shapes can help to communicate the form that the body will take and the possibility to change its dimensions: Expanding or condensing the form in size supports modulation and changing the level supports an increased movement range. Many examples of this have occurred, and the one described here came from a visually impaired facilitator, Holly, in a workshop with only visually impaired children (see Holly's point of view in Chapter 8). Holly described a movement we sought to do, which would then be used in a partnering exercise. She used words to inspire their imagination, assuming that they were all familiar with holding a ball and moving with a ball.

Holly asked the children to 'move your hands as if you are holding a large beach ball between them' (she does it to accompany their timing as they start moving and I move around keeping watch and stepping in gently in case of possible collisions).

Holly continues: 'Move it around the front of your body', she accompanies them and when they bring their arms too close to the chest she asks them to inflate it fuller to keep the shape. This idea has been taken further in other workshops and we also use real soft inflatable balls or invite more imaginative movement: 'Move it up in front of your face, or down to your knees [...] imagine the ball gets smaller or bigger so that your attention follows the movement you make.' In this way, we pick up on movements that participants are making and affirm them. We also invite children to describe what they are doing: 'I have a ball, I am moving it over to the side' and we may add 'to the right side' and encourage a continuation right around to the back space and invite them in turn to share with others what they are doing with their ball. In this way the members of the group who feel confident with words describe what they are doing and so others can do likewise or continue with their own exploration. Then we stand back to back and try to roll around the circle of the arms that the imaginary ball creates, to face each other, our feet walking us around the rolling point of contact. Sharing invites the sense of a group process and generosity with ideas supporting creativity and ease in being with one's own movement whilst others do theirs; it underlies the freedom to move with continually shifting points of contact with another.

There is a visual or inter-sensory aspect to this as the description evokes a visual sense, but, once incorporated, the imagery is used to entice and invite movement exploration. Images from day-to-day life serve the purpose of being familiar and can be then used to expand the movement experience beyond the usual day-to-day range or expectation. This aspect also welcomes differences – by working to avoid disadvantaging anyone and so avoiding resistance or an inhibited sense of self. Feeling different can inhibit body awareness and cause retreat from the inspiring terrain of facilitated movement into the terrain of disability, where a visually impaired person is effaced and may then self-efface their able, but different, bodily sense. It may be helpful to cite examples of *non*-facilitation that are unhelpful and that efface the visually impaired participant. 'Find a place in the room' – the visually impaired dancer who doesn't know the room cannot orient to 'find a space'. Instead, guide them around the room first, and then across it. Then ask them if they can find their way to where they would like to stand, or towards your voice where there is a space and then let them know where they are in relationship to others, for example in a circle, and the others can say their names to indicate where they are.

'Choose a partner' – the visually impaired dancer may not know where the other dancers are and cannot see who is moving towards or away from them, so they wait in discomfort, and if there is no response then the sense of exclusion, shame and disability sets in. It helps to propose partners of people who are more or less the same stature and to change partners regularly.

Forming a sense of a group is an important beginning, generally done by initially forming a circle to create a sense of community – this is another orientation task most easily done by holding hands, forming it together and then dropping hands. The social significance of choice is important and requires facilitation. The name circle and other meet-and-greet exercises are essential in a group of people who are new to each other. Some other positive examples:

1. Everyone is given a number, one or two by going behind them and tapping them on the shoulder saying '1', then to the next person '2', the next person '1', etc., until everyone has a number, then all number 1s turn to the person on their right, and all number 2s to the person on their left. This is a pragmatic solution if the difference in size does not matter. If the group involves sighted and visually impaired people then they can alternate when forming the circle.
2. 'Move in this space and when you find another person stand back to back with them. Feel the difference between your backs and the length of your arms by putting the backs of your arms together, palms facing forwards and moving your arms to full stretch out to the sides. Initiate movement from any point of contact and see what happens'.
3. After a short discussion/check-in, one partner stands with arms open, and the other stands behind them and places one hand on their partner's lower back who then moves from that place. The mover is encouraged to notice how much information they get about their body moving from this place. The roles are then swapped before a chat about the experience.

The ability to transition from one moment in space to another and from one level to another depends on finding the pathways through first-hand experience in an environment that is safe and supportive.

Movement observation – An interpersonal process of interaction with peers

Self-observation and reciprocal peer interaction are a means to answer questions and to probe deeper into an experience (Vygotsky 1978). In all the relational tasks, whether it is moving into a stretch, lying on the floor, or moving with another person or the group, there is a critical balance between giving space and time for the participant's exploration and for the teacher to assess where each person is relative to the task – to assess the support they may need to manage it. There is an aspect of observational assessment in each teaching point and an aspect of immersion in the experience for the participant. As a witness of the participant's experience, the teacher has to carefully assess the evident level of achievement or engagement, to provide support and further information specific to the participant.

The teacher and facilitator may well work directly with a participant in order to sense and feel where there may be a gap in understanding or to demonstrate more clearly and with more detail according to the questions or issues that arise in the process. This role of accompanying the participant's experience is key to the inquiry.

Clear verbal instructions and descriptions are essential to this process. The teacher mediates the learning experience with the use of verbal description to support the participants to notice if they are with the task or digressing, an awareness that brings more autonomy. Hearing movement language also supports learning to communicate about movement and how ultimately to be more aware and self-directive, more effective and more present. Simultaneous description is used to describe the movement as it is being done and also indicates how to move and navigate the space together – such as 'let's take a walk'. This narrative accompanies the movers as they move in space and provides information about the quality and form of the movement that is the current focus. The facilitator will describe what is going on in the room as the process unfolds, such descriptions enable the visually impaired participants to know what is going on around them and validates the activity. It is a means to steer the room in a particular experience and to feedback on what is happening.

Description is also used with demonstrations and is repeated as the movers assume the task so that repetition of the information enables them to follow the words as well as assume the role they have taken accordingly. In these moments, a participant may need to stop and rest, to digest the information that their body has absorbed, or in some cases, to stop due to discomfort. Even if this does not appear positive it is an important and significant moment to acknowledge and provide reassurance, as such moments are equally important to learning and development, and being true to oneself.

This mix of objective observation and reflexive introspection provides valuable information to work with and progress from. Fundamentally, being aware of feelings or mood-states is being self-aware, and then naming them brings emotional intelligence and the ability to monitor and regulate such states (Goleman 1995: 47). Access to both subject/object perspectives provides a balance to support self-reliance, maturation, integration and progression towards a deeper understanding of the intuitive body. There we discover the primal inherent interrelationship of emotion and motion, of self and movement awareness, assisting the realisation of intersubjective relationships that may not otherwise form.

I have introduced these key aspects and modalities to clarify the intention of the praxis and its outcome and to bring the story of this work up-to-date. Many sighted and visually impaired participants have contributed to these concepts during the workshop experience, and company members also in our meetings continued explorations and training sessions. It is time now to look more specifically at the methods and processes that evolved from the legacy of Paxton and Kilcoyne, which I have categorised as 'modality-specific approaches', a concept I took from research into the integration of visually impaired people, undertaken by Heller and Gentaz (2014: 57). With all these methods the experience builds the sense of the body as a resource for freedom to move, for movement exploration and for greater self-awareness and self-definition. CI provides the movement vocabulary for each person to use as a basis from which to move freely and in ways that are not available alone. The rolling point of contact becomes the point of initiation and departure, of support and leverage, of weight taking and indicating direction.

Modality 1: The lower six inches

Rationale

The ground under our feet is the supporting surface from which we found our way as infants; from lying to lifting the head and pushing ourselves up to come to

FIGURE 4.1: Rolling modality 1. Lower six inches.

FIGURE 4.2: Rolling modality 2. Finding pathways along the floor.

sitting, to hands and knees, and finally up to our feet. Usually, we do not have to be shown how to roll or lift our head, or how to come up to standing. However, as the developmental process is innate it can also be tripped up or impacted by the intervention of parents who, for example, place a child in sitting position too soon, or by the limitations of the environment in which we grow. If an infant is placed in sitting, she has missed the significant stage of rolling to side-lying and finding her own support to sit – this requires support for the head 'righting' to come from the whole spine and from rooting the tail, as well as pushing from the arms and levering from the feet and toes to fold in the legs. If we do not spend enough time on the floor or a horizontal surface prior to moving away from the earth, the body can lack the core movement of the spine and torso, and the stability of the limbs to weight bear at the hips. Likewise, if we do not learn by ourselves to come up to standing, we also miss finding the pathways back down. These movement pathways are primed from the contact that all parts of the body have with the floor through the skin, from the side or palm of the hand to the toes when the belly lying, or from the heels to the head when on the back, and so on.

From early pre-natal development, movement is a reflexive response to internal stimuli, to changes in position relative to gravity and to tactile stimulation of the skin. Touch is considered to precede an ability to respond to other sensory stimuli, such as taste, smell, light and sound (Bainbridge Cohen 2012). As previously mentioned, touch and movement are evidently interconnected, and physiologically motility is a form of innate intelligence (Siegel 1984; Dymoke 2019). The body requires stimulation to prime the available reflexes and requires movement experience to learn and enable our actions to be effective (Bainbridge Cohen 2012). Neurological disability impacts this process, as do negative experiences occurring in the process of learning to become a moving being. Our motility is a language that evolves in patterns and follows pathways that connect the mind and body, the centre to the periphery and the self to the other; it is essential to ontogenesis. Hence the significance of working in this way to return or regress is to rediscover these early, primary reflexes, their underlying support for associated developmental pathways and patterns, to reinstate a baseline response and potentially increase agility.

The reflexes and gravity

Primed to move along the earth, our physiological reflexes underlie the movements we make and are stimulated directly from tactile stimulation from; in this case contact with the floor. The reflexes are involuntary movements in response to a stimulus, providing the neurological pathways upon which finer and more intricate movements are established, and maintain our ability to respond and move without thinking, i.e. to survive. The whole body lying on the floor enables a state of rest on the one hand – with the pull of gravity drawing our weight downwards, releasing the effort required to maintain uprightness, enabling the tissues to release, for the breath to pass through like a wave reaching to the tips of the toes and into the underlying surface of the earth. On the other hand, if there is an underlying sense of imbalance in the body, of distress or discomfort, it is extremely hard to find rest and ease on the floor and the body needs instead to move, to discharge the stimulus and find some other sensory support or modality of being.

Lying on the floor is a place of return, of returning to the self and to homeostasis – a place for recuperation and rest. Lying enables maximum body contact with a surface, so the fluids can flow along the whole body without the downward pull to the feet that comes with standing. With the support of gravity, the fluids flow between tissues and cells, and in this position, the membranes and layers of cells can 'breathe' and detox. In a state of rest, other sensory stimulation from the environment can be quietened to lessen or annul the need to respond, to make voluntary movement – it becomes the place of 'do not disturb'. Coming into movement from this place involves awakening the senses, awakening the movement responses and rising away through rolling and then levering through the upper limbs to lift the spine up headfirst. The spine is the central axis, with the digestive tract and the main vascular system channels in front, and with the viscera, which are protected by the ribs and pelvis creating the container of the torso. Rolling this container around requires reflexes in the neck and lumbar regions that trigger a turning of the body in response to the turning of the head; turning the head initiates rotation in the neck and turning the pelvis initiates rotation in the lumbar spine. In either case, this rotation can bring the whole body into rotation or sequence through segments. Ease comes through practising the movement with initiation from either end of the spine and from either right or left sides.

With this information at hand, spending time lying down is valued differently to the assumption that this is a passive place, and that there is nothing to learn here. On the contrary, once these reflexes have been experienced, much is revealed to each person about their state of ease and connection to their body, and for each to revalue or come to understand them as a means to self-assess as well as self-regulate. Having spent time with these primary reflexes the tone of the body adjusts, as the nervous system discharges any latent effort or stress in the somatic system and the interoceptive state of the viscera also has time to operate with more blood flow to the system. From this place of rest, the body may be more available to new experiences and able to engage in movement explorations and research – having cleared sensory-motor pathways this place of returning to the self can be an attentive place to start.

The horizontal surface of the earth provides a safe environment in which the body can learn the interconnected sense of all the parts in relation to the spine, the axis around which it moves in rolling and with which it progresses headwards to the vertical (Bainbridge Cohen 2012). Having explored and embedded this initial sense of self and sense of a body 'schema' (Schilder 1970; van der Kolk 2014), these patterns remain in the body and are a core support when ambulating on the earth as a bipedal being.

Lying on the floor

Each modality refers to the sphere of movement that the body is engaged in, including its position relative to the pull of gravity. It includes the sensory affinities relative to this 'sphere of operation', such as touch and kinesthesia, that stimulate awareness and the support of the underlying reflexes upon which integrated movement is built (Bainbridge Cohen 2012). I start with the purpose behind the activity of lying and then refer to my experience of facilitated sessions, to note the key aspects and variations that can be offered, and acknowledge the innate need to change position whist lying.

Themes: Awakening awareness of body as a whole through connecting to the 'earth' and gravity through touch and sensory feedback (proprioception).

Facilitation starts by introducing what is going to happen, for example: 'We are now going to find our way to the floor and lie down. To do this we will spread out in the space and then find our way to lying on our backs'. It is important to be clear beforehand that there is enough space for all to lie comfortably. It helps to 'mark' the space by asking the participants to stretch out in all directions once lying, to feel the space they have around their body on the floor, which may involve meeting others around them. This inevitably normalises the event of making contact with another person; some automatically say hello and their names when they meet another, others just continue on, and it becomes more familiar as a group exercise. Be aware of anyone having difficulty in getting down on the floor and coming up again, or any risk of falling.

Facilitation focuses on body-mapping in order to have the sense of interconnectedness of the parts to the whole. The focus on sensing and feeling each position and body shape, through physical contact with the floor, brings awareness to the skin and the tactile sense, attuning to the sense receptors that bring awareness of temperature, texture, pressure and movement.

Whilst lying, information about the experience is provided by the facilitator who guides the experience of connecting or 'bonding' to the support of the floor – to the sense of yielding, softening and letting go of tension. If someone is unwilling to lie, they may choose to lean back or sit. Bring focus to the breath: generally, when lying, the breath slows down over time, as the body rests its weight on the floor. Invite participants to breathe in through the nose and out through the mouth, and to lengthen the out-breath (exhalation).

I lie

Version 1. Whilst lying I explore the idea of sensing the shape of my body on the floor – this requires attending to tactile feedback in each position. After a while, one side of my body wants to change position to balance the tone. I follow the invitation to turn onto my back and open into an extension, and later continue over to the other side. I note the changes in position and my tendency, like others, to flex the limbs when side-lying. I then turn to lie on my belly to see if I have a tendency to open into extension there and notice one side remains more flexed than the other.

Variation 1. Come to side-lying and introduce full body flexion on the out-breath (gathering around the navel area) and opening into full-bodied extension with the in-breath. This is a way to modulate the tone of the whole body, to regulate down any states of tension.

Version 2. This version is effective for participants who struggle to relax on the floor. Try using the image of lying in the sand, so the body feels softer on the underside against the floor as it turns. This invites a sense of yielding to gravity or softening the body tone, with a sense of following the shifting weight as the body moves in and out of position.

Version 3. Try using soft small inflatable balls to support parts of the body that feel uncomfortable, such as in front of the belly when side-lying, under the head, or under the knees in back-lying, etc. (see Figures 4.1 and 4.2).

Modality 2: Rolling (Figure 4.3)

Rolling is primary to our mobility as infants and so I refer to it as the second modality – following lying, this is the starting place from which the body moves progressively upwards to the next sphere of being: sitting. I start with the context of the roll, then describe 'my roll' and then the process of facilitation and the different types of roll.

This modality was referred to as 'the lower six inches', the level closest to the ground. At this level the experience of movement involves the turning of the

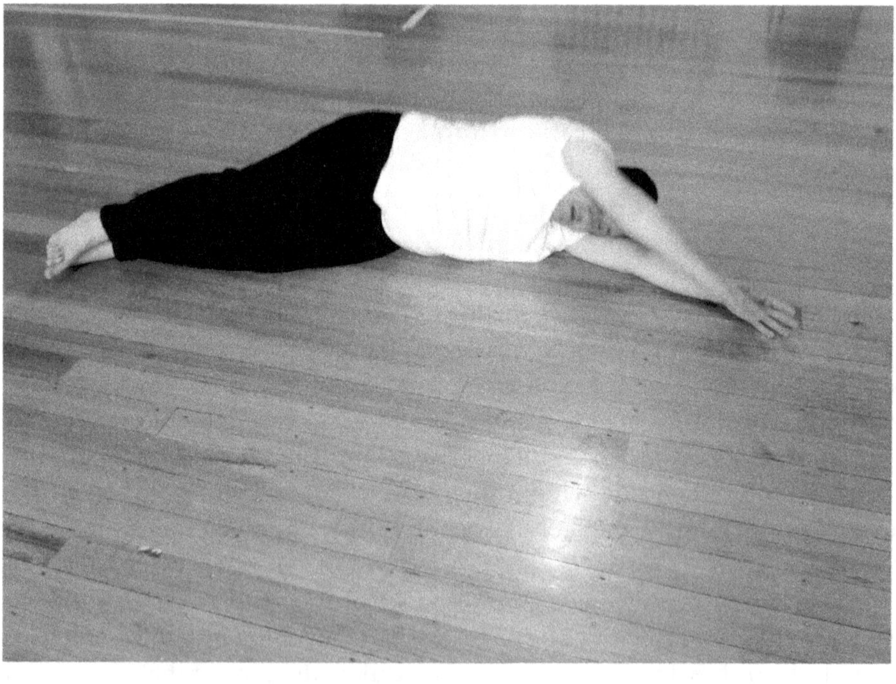

FIGURE 4.3: Crescent roll, connecting to the whole body.

head and torso from back-to-front lying or front-to-back lying, turning over one side, and some may experience the sense of falling in this transition. Rolling tonifies the core muscles along the spine and abdomen to prevent the sense of falling onto the floor and enable the momentum to connect and integrate the whole body. At this level, touch and movement are the primary senses that work together with the vestibular system to establish proprioception of the body in contact with the ground, the earth, from which we will soon depart to explore wider horizons, and to which we return to rest. The patterns experienced in these modalities are returned to on higher levels and surfaces but with less fear or possibility of falling due to unfamiliarity, as the practice on the floor retains the movement pattern available to be recalled or accessed unconsciously. I recall my exploration of rolling to illustrate how my idea based on previous childhood experiences evolved.

I roll

I experienced moving from a place within my body to roll against and along the floor and how this integrated my *idea* of movement with my *perception* of my body in movement – it established an enlivened and animate sense of body awareness. I have heard about rolling, and I have a sense of what it is, but I am not aware of the different ways of rolling.

Lying on the floor gives me a sense of my body surface, size and weight. As I turn from one side to the next, as I do in bed, I feel my whole body moving. I learn to roll my body by starting with a turn of my head that continues down my spine and I continue, my arms and legs lengthening as I arrive on my back or belly and flexing at the knees and elbows as I arrive on my side. I notice my body does this on its own and sometimes I knock my knees and elbows on the floor if I go too fast. I try to go smoothly with the understanding of allowing my limbs to lengthen on my back and front and actively keep them outstretched as I roll on my side and continue rolling like a tree trunk across the floor. I may roll into another person but we respond and roll away again, or just pause and rest.

I learn that I can push myself over using a hand or foot on the floor to lever myself away from the floor and then to roll around to the other side. I can also reach over to the opposite side, reaching with the right hand over to my left, feeling the spiral through my body till I have rolled over from my back onto my left side and then onto my front.

I learn that the flat surface of the back and front of the body means I have to engage in turning around the underside of my body that is on the floor and lifting up the side that is moving off the floor. I can trust the floor to bear my weight and I feel lighter as I am moving compared to when I am just lying on the floor still.

After a while, this movement feels familiar, like I have been here before, and I start to feel at home in my body and appreciate the sensation of movement travelling through me and taking me into space.

As I roll I feel the momentum of my weight and this allows me to continue to roll if I wish. My weight falls into the floor as I roll from my side to my front, and, like a wave on the beach, the momentum continues so that my roll continues along the floor till I stop. This sense of movement in space is very liberating and I take care not to roll too far or too fast as it over-stimulates my inner ear and I can get dizzy or nauseous. The sensation of momentum, of my weight falling and catching itself, stimulates a space–time sensation that opens my mind to perceive the possibility of continuity and timelessness.

Rolling provides me with a way to connect to my whole body, and in rolling I perceive my body moving and feel more fully connected. It is like I have crossed a bridge into unknown territory and had lots of surprises and discoveries that still remain with me, like a mapping of my inner being, a sense of body awareness and of knowing what that means.

Facilitation relative to CI

These first two modalities create a baseline of connection to self through the body, and although we may expect such states of being to be common to all, there is still a likelihood of differing responses and individual 'histories' coming to the surface. From this baseline experience, it becomes part of the 'work' to share both the physical aspect of 'doing' and the affective aspects that may stimulate memory and enliven a felt sense or state of consciousness which may not be familiar. Some people are not keen to lie on the floor, they may come to sit or keep their head lifted in order to feel secure. The purpose is then to invite the action of rolling – to initiate movement from the discomfort of lying in stillness and in so doing to introduce levels of coordination and alignment that may not yet be available as a resource. In discovering purposeful movement, support for other habitual states enables freedom to move through them. It is the moments of verbal interaction in the experience of partnering that the bridge between body–mind and movement is made and learning takes place.

At the start, as with lying, this primary and nurturing place supports the mover to connect to the body and all its parts, to have a sense of trust and agency. To roll on the floor together with others can be a liberating experience. Starting with a basic roll like the 'tree trunk' roll, of the whole body in one piece, there is no sense of right or wrong. Each person executes the roll in their personal style and in their own time, to discover certain idiosyncrasies, their level of physicality and confidence in moving. The facilitator can then attend to tension or patterns of holding and invite more extensor tones to lighten the body or prevent knocking the knees and elbows (see Figures 4.1 and 4.2).

Having explored the basic roll, other types of rolling can follow over the next workshop sessions or be interwoven with coming away from the floor. The purpose is to continue mapping body shape, the whole-body form, and to re-emphasise the centre–periphery connection.

The facilitation will progress to specific types of roll with simple changes to the point of initiation. These, and the dialogue through hands-on partnering, support the mover to sustain the body shape that forms the pattern. The progression can be indicated through a clear demonstration of the steps and stages of the roll, using hands-on prompts to clarify and guide away from loss of form or ineffective patterns towards more integrated and effective ones. The key aspects, or teaching points, remain consistent to the experience; letting go and releasing into gravity; turning the whole body around the axis; taking time for the vestibular system to connect with proprioception of the form and being with the organisational baseline of the horizontal plane. Key to the roll is the rise and fall of the centre of gravity and the momentum generated by this movement, as the initiation from the centre connects out into a full body pattern. The process is a cycle, starting with the stimulation of primary sense receptors in the skin, joints, joint membranes, tendons and fascia, to generate primary reflexive responses, which involve muscular contraction and lead to a change in position. The initial movement pathway evolves into a formalised, voluntary and purposeful movement – which relies on practice to attain a sense of organisation – which in turn relies on this body–mind feedback system for improved effective action to be realised. Eventually, the movement becomes integrated and fluid with ease and enjoyment – it becomes part of a movement language that informs and embellishes movement on other planes, like the rolling point of contact.

Facilitation involves providing a focus for further exploration and the introduction of other types of rolls, or qualities of doing them. In each case the focus is explained whilst being demonstrated, and this takes place for visually impaired people by tracing the mover with their hands-on key landmarks such as the centre and peripheral points, and by tracing down the spine to feel the axial formation.

a. With a focus on the underside of the body as it turns and meets the floor – such as starting side-lying with one hand on the partner's belly and another on the front of the limbs to feel the body turning over that part and meeting the floor. With a focus on the back side as it rises away from the floor, up onto the side that is moving into the floor, giving a sense of sidedness rather than front vs. back.

b. Focus on the centre of gravity – turning around that. Feel the centre drop closer to the floor on the belly and rise up higher from the floor on the side, and again on the back and next side. The centre takes an oval pathway and momentum comes from the shift in level and increased gradient from the short to the long side. By placing a hand front and back you can trace the flatness of the body on the belly or back, relative to the height of being on the side.
c. Focus on the spine, the sense of an axis, with one hand on the sacrum and the other in the middle of the spinal column on the back. The front of the vertebral column is deep to the centre of the body in front of the spinal cord and behind the organs.
d. Focus on the round cavities of the skull, ribs and pelvis. Roll around the shape of the skull, ribs and pelvis. They are all different sizes and shapes, the head being the deepest front to back, the pelvis or rib cage being the shallowest and widest.
e. Focus on the form – e.g. a spiral roll. Initiate distally from a hand reaching across the upper body in the direction of travel, this turns the upper body first and a spiral through to the opposite limb results. The same can be achieved from a foot. The roll proceeds around the outside of the limb, leading from the fingers and is very effectively traced by a partner with their hands. The same can be done from the feet, taking a foot and tracing a spiral to the opposite hand.
f. Focus on curving the spine laterally into a broad 'C' curve flat on the floor (see Figure 4.3) with the limbs extending this shape, so the hands and feet are also to one side of the head and tail, forming a crescent shape. Always maintaining the hands and feet to one side *and ahead* of the navel, the roll continues onto the belly, the other side and back onto the back. The inner curve of the crescent is the shorter side of the body which faces the direction of travel and the longer side is the furthest away. The opposite can also be done – rolling towards the longer side. This pattern requires hands-on tracing to help maintain the crescent shape for longer as the tendency is for the crescent curve to flatten out on the back or belly.

Interaction and engagement with a partner

Whilst this 'puzzle' can be done independently by sighted people, facilitation helps the mapping of the positional changes. In a Touchdown Dance workshop, the sighted partner does each roll with a visually impaired partner tracing the pattern as described, pausing with the movement when it does not feel clear. They then swap roles and pause again or 'rewind' when necessary. This dialogue is peer interaction – it involves social skills and mindful verbal exchange during the physical interaction. This means that fears and inhibitions surface in the process, and with mutual support, it is possible to learn to override them – through formalising the movement and feeling it concretise, yet paradoxically become more fluid, these fears and inhibitions are moved in the same way. The partners become each other's research participant and their own teacher-subject. A shared sense of achievement takes place at whatever level is reached, with a feeling of what is now achievable, before adding in more information or identifying a point of focus to assist them to go further.

For some people, a particular type of roll can be more of a challenge or an alienating experience, for example for those who have a negative body image, whilst the same roll might be a positive and affirming experience for those who do not. Having experienced the prompting and support to accomplish the roll, the effect can be healing and transformative as body image and movement experience come together in a reciprocal and mutually affirming way (Gallagher and Cole 1995).

Modality 3: Back-to-back sitting (Figure 4.4)

Themes: Feeling confident to say 'yes' or 'no' to taking weight; initiation and following; taking responsibility; listening; tuning to another's physical strength; differentiating ability.

FIGURE 4.4: Back-to-back – variation 1.

This partner work involves relating to others through movement and feeling the potential of a duet form. The facilitator needs to introduce the concept of the under dancer and the over dancer; the under dancer will be taking weight and supporting the over dancer who learns to give weight efficiently. It is also advisable that two people of similar size and stature team up for this exercise, where possible.

In these exercises, the ability to yield to a surface such as a floor or a partner enables a sense of the whole body in movement, and further movement can then be initiated in succession from a place of softness and ease rather than tension and withholding. The back-to-back exercise establishes a realm in which sight is not useful, regardless of who may have the vision. The point is to sense and feel through body-on-body contact. This exercise is done early on in the workshops to establish a baseline for communication through touch and movement, and an acceptance that both can participate in the same exercise whilst having very different or similar flexibility. This exercise starts from a concrete base position and returns by reversing the pathway back to this base – i.e. sitting back to back. It then develops into variations that progress from the same starting position and return again.

Whilst the first modality involves the sense of 'yielding' to the surface of the floor, this modality involves the support of the back surface of the partner as the stable surface to return to, and the floor as the surface to move on. It begins by finding a good 'upright' sitting position so that as far as possible both are sitting with the pelvis under the ribs, with the weight balanced across both sides, and the spine erects with the lower back and neck curves leaving a space between the surfaces. The head rests on the top of the spine, perpendicular to the floor and facing the horizon. The legs are extended if slightly flexed at the knee when the muscles are tight. After a few minutes of finding the optimum position, there is a quiet moment to sense the landscape of one's own back against one's partner's and to notice the movements taking place in one's own body and theirs. The facilitator then indicates that one person folds forwards over their legs, slowly, taking their partner 'with them'. This means that if my partner folds forwards I allow myself to extend backwards over their back, my head may rest in the curve of their neck or between their shoulders. The weight of my body rests on their back and we breathe, with each breath we may hesitate to 'let go' but the facilitator encourages us to extend the out-breath over time so that the body softens and the one underneath deepens into their forward fold and I, on the back, provide the weight that enables that to happen. If my partner cannot go any further or feels it is enough they initiate the return to sitting and I comply, we return to sitting up and feel the change in our whole body tone. I then fold forwards over my legs, taking time to ensure I do not leave my partner 'behind'. In this role, I follow the same instructions and return in the same way.

Following the experience of yielding to the support of another person, we discuss how it felt. For some people, whether visually impaired or sighted, it is challenging to move backwards into space, particularly if this is not a familiar movement. Allowing the head and body to go backwards triggers physiological responses to hold oneself up to prevent falling, so the act of trusting the person behind takes time and a gradual descent is important. The displacement of the head backwards without support will activate reflexes heightened by the fear of falling. It may take several attempts at going a little way backwards, then returning to sitting up so that the cooperation of the person underneath is established, as is the sense of listening and receiving the weight of the partner on top. The under dancer provides the support and so goes at a gentle pace to avoid the over dancer feeling afraid.

The under dancer, the weight-bearing person, also needs time to trust they can take the weight of their partner as it comes into their back as they fold forward. The key instruction is to maintain as much surface contact as possible so that the weight dissipates over the wide surface of the back, along the whole of the spine, and down into the pelvis and floor. So both the person yielding their weight to the supporting partner and this supportive partner need to explore and experience their own limits and communicate them. This can be done in movement, as the one underneath will pause their descent forwards if the weight gets too much, using their arms for further support of their upper body by putting their hands on the floor under their shoulders, to either side of their legs. This added stability assists them to return directly to sitting up, using a push to come up through the back and arms. The over dancer can feel this movement coming up underneath their back and will then respond by coming back up at the same time, following the initiation and using that support.

When a limit is reached for either partner it becomes evident by stopping – it is a felt sense, like saying 'no' to more. As in any dialogue, it must be possible for those involved to stop at any point; so either can communicate to their partner, without any judgement or protest, when it is enough. The supporting person has the responsibility for their own limits and their partner's safety and should not remove their support suddenly from under their partner but return to sitting, where both return to their own support – sitting on their pelvis – whilst remaining in contact.

Weight exchange

If the weight has become too much for the under dancer to push back to sitting, it is best for the one on top to use more effort to sit up, rather than to separate from the one underneath. In any event the person on top must not push their weight onto the support but allow it to be 'taken', to follow the one underneath, and finally to let go when it is established that it is OK to do so. The pushback from the one underneath supports the return to the vertical, signalling to the one on top who then complies fully, aiding the joint return by following their initiation.

Reciprocal support

These subtle shifts provide key non-verbal signals that establish the movement relationship and a sense of mutual responsibility. Through the physical contact and regardless of not being able to see the other, each is attuned to the movement that they both experience, to the idea of their specific roles and to the idea of support and orientation to the support. As they are both seated on the floor, there is little chance of injury with the centre of gravity only a few centimetres from the ground.

Expanding and condensing, using gravity to go further

The person who moves back over the support is lengthening away from their centre and over the centre of the person underneath so that the weight of the body on top moves through the support and into the floor until they come back to the vertical and take their own full weight again. This information is given by the facilitator to provide reassurance to the support person, and for both to realise that gravity is on their side. Gravity is key to how we organise movement towards and away from the vertical plane, whether into the front or back or whether side to side. It is also important to understand the concept of the centre of gravity in the body and how to organise around that.

The kinaesthetic aspect of the embodiment process is where movement learning takes place, and the density of weight in the physical contact creates both the sense of support for ease and trust. This is why such weight-bearing exercises are important in the early stages; the body gets to sense and measure how much weight is useful, beneficial and feels good. In this body-on-body contact the movers receive a sense of their own body schema, its size and shape are revealed against another body. Such dimensions and physical differences are not otherwise perceivable to many visually impaired people.

Variations

Once mapped and clear, the under dancer becomes a base of support. Other versions of the back-to-back position open up as the dialogue between the dancers becomes more trusting of managing weight and the role exchange. From this position it becomes possible to move further into the supporting role and in other planes, to explore variations and more mutual roles.

Forward/back. *Stage* 2. This version of the exercise is possible for those flexible enough to bend forwards over their legs. It should be progressive, depending on each person, and as it involves full-body support it can be helpful for anyone with stiffness and previous back injury if done with care and comfort. The under dancer reaches a comfortable forward bend over their extended legs, their hands placed under their shoulders to support the weight of the over dancer, who is extended backwards over their back with their arms extended to take the upper back into further extension. The over dancer can bring their knees up and place their feet on the floor close to their pelvis. This lifts their pelvis up off the floor and by pushing from their feet they can slide headwards up the back of the under dancer to lift their pelvis up higher. With the arms extended over their head and reaching into space over the under dancer's head, their body is long and the feet are the only part in direct contact with the floor. The whole of the over dancer's spine should be in contact with the under dancer's back to ensure weight is equally dissipated through the whole surface. To return to the base position, the under dancer initiates a return to sitting up and the over dancer slides back down, returning their arms to their sides and, once sitting on their pelvis, straightens their legs out ready to reciprocate.

Side-to-side (Figure 4.5). Having done back to back with weight-bearing exchange, the next starts from the same sitting base position. The arms are then

FIGURE 4.5: Back-to-back variation – side to side.

lifted out to either side and extended from the shoulders. The palms of the hands face forwards with thumbs up, and the back of the arms rests against the partner. It is important to maintain back-to-back contact, to move at the same time to the same side, i.e. one will go to their right, the other to their left. Then tilting the head to one side (both in the same direction) the neck and shoulders follow into a side bend – so that the arms on the shortening side move towards the floor. In this side bend, keeping the backs together, the little fingers will eventually touch the floor, then both return, reversing the pathway, leading through the spine, then shoulders, neck and head last, to come back up to the vertical to sitting in base position again with the arms out to each side. The return may have to start from the base of the pelvis, returning as above, along the spine till the arms return to a balanced extension to the sides and the head returns to the top of the spine.

This is then repeated to the other side. Once this form is established, the partners explore moving a little further along the floor, sliding, along the underside of the arm till they both lie fully on the underside of the ribs and pelvis with the leg and arm of the other side lifted off the floor. Coming back up requires maintaining back-to-back contact with the initiation from the tail and sitz bone on the base of the pelvic half that is on the topside, reaching down towards the floor and then the rest of the spine coming progressively back up to the vertical with the head coming up last and the underside of the body lifting up off the floor sequentially from the upper arm to the little finger.

Doing this back to back gives the opportunity to feel the progression of the movement through much more of the body surface and to map the body position and the pathway into the floor more progressively. Lateral movement of the spine may not be a familiar movement and the progressive pathway in and out of the floor provides an opportunity for both movers to orient and organise their movement together with the support of the other, gravity and the floor. The key challenge is to orient to the front without turning around the spine. It is essential to remain front-facing so that the movement flows along the underside of the arms to bring the body to side-lying, and the return is smooth, with a need to override the tendency to lift the head up first rather than to rewind the pathway and have the head come up last.

Twist (Figure 4.6). This variation is spiralic in orientation and so less strenuous on the back. Starting back to back with the arms out again, and legs wider apart, both turn to one side and face or look at the hand that is leading the twist, and so going behind. Then return to the mid-line base sitting position again with the head coming back last and repeat to the other side.

Twist Stage 2. Repeat the same twist and facing in the direction of your partner's feet, extend your arm along the ground next to the outside of their leg and slide down along the underside of your hand and arm to lie adjacent to your partner,

FIGURE 4.6: Back-to-back variation – twist.

with backs together. Return by leading with the upper arm to reach around the back of your partner, coming up to sitting with the head coming up last.

Twist Stage 3. This stage brings in the under and over dancer roles again. Continue the above twist but one partner chooses to roll over onto their belly and the other keeps their back-to-back contact whilst swivelling around to lie across the middle back of the under dancer, still back to back. The over dancer extends their body, arms and legs fully, so the hands and feet contact the floor, and neither is in a twist anymore. To return, the under dancer initiates a spiral roll and pushing with both hands spirals up to sit and the over dancer follows to sit also.

Surfing (Figure 4.7). This is not the same as crowd surfing! In the lying position described in Stage 3, the over dancer lies across the back of the under dancer and places their feet on the floor ready to push and reach through their arms, and ready to extend their body over the under dancer. The under dancer starts a spiral roll in the direction of their partner's head. The over dancer extends their arms over their head with palms up so they can slide along their arms onto the floor as they are taken headwards over their partner's body. The over dancer takes a ride in the direction of the roll, surfing over the rolling body, sliding along their extended

FIGURE 4.7: Back-to-back variations – surf.

arms and holding their head up off the floor. The surfer can go off the other side or pause before their pelvis is fully over, so rolling back is possible. The heaviest part is to return with the weight of the over dancer's torso across the back, so the over dancer gets ready to push with their hands over their head as the under dancer initiates the return.

The under dancer can stop at the point they feel the over dancer's centre pass over theirs and then reverse the roll to take the over dancer back. The over dancer sits up when their pelvis comes over the side, the under dancer continues to spiral roll to come onto their pelvis and sit back to back again.

Caution

Be aware of tightness at the back of the legs (hamstrings) which will mean that some people will want to flex their knees in the forward/back variation. Invite them to go as far as they can without bending their knees to feel the gentle lengthening of the back of the legs – without forcing so discomfort is avoided. The idea is that the weight of the partner's body on the back supports the gradual lengthening of the back of the legs.

Tight muscles in the back of the legs can prevent the lower back from flexing forwards without bending the knees. Also tight muscles from the front upper thigh to the pelvis and spine limit extension of the spine and prevent an upright posture – so allow the knees to flex so that the hips flex also as above.

If there is an imbalance in the mobility of the two movers, ensure the one with greater mobility works to the level of the one with less, as they will still benefit and the one with less will not get injured.

Be aware of getting a balance in size and body weight in the pairs and also not to go into the surfing pattern with people who have back problems or if the over dancer is not able to monitor how to extend their weight over their partner's body.

Surfing involves the rolling partner facilitating the other to have a 'ride'. Both feel their body is accepted and have a positive perceptive sense of their body image – both should feel trusted to carry the momentum of the surf by rolling underneath as the other 'surfs' over the top. The facilitation focus is on the initiation of the one rolling underneath, whilst the one on top initiates a push from their feet and extends fully to be lighter and to slide more readily along the floor – in order to lessen the load and enable a more exhilarating and faster surf. This moment of listening for the impulse is not visible, it is a tactile sensation found by following the idea of the support initiating and the mover being ready to 'go' and to do what is needed to enable the sum of both movers to create something smooth and functional.

Modality 4: Stand on 'all fours' – The low 'bridge' or 'table'

The sphere of operation and spatial orientation of this modality is higher up than when sitting, as the under dancer and over dancer roles become more specific and collaborative in order to develop from the back-to-back sitting on the vertical plane to a horizontal one, where more movement in and out of the floor is possible.

The 'low bridge' is a structural reference, with one partner providing the 'bridge' for the other to move over, around or under. The hands and knee position or 'all fours' is a familiar place for the body in childhood but may be difficult for adults to resume as it requires an agile back and confidence to take weight on the hands and knees. With this in mind, the previous modalities are essential starting points for the experience of weight bearing using the full expanse of the back and the use of the limbs for the mobility of the supporting structure. Being on 'all fours' is a mid-way point towards standing and enables the movers to explore different ways of using the support to move in ways not possible alone, whilst being close

FIGURE 4.8: Low bridge from a squat.

to the floor. Facilitation of the baseline position is important again, as it becomes the place to move into and out of when the partners are more aware of the possibilities of each role and how to be cooperative. The under dancer and over dancer roles are very clearly defined.

As the exercise involves remaining on the knees, which can be bony and discomfort can be felt with the weight bearing, it is best to use mats or to wear knee pads that are used in sports activities or martial arts.

It is also advisable that two people of similar size and stature team up for this exercise, where possible.

As a sighted mover I sense and feel my shape and I see my body shape, looking down from the front to my hands, knees and feet. I have to sense and feel the joint angles at my knees and shoulders to find the 'box' position. I then extend my neck so that my face is parallel to the floor.

As a visually impaired dancer, I 'see' my body shape proprioceptively, through the movement of my body into the position. I follow the facilitator and the guidance of my partner assisting me to position my hips over my knees, find the 'box' shape and face the floor with weight equally on each hand or knee.

FIGURE 4.9: Low bridge – all fours.

Initial baseline position

Coming onto 'all fours' can be done from sitting, by turning to one side with the leg flexed and the arm extended. From there come to 'stand' on the hand and knee of the flexed leg by lifting up the torso and bringing the other arm and leg in underneath to stand on all four limbs. The knees are flexed, placed hip-distance apart, the top surface of the feet are on the floor and the toes are not turned under. The arms are under the shoulders, shoulder-width apart, so the spine is 'suspended' between the limbs and elongated from the head to the tail.

Standing on 'all fours' in this exercise needs structural detail to make a box shape that is strong but flexible. The upper surface of the back needs to be as flat as possible, so encourage a long spine with the pelvis and ribs horizontal to the floor. The head is on the top of the spine, the face facing the floor and continuing the length of the spine. The hips are balanced on either side of the spine and perpendicular to the thighs, the knees are under the hips with a 90-degree angle between the upper and the lower leg. The lower legs and top of the feet are also on the floor to provide a further weight-bearing surface. The shoulders are also balanced on either side of the upper spine, and the arms are perpendicular to the shoulder joint – straight but not 'locked' or fixed at the elbow so that they can flex outwards if needed. The hands are directly under the shoulders with palms down on the floor, fingers spread to provide a large surface area to support the weight of the upper body.

Focus of facilitation

This box shape is called a 'low bridge', or 'table', it is a stable structure for the over dancer to move over, one that can also move when required. The stability of this structure is felt by bracing the muscles around the joints, such as the elbows. The possibility to lower the 'table' means flexing the elbows and hips to fold down towards the floor. This is necessary if the over dancer feels too high or unsafe and wants to come closer to the floor to find their way off. Being on the hands and knees is not a fixed position, once it is clear, time is taken to explore movement reaching forwards from the top of the head and back from the tail, rocking back and forth over the hands and knees to feel along the spine. This also helps to 'warm' up the weight-bearing joints of the wrists and knees and to feel how the weight shifting can spread the weight into the floor through different surfaces. Coming back from the tail brings the body back over the foreleg and feet and shifts the weight from the knees. The lower leg and feet are also weight-bearing surfaces to spread the weight further into the floor. This can also be done sideways, shifting the weight from side to side to experience flexion at the elbows and hip joints, across the hands and lower leg/feet. These are small movements shifting the 'table top' in space rather than the table itself.

Weight-bearing limits – Caution

If the weight is too much for the support the answer is to know how to go to the floor with the weight safely, by folding at the joints, and avoiding strain or collapsing.

If there is difficulty sustaining this position, it is possible to take frequent rests by coming back to sitting or lying. It is recommended to take time to move back and forth from sitting to all fours, particularly for the visually impaired partner to feel this pathway is clear for them to manage autonomously. Then when they are the bridge for their partner they have already explored the transition back to the floor, which is necessary to take the weight down.

Explore partnering that involves very little weight sharing first: the bridge moves over their hands and knees, folding up and down, and the partner follows with one hand on their lower spine, the other over the crown of their head. The over dancer then places both hands on the shoulders of the bridge and compresses down into their hands for a few seconds, checking if the weight is comfortable. This can be repeated, increasing in duration with rests in between. Then the same with hands on both sides of the pelvis and compressing down into the knees.

Summary: Work in pairs of equal size/stature as far as possible and take time to assume the correct table position for taking a partner's weight and moving effectively down with the weight if needed.

Be cautious about the stability of the table before taking weight from the over dancer and before the over dancer gives their full weight. Vignettes of the different variations of the low bridge follow to illustrate.

How to offer the best support and to still be mobile? The stability of the bridge is important to maintain an effective support for the over dancer. Mobility is required for the transition from the bridge to one side or to lower to the floor and continue from the bridge to further low-level interactions. For ease and flow, the over dancer self-supports to lessen the weight.

Alignment principles for optimum support; when 'stacking' the bones over each other they should be aligned to the direction of the weight, and this avoids straining the joints and muscles. Weight bearing has benefits, as weight passing through the bones stimulates bone density and strengthens the bones; however, the over dancer has to take their own weight again on their hands and feet to assist when needed.

Making the bridge flow

I recall making a low bridge by assuming a position on all fours, with my hands flat on the floor and fingers extended directly under my shoulders. I recall my knees are at a right angle, directly under the hips, the top of my feet flat on the floor, soles upwards and the toes extended (see Figure 4.9).

My visually impaired partner traces my body shape along my spine, the back of my pelvis to my knees and from knees to my feet to check my feet are flat. Then from the back of my shoulders to my hands. We talk it through, and I feel my body shape and adjust from the feedback I get from his hands. His attention to my hands points out a slight imbalance of weight distribution on the outside of my left hand which I had not noticed. I then notice I have more weight distribution in my right hand and adjust my shoulders to balance it out. We swap roles and visual information assists my hands to tell him what I notice. His hips are tilted forwards giving more lumbar curve and creating a dip in his alignment. With my hand on his lumbar spine, he lifts up and feels his abdominal muscles working in the front – and his breath releases down into the pelvis. He feels more stable as a result. He also drops his head towards the floor, so

I suggest lifting his head to feel his ears in alignment with the horizontal and his face to the floor. He lifts up higher than needed, shortening the back of his neck. He adjusts his head back and forth and releases some tension in doing so.

Low bridge variations – Illustrations from partnering experience (see Figure 4.8 and 4.9)

Bridge 1

My visually impaired partner assumes the all-fours position, and we take time to verify the box shape. I assume the same position on all fours alongside him, in contact shoulder to shoulder and hip to hip.

My visually impaired partner explores the starting position, both of us on all fours, his right side against my left side. He lifts his centre up by extending his arms and legs under him. Positioned close to my left side, his adjacent shoulder, ribs and pelvis touch mine, he leans into my support with his right side and starts to roll to place his back onto my back. He slides off and has to regain his own support back on his hands and knees. It is good practice to find the pathway on and off so he can resume his self-support. After a few attempts, he manages to register the height of my back and lift up just high enough to place his adjacent side above mine, to then roll onto his back on my back, shoulders to shoulders, pelvis to pelvis and his spine over mine. I feel his weight drop into my hands and knees, and lower slightly as he makes an adjustment to continue over to come off on the other side. He reaches for the floor with his left hand and foot but finds difficulty continuing the roll for fear of falling. We take it slowly, I lower my bridge taking my centre down closer to the floor so that he finds his own support on his leading hand and leg (his left side) and then comes over onto his other hand and knee. I realise with him how the back-on-back position is a challenge and that it is unfamiliar to be off the floor. It is a challenge to feel confident in this position, to understand how high off the floor he is, but this is helped by having done the previous back-to-back sitting exercise and the twist.

For those with different heights, when the bridge is too high, it has to lower for the over dancer to come onto it, then rise up to the box position if the over dancer is confident enough to be off the floor and not too heavy, if so lower again to let them off.

Bridge 2

As the 'bridge', my partner takes the under dancer position, as the over dancer I place a hand on the top of his head and one on his sacrum. He moves forward from his head towards my hand, and leads back from his sacrum and tail towards my other hand, feeling the shift from the supporting hands to his knees and lower leg/top of the feet against the floor. This establishes a sense of the spine and its alignment between the head and tail. Keeping one hand between my bridge partner's shoulders, the other on his sacrum, I, the over dancer, place my centre (below my navel) onto the mid-lumbar part of his back. Taking my arms away so that my arms and shoulders are over the far side and my pelvis and legs were on the near side. It is possible to touch the floor with my feet and hands and explore taking them off, to be like a seesaw across his back. I then try to balance on my centre with my head, arms and legs all lifted off the floor, and my spine in full extension like a skydiver.

Caution

Watch that the over dancer does not push themselves onto the bridge from the side but places their centre on top of their partner's.

There are different ways to come off the bridge. Once on the bridge, the over dancer can explore gymnastic positions, such as a supported handstand, by placing both hands down to the floor and reaching up with the legs and feet. To come down, lower the legs, gently going back onto the feet, so the head and arms come up, and come off on the feet side. More advanced options come as the under/over dancers gain experience and change roles.

Bridge 3

Starting as in bridge 2, as the over dancer I place my centre, belly-down across the back of the bridge a little lower towards the base of his spine/sacrum, I initially use my hands and feet on the floor to keep my weight off, and from there explore giving my whole weight, so he can agree to take my weight at that point. This is usually with the weight flowing directly into his knees and into the floor. I try to roll up the bridge towards his head. This is the crescent roll I had practised on the floor but now over a bridge. It is easy to get disoriented rolling from the side curve over onto my back. It requires a supple back to remain in contact with the floor with my hands and feet. Inevitably my hands come off the floor and I move more quickly onto my other side, then I am at the top of his back and have to stop and come off to my feet. Neither of us can 'see' what is happening but he can feel where I am; he lifts his head up when I have reached the top of his back. I can then start to roll back down towards his sacrum, at which point he ensures that his toes are not tucked under but flat on the floor in case I roll off the end onto his lower legs. This is avoidable if the under dancer responds by lifting his pelvis to signal the end of his body or says so directly – but if I roll slowly I can stop at any point and come off, or my partner folds at the knees and elbows to bring his pelvis down to his heels and both of us come down to the floor.

The interaction of the dancers ensures that both are involved in modulating the relationship to optimise comfort and movement range. As the over dancer I modulate the weight I give to the bridge by using my hands and feet on the floor and by keeping moving, rather than hanging for too long. My partner, the under dancer, modulates the height of the bridge by lifting or lowering as necessary so that I know where the floor is and to safeguard us both.

My visually impaired partner was initially heavy across my back, so I had to engage my abdomen. He took his own weight when he could but as he rolled onto his back I used my arms and body to come up underneath him to give myself more active support and a firmer surface for him. I lifted my head at this point as my back is shorter than his and he decided to come off onto his knees. He realised it was the crescent roll and felt toned and stretched. It felt like a bridge 'too far' at first but he tried again and it came more easily.

The wave

In this version, three people placed side by side on all fours make a wide bridge to support a person lying across them, floating them by making a wave-like movement with their backs. This triple bridge transitions slowly by lowering themselves down folding back onto their heels, into a child's pose so that the over dancer can roll off onto the floor. This version is a trust-based exercise and another way for visually impaired participants, in particular, is to explore being off the floor and measuring the difference in body shape and size.

Familiar places – Integrating different modalities

It is helpful to recall already familiar structures and to build from familiar places. I remember the experience of rolling on the floor, of feeling my weight pour onto a surface. When I come to work with a partner on all fours, this time the surface I am moving on is of another moving being who responds to my weight, taking and meeting my weight so that I can move freely in a way that I cannot otherwise do without the bridge there.

Transition from back-to-back sitting to low bridge (see Figures 4.8 and 4.9) – in back-to-back sitting, one person folds their legs at the knee, knees perpendicular to the floor (so knees up) and brings their feet close to their pelvis. Reaching forwards over their knees with the arms and head, their spine curves forwards. Meanwhile, the other partner leans back on this support, brings their feet in close to their pelvis and pushes their centre up so that they are sliding up their partner's back and no longer sit on the floor. This momentum translates through the under dancer, who then lifts their centre up and reaches forwards to come over their feet into a deep squat and then onto hands and knees, taking the over dancer with them, whose legs now extend and their back is extended over the bridge – pelvis on pelvis. To come back the under dancer folds back carefully into a squat and to sitting and the over dancer comes back onto their own feet to sitting, to lessen the weight on the under dancer's knees.

For the over dancer this feels like taking a ride into the 'back space' and is very helpful to tune into the sense of 'falling' or moving backwards, headfirst with the arms reaching in the same direction.

Modality 5: Lifts – Pathways into space and back to the floor

Front to back lift: It is helpful if the partners are not so different in height and build. The under dancer stands in front of their partner, turning their back to them. The over dancer approaches from behind and the under dancer bends their knees as the over dancer reaches up with their arms and places the front of their torso against the under dancer's back, who bends forwards, taking the over dancer onto their back and by straightening their knees they lift the over dancer up off the floor. The centre of the over dancer (navel area) should be slightly higher than the centre of the under dancer so that the under dancer can lever them off the floor and distribute the weight over their back and evenly into their legs. The under dancer supports their upper back by placing both hands on their knees, arms straight to act as a brace.

The visually impaired dancer may find leaving the floor a great challenge as they then have no idea how high they are from the floor or how to get back down. So the demonstration has to show this clearly, so the visually impaired dancer traces the movement with the support of a sighted person, who indicates the gap between the floor and the feet of the over dancers – which is rarely more than a few inches.

Back-to-back lift (see Figures 2.1 and 2.2): This is the same as above but standing back to back, touching and finding mutual support. One decides to be the under dancer and bends their knees and to lowers their centre relative to the over dancer's centre, and slides down the over dancer's back to sustain contact. The under dancer bends forwards with the over dancer keeping their back in contact so the under dancer takes their weight. The under dancer scoops the over dancer up by levering their pelvis under the over dancer's, to lift them off the floor.

Other versions: Rather than back to back, the under dancer turns 90 degrees so they stand in profile to their partner, then bends over to have their hands

FIGURES 4.10: Aikido Arm – turning.

on their knees and the over dancer goes across the under dancer's back and is lifted up, they can then roll up the back as if on the low bridge and slide off onto their feet.

Front lift: In this lift, the person lifting (under dancer) stands with their back to the over dancer and bends their knees to lower their pelvis. The over dancer rises up on their toes, extends their arms up over the shoulders of the under dancer and brings their pelvis to the level of the under dancer's lumbar spine. The under dancer bends forwards, taking the over dancer with them onto their back and then straightens their legs to lift the over dancer off the floor (just a few inches). A visually impaired over dancer can still feel anxious about how high they are and be invited to reach back down to the floor with their toes, so they find it as the under dancer stands up. This ensures the over dancer remains as actively involved as possible in all stages.

Variation: The over dancer lies across the under dancer's back, as in the low bridge, and rolls up (doing the crescent roll) and comes back onto their feet. The ultimate end is to roll up onto the shoulders and into a shoulder lift.

In all these lifts, it is important to return to the rolls and body schema experienced on the floor, in the spiral rolls and aikido rolls for example, and the ability to extend the body into space to be lighter – to feel the connection from the centre to periphery.

The aikido arm principle is worked into the teaching in various ways as a means to connect from one hand to the other across the shoulders or chest, providing a surface to move along as with the back-to-back exercises (see Figures 4.5 and 4.6) and in Figures 4.10 and 4.11.

Shoulder lifts: The under dancer faces the over dancer and lowers onto one knee, offering the over dancer the top of their shoulder with the arms flexed at the elbow and hands ready to take the weight of their legs and lift them up to shoul-

FIGURES 4.11: Aikido Arm – moving into space.

der height, so the body is like a horizontal plank. The over dancer reaches up and places their centre on the shoulder as close to the neck as possible so the under dancer has to incline their head away to the opposite side. Having tried this once or twice to test being able to take the weight, it can be tried from standing – the over dancer walks towards the under dancer so they move in confluence and both centres lift up at the same time, and the under dancer can walk while carrying them (see Figure 6.2).

Caution: Watch the weight on the shoulder and protect against neck strain by avoiding falling onto the shoulder or taking the weight for too long – moving with the weight is lighter than standing with it. Full body extension is important to prevent 'dead weight'.

Once all these modalities have been explored and discovered, they become integrated into fluid improvisation, moving in and out of contact, sharing and exchanging support and mover roles. We work with this range of modalities, drawing on the skills base also in our preparation for performance devising and studio practice as research as we integrate them within a creative process and interweave with other forms of improvisational techniques according to whom we work with, as the next chapters explore.

5

Workshops: Our Partnerships and Projects Since 1994

This chapter provides insight into the contexts that Touchdown Dance has been working in since 1994 when I took over direction of the charity from Paxton and Kilcoyne and continue to do currently. The work with children and young people became a priority as many adults asked 'Why did I not get this when I was young?' and more so when improved freedom to move was tangible, visible and stated in feedback from participants, family, staff and peers. The purpose is to indicate the benefit of movement work such as CI and the scope of interdisciplinary practices that it provides within a care-taking, inclusive environment. As with adults, the work proved to enable visually impaired individuals of all ages, to have attention to specific limitations, to find ways of opening to new movements, concepts and relationships with others and space, that were otherwise lacking. Clearly these projects required administration, fundraising, project management and much preparation and monitoring, but as we established partnerships with other charities, institutions in education and health settings, we have sought to maintain these to reach those hardest to reach members of the community.

Over many years, our workshop practice has grown from the initial partnership with the Royal National Institute for the Blind (RNIB) to other organisations LOOK UK (for visually impaired youth); SENSE UK for sensory impaired (all ages and abilities); Henshaws Society for the Blind (with visually impaired children and some siblings on family fun days) and more recently with Action for Blind where we work with young children whilst parents met for peer support. It became imperative to reach children and locating groups was a challenge so we worked in schools for learning disabled children and mainstream schools, offering bespoke projects as well as operating as core partner for the national program of arts in education, Creative Partnerships. More recently we have delivered specialist input with grants from BBC Children in Need and delivered workshops in social and cultural centres or venues with funding from the National Lottery.[83] We have also had research and development awards from the Arts Council of England and grants for touring productions with workshop residencies in venues and dance agencies. We have delivered staff inset or training days in schools, arts organisations and institutions and presented at events at home and abroad. Internationally we have worked in partnership with LOOK UK (in Belgium) and with the British Council (in Spain, India and Sri Lanka) and travelled to perform or participate at events in Germany, Australia, United States, India, Sri Lanka and Portugal.

From the session notes I have kept dating back to 1997, I notice a sense of continuity since my first experience with deafblind adults in Denmark. The notes include participant, family and staff feedback, imbued with a sense of space and extended time for detailed self-exploration, an awakening to self. There is excitement in response to enticement and to the experience that offered the possibility for self-development, for moving beyond the habitual bodily sense and the normal movement range. I recall a sense of 'no going back', like the 'hatching' metaphor depicts (see appendix 2). On all occasions we operate with the same 'ethic of care' (Etherington 2007: 604) and pedagogic ethos, using the principles and person-centred approach in each setting. With the inevitable variation in participant's experience, each workshop offers potential and a wealth of learning to be shared by all. The application of CI was developed further with the addition of the developmental and somatic movement education approaches from Body–Mind Centering® (BMC®), theatrical interactive games, with the use of props, sound, voice, movement metaphor and by stimulating the imagination through play. I have selected examples that were well documented and consider the impact of the work on development, mobility, special orientation and communication skills. Where possible, I provide the recorded feedback from participants and reference the core foundational principles given in previous chapters and other theoretical underpinning that is available in research undertaken in the field. I begin by going back to 1998 and continue forward from there to more recent projects in education, I continue with youth in community settings with LOOK UK and finally I revisit a workshop series with adults from 1999 in Manchester.

Children (Figures 5.1 and 5.2)

Workshop provision for visually impaired children and those with additional needs

The first example given here is taken from one of the first workshops delivered to visually impaired children at a school in Bradford in 1998, where we worked together with two groups of partially sighted and blind children, one group of five children aged 9–10 years old and the second group of six children aged 12–14 years old. We were given basic information about the visual ability of each child

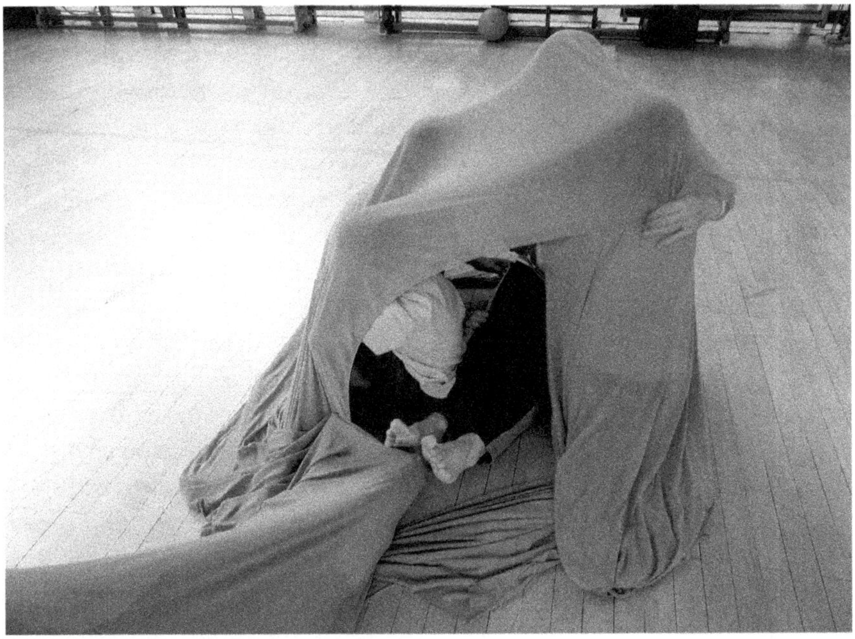

FIGURE 5.1: Workshops – visually impaired children under fabric.

FIGURE 5.2: Workshops – visually impaired children under fabric.

prior to attending, indicating if they had any useful vision and mobility issues. This information provided an idea of the 'base-line' ability from which to plan the session content and forms of interaction to uses. I review the first group session in which two children were partially sighted, two others had 'some useful vision' and with the fifth that meant three were 'registered blind'.

The space was a room with a vinyl floor, cleaned and airy as requested, with no furniture other than chairs at the edges. The first group entered in PE kit, I asked for us to 'make a circle' which was not a familiar task. The circle shape became more established after we shared our names, then I used some background instrumental music for a warm up. This involved moving the hands in the full movement range using description such as clasp and fan the fingers, rotate and turn the hands up ('happy hands') and down ('sad' hands). This playful sense was to awaken feelings and to see how they responded to emotional content with the movement instruction. Continuing to the elbows and arms at the shoulders, we moved the hands through space around the chest, in front, to the sides and behind, then with arms extended. During the first half of the session, I assessed the level of attention and response and grasped an idea of each child's ability to understand verbal instruction, spatial perception and their receptivity to explore bodily movement.

Breathing in whilst lifting the arms, and out whilst lowering them followed, inspired by yoga and Qi Gong (a Chinese approach to balancing the energy flow in the body) to address postural balance. The baseline stance involved standing with weight equally balanced over both feet, feeling the weight meet the floor through the underside of the feet (inspired by Paxton's 'small dance') and the whole body rising upwards through the top of the head. This brought attention to the feet and awareness to the whole body 'schema'. Slow pace and gentle flow were significant at the start to ground the energy and find a sense of stasis – as the energies were mixed; one reticent, another elevated and excited.

Holding hands in the circle, we went onto tip-toes to balance together, finding imbalance to begin with and using the circle to find support for standing on one leg, parallel knee and ankle 'bends' and so on. The hand holding indicated that the group was happy to make physical contact. An alternative (used in group 2) is to put hands on the shoulder of the neighbour but hand holding worked well in this younger group. Then we ambulated around the room with one leading the chain and passed under the arch created between the last two. I changed the step pattern to two steps down (knees slightly bent) and two steps up on the balls of the feet. Then we broke the chain back to the 'circle' and did variations, two steps and then a turn. The purpose of this was to see how confident they felt to move into the space, to orient to the room and take the lead.

Another exercise involving spatial orientation was done in pairs, with one partnering me. We did an imaginary journey around the space, finding our way as if on a rampage through an imaginary landscape, describing it as we go – here we have to climb over a rock, come down onto hands and knees, sit down and have a break by a river, lie down and roll on the grass, come up to walking again and then change roles. Each chose their favourite moment to share which we all repeated and linked together in a sequence.

Then back to the circle, now lying on the belly, we place hands palms down, next to the shoulders. Gently lifting the head up to listen to my voice, extending the neck, we lift the shoulders off the floor and rest on the forearms, sphinx like, pushing backwards/feetwards from the hands. This brought effort and energy into the upper body and some could slide on their bellies. This BMC-based® developmental exercise enhanced co-ordination and a sense of agency for each child. Followed with rolling from the belly to the back and over onto the belly again, they felt a sense of balance on all sides. Sequentially gathering in the limbs around the navel and opening out into space, brought a sense of coherence to the whole bodily sense. Back up on to our feet we did open and close again but standing and using the breath.

I chose Nigerian JuJu music to invite free movement in response to the strong rhythms, the instrumental and vocal melodies and incorporated their movements in a sequence. Then back to a standing circle, we repeated the hand dance we did at the start to 'cool down' and share feedback comments.

This age group was comfortable with each other and happy to work together. They responded well to unlocking the imagination in a playful interaction which took them into movements that were unfamiliar and unusual to them; from sensing the flow of movement to changing direction and self-directive orientation in the space. In the handholding and in the partnering, they learned to accommodate a comfortable pace that their relationship determined and created a more concrete sense of trust. Inherent to these spatial relationships is the opportunity to work to the group rhythm and to their own, recalibrating their sense of potential to relate to others in self-movement.

I provided the form and each child found his/her way to author their own content in terms of movement and pace, spatial orientation and timing of transitions, feeling a space to breathe, to let go of the need to repress energy or emotion that conformism in school often requires. We didn't need to use objects as physical contact and movement with each other provided enough interactivity. The core principles established a conducive working relationship, awakened curiosity – and a sense of surprise. Follow up sessions went with ease as they had not expected to do such interesting things and 'dance' became demystified – a possible 'I can' replaced a sceptical 'I can't'. The group energy was positive till the end, and the teacher shared that she had seen them all do things she had not considered possible, in reference to the sedentary and ambulatory limits usually imposed on their day-to-day schedule.

The second group didn't do the hand holding chain but more partnering work, trust-based exchanges, sitting back-to-back and leading and following in an imaginary 'jungle' environment which brought talking/descriptive dialogue into the experience, and some sound effects.

Analysis and observation

Key to the outcome of the sessions was the stimulus and enticement into movement with reassurances of safety. Each child was acknowledged by myself and the group, both seen and heard, whilst co-participating. The focus was on the whole-body experience, supported by developmental movement theory and practice to inform the assessment and analysis.

First, I hold the whole space and guide them progressively from arrival in a circle to orientating to the space, to each other and to their self/body schema. Evident to the structure and content, the style of the facilitation, is the ethic of care – the attention to safety, access and equality, both prior and during the session. In the warm up, the gentle invitation to follow me, to be co-operative, and attentive to self, enabled me to 'assess' or 'calibrate' the availability of each individual to participate in the experience and in the group. The sense of containment brought about by this ethos enabled me to deepen the realm of the exploration, to travel through the three dimensions, from being upright on the vertical to moving forward and backwards through space, to lying on the floor, organising the sense of the spine lifting and rolling gently around the axis (Bainbridge Cohen [1993] 2012). The influence of BMC® approaches is apparent and appropriate. In the study of the body and ontogenetic development, Bainbridge Cohen identified the significance of the use of imagery in traditional learning, and how this is enriched through finding the 'kinaesthesia of it within ourselves' (1993: 64). This follows with any perceptual understanding of the body – any idea becomes more vital to the 'self' if it is sensed and felt in its original form – i.e. experientially. Bainbridge Cohen distinguishes these two learning styles as different 'mind' states – the one being the attentive mind conscious of the task at hand and the other being the mind of the material, the different types of sensation and feeling responses that attune to the bodily sense, to nurture new insights and future expectations (Bainbridge Cohen [1993] 2012).

Finding the hands and feet: Based on the fundamentals instilled in the work by Paxton and Kilcoyne and furthered in my own research through BMC®, this exercise aims to establish an initial 'grounding' state in the body, for effective channelling of latent energy. We generally start standing, moving the hands at the wrist as the familiar gestural, tactile and motile part of the body. The hand is peripheral to the upper limb and visually impaired people may hold a cane to connect the body to the ground in front of them or to meet the unseen world that approaches as they move into it. Without sight, these children may have missed the significance of hand to eye co-ordination usual to infancy. They may have missed the visual tracking of an object in space and the response of grasping it to then bring the hands together at the mid-line to explore it bilaterally. This response underlies the orientation of the body–mind, and unifies the tactile, kinaesthetic senses with stimuli from the other special sense organs (hearing, sight, smell and taste), to provide a sense of coherent self-perception (Bainbridge Cohen 2012; Kestenberg 1975; Sheets Johnstone 1999, 2010). Developmentally the hands are initially independent and then in joining together, they conjoin the two sides of the body.

Finding the feet: The usual developmental milestones in which infants come up from sitting to kneel standing and up to standing on their feet may be a challenge to visually impaired infants and children to find, with lack of sight to locate furniture and objects to lever into to bring their weight up. Without visual stimulus, the need to crawl, to come up to stand and walk around in the world is lacking and the movement range of the lower limbs may be limited. Cautious parents afeard of accident and injury may disencourage autonomous exploration due to lack of insight into how to do so safely.[84] Paxton and Kilcoyne (1993) researched the apparent underdevelopment of the muscles of the lower limb and its manifestation in lack of motility of the toes, of the feet, ankles and forelegs. These gaps are caused by incomplete development due to the general tendency to discourage moving which other research findings concur with. I consider an example as it justifies the importance of the developmental approach of our input for children at any age.

Underpinning research findings

The impact of movement exploration on the quality of life of visually impaired children has been evident over time in the Touchdown Dance workshop research. This became another motivator for working with children and has since been confirmed by more recent research into the negative impact of physical *inactivity* on 'locomotor dysfunction' and ultimately on the quality of life of visually impaired people. Research undertaken by Julie Jennings (2009), as National Development Officer for Early Years at the Royal National Institute for the Blind (RNIB), looked into early intervention protocols and child-centred team working.[85] Jennings suggests the need for improvement in communication between screening and referral systems and notes a faster more efficient service for hearing impaired than visually impaired children which impacts on lack of input from very early on – the most critical time. The National Service Framework (NSF) outlines the benefits of early intervention for infants 'Delaying early intervention can result in irretrievable loss of function or ability [...] or the intervention being less effective.

Early intervention has a positive effect both in terms of promoting development, and minimizing decline or regression among children with developmental difficulties' (Jennings 2009: 23).

Dale and Sonksen (2002) have argued that it may be possible to minimize the risk of developmental setback that has been identified in some young children with profound visual impairment if babies with a visual impairment are identified early and intensive health and education developmental support is provided in the first two years of life, including visual promotion (for form vision).

> Part of effective early support is raising awareness of the potential developmental, emotional and social implications of childhood visual impairment across agencies, to ensure prompt referral from Health to Education services. The early involvement of visual impairment education services can support parents' in promoting their child's early development and learning using the Early Support materials, and in particular the Developmental Journal for babies and children with visual impairment.
>
> (Dale and Sonksen 2002: 23–24)

Following these early workshops and due to the brevity of the session, I learned to avoid one-off sessions and to deliver a series so that progression could be monitored and further benefit experienced. This would also support staff to witness and discuss the individual and group impact. Over time, the sessions rarely went as originally planned in accordance with the ethos of supporting participant voluntariness. As the flow of material followed the participant interest and response to the proposals, participant inclusion was assured and relationships recalibrated. This meant that they were very rarely dissatisfied. If so, it would generally be due to complex underlying circumstances or conditions, such as trouble at home, ingrained lack of self-esteem or inherent problems with social engagement and trust caused by a learning disability or Autism. Occasionally a touch averse participant or touch sensitivity would also need to be addressed through adjustment of the task or type of touch and would generate new ideas to add to the options available. Touch aversion could also be a reflection of personal circumstances or of the culture imposed by the school. As the next example indicates, touch-averse members of staff learn to appreciate the conducive peer relationships that physical contact engenders in CI, and note its merits as an educational tool and as a form of communication that levelled the uneven ground for blind children.

St. Vincent's school

We have delivered many projects at St Vincent's school for visually impaired children, some who have additional needs. Since 2015, we have delivered workshops there due to grants from BBC Children in Need. Over these past 5 years, we have been delivering workshops and recreational activities at the school with support from staff. We have worked with a group of 11 children and as numbers diminished due to the pandemic, we have worked both one on one or with small groups.

We undertook to meet three 'differences' that we measured through observation, feedback activities such as constellation exercises, and participant led discussion; *improved sense of positive self-identity; Increased use of expressive language; More confident in engaging in group activity*. As I go on to explore, these differences were not necessarily met by all of the children, or to the same degree, however the overall benefits were clearly identified, visible to other staff, and reflected in their maturation and school progress. The Headmaster stated that all the children needed to 'do Touchdown Dance', such was the difference in character formation, gait, self-confidence and maturation. Initially the group included children from different year groups who benefitted from the social engagement inherent to each session and felt more integrated in the student community. Individually they gained in physical awareness as we introduced ways to explore their body image, spatial orientation and mobility, physical and verbal interaction with others.

At first, compared to sighted children, we found reluctance to access this movement-based activity. As with many adults, we found low self-esteem and disinterest with underlying repressed emotions, we found restraint and withdrawal in many of the children in their reluctance to move or in their resistance to explore alone. These issues were successfully addressed particularly in the interactions that involved role change, having a variety of small group activity, sharing feelings and achievements. Some of the children who lacked independence in walking around the school became able to find their own way with ease and this supports the aspiration of the headmaster who wishes for all the pupils to continue into a job or further education beyond school.

Our first group drew on a range of ages across the school. Eleven children were referred to the group for a weekly session of an hour of non-curricular activity, for the purpose of nurturing self-directive movement, the imagination and communication skills, the expression of feelings and emotions, to enhance self-confidence/esteem, social engagement and body awareness. Often referred to as 'body-language', the majority of visually impaired children lacked physical responsiveness and tended instead to focus on listening, it wasn't part of the school experience to learn about being aware or attentive to posture, comportment or ease with, and adherence to, the visually dominant etiquette, and so to conform to the exclusive social normality. With such a range of ability and personality, this was not deliverable within the curriculum – the gaps that were needed to be filled were developmental and primary to intellectual education. The best would be to progress through infancy and childhood into puberty with a greater balance of emphasis as the research referred to above indicates. Touchdown Dance has no ambition to do this, we can only contribute and offer support and inspiration, as well as guidance to the children, the school and to hope the individuals transmit this within their own lived experience into a changing society.

The first session

After gathering together sat on the floor, Paula Hampson and myself formed a circle and had our first group check-in. Some children sat upright, some slumped forward, some more reticent as they didn't 'want to do dance'. We start with this friendly get to know you circle, asking for a name and three words to identify things they each like to do. Saying one's name is easy the three things hard for some – inspired by each other the trend is most often to do with technology, screen time, occasionally listening to music, more rarely to sing or rap, or do anything physical, most often walking a dog or going shopping. One or two may have done dance before, and their enthusiasm is tapped into later. We talk a bit about what we will do together, how to work together, to support each other to feel positive and discuss ground rules. At this school, these were easier as we had no one with 'anti-social' behaviour or social engagement issues, to be concerned about. In other situations, we would have a chill-out zone with fabrics and a ball that one of the three of us would facilitate if needed. At this school, only one or two children had useful vision, and most had other impairments or mild learning disability, so the task would be to progress at a pace that could carry them all, and those of us with sight would hold the others safely from collision. The ground rules were about safe care and care of each other rather than 'rules' to manage conflict. In this group, the members did not know each other well and so peer pressure or interpersonal tensions were replaced by personal limitations for most, and for one, over enthusiasm. We all agree that we will make movements and explore new ways of moving and make our own dances.

After the check in, we mapped the space, a large gymnasium, by walking around it and finding our way to the edges. This happened in solo, but we accompanied when needed to guard against collision. We wished to allow as much autonomy as was safe to do. From the edges, we called out our own names to each other across the space. Then we walked back towards the centre to find each other and make a circle. We asked 'who did you feel close to or far away from?' and they all responded differently. One noticed: 'we can feel how big the space is!' To bring awareness to their bodily sense, we do deep in-breaths with hands on each side of the rib cage. We use expansive arm movements to amplify the sensation of inflating a balloon and reduce their amplitude as it deflates on the outbreath.

Mood and movement

Each took a turn to do a movement that expressed their mood, having done it they described it and we all did what we understood from the description. We complimented each person for their interpretation as they couldn't see to cross reference with each other. The following examples are in the present tense to be more succinct: **E**, slide steps to his left and flicks his head; **C** touches her knees and reaches up; **O** touches her ears and above her head and then reaches to touch the floor and jumps back up; **L** doesn't know what to do and just brings her face to her hands. **Lt** skips two paces sideways and back again.

Warm up

The purpose is to 'wake up and shake up', to orient to the self, each other and space, using self-touch and/or objects. The idea was to awaken awareness of the sensation of movement, (kinaesthesia) by moving each body segment. A gentle sensory start may entail rubbing the hands together and then placing them on the head or face and to work down the whole body. Then to press the hands on the top of the head and release after a few seconds to feel the rebound through the spine and the spine lengthen.

Feeling uncertain L asked to sit back on the bench and sat head down as if just waiting for time to pass. I gave her a small inflatable ball, she says 'I feel so tired, I can't concentrate [...] I need to sit out'. Paula continued the warm up and I showed L how to use the ball in her hands, and on her own body – to squeeze in, slap it and roll it on her body surface – she liked that. I then showed her to have it behind her back, to lean into it against the wall, to support her to rest. This helped her to sit more upright and set up another 'rule', that is, gave her and others the permission to opt out. Later in the class L joined in again, illustrating the principle that respects the right to say no, to negotiate alternatives. L's need in the moment was met; after witnessing from the edge and still included, she realised what it was we were offering and refound motivation to participate.

The session continued with stepping back out into the gym, taking short steps, then longer, fast and slow, pretending to be 'on the moon', then skating/sliding (inspired by E), walking on hot coals, then on soft wool. We all moved together. This was followed by taking an inspired risk and asking them to recall the movements from the start and doing them in a new, different 'quality'; O did her sequence like a 'leaf in the wind' (her words), C did hers 'spikey', E did his slower and with sharp emphasis. We then did 'freestyle' to music for seven minutes, all improvising and at times partnering or exchanging, particularly with C and others with no residual vision.

The mood scale

In the following weeks, the check-ins continued in the same format, they could share three things they had done since the last week, and then we added the 'mood scale'. I devised this as a physical constellation of each child's mood. It usually involved using a yellow ball at one end of the room (for sunshine, feeling good) and blue ball at the opposite end (for feeling 'blue' or sad), and each person takes a place on that line according to how 'happy' or 'sad' they feel. We also invite them to say how they feel, to find other words and unveil other emotions or felt states. As they couldn't see the balls, In this instance, we use a large piece of fabric that we all hold with both hands, standing around the edge of it, pulling it flat so we feel it as far as possible on the horizontal. We then lift it up or down according to how we feel – up high if we feel 'happy' or 'up' and down low if we feel sad or in a 'low' mood. This three-dimensional connection enabled the group to feel their differences and to sense the mood of the group that was otherwise unacknowledged as most could not see each other's faces, demeanours or postures, and missed the possibility to open up emotionally to each other. Repressed emotions were apparent to us and emotional illiteracy was also apparent as we explored other words for 'happy' and 'sad' or 'excited', and over the weeks new words surfaced to show they had been thinking about what to say; 'ashamed', 'unsure', 'upset', 'exhausted', 'boss', 'not good', 'so happy I want to announce this in assembly'.

Finding self-touch and self-movement

Whilst breathing consciously we asked them to put a hand on their own belly, then on the lower back, into moving and brushing down the body at the same time, to waken up the 'skin'. Then swinging the arms from the shoulders, not like windmill arms but loosely, around the body and up over the head – brings in the torso. This is inspired by release dance and martial arts as a means to feel the body weight move and bring awareness to different body parts.

Finding gravity – Releasing weight

In twos, one partner lifts up their partner's arm and drops it, so they have to try not to hold it up. This is fun for some but bemusing for others who can't let go of holding or controlling it. On mats, we lie down and lift up a leg, feel its weight and replace it, then the other, then both. We lift one leg and cross it over the other to roll over onto the side or belly. Then in twos of the same gender one (A) lifts up their partner's (B's) leg just an inch off the floor, B tries to release its weight as A spontaneously drops and catches the weight without it touching the floor.

This introduces trust in a simple way and leads to other 'trust' exercises in partners. In partnering, they tended towards choosing the same gender and we monitored height/size differences as far as possible and introduced the ideas in a progressive way. Standing facing each other, arm length apart, they find each other's hands and put them together, palms to palms and elbows flexed. Following instructions, they lean towards each other without stepping, to feel the sense of weight arriving in the hands. Each pair feels how to support each other and then push back equally from the hands to stand on their own feet again. This is repeated several times, and any imbalance becomes evident and usually fun. We contain the possibility of falling as they are more likely to lean in towards each other and enjoy that proximity, than to want to push away. Some however learn to push away so hard they have to take steps back and enjoy this lunge and vestibular stimulation. The idea of agency is emphasised as they then have to find each other again.

Variation – one pushes and the other walks backwards with the momentum rather than pushing back. When the impulse ends, they reciprocate – this is very popular and a good way to introduce walking backwards. Positioning in the space to avoid collision is essential.

The challenge of this builds on the gravity work and postural work that we warm up with (the breath and self-touch exercises). O found it difficult to maintain her own 'base' and wanted to step in, she struggled to work in synchrony with her partner, to mirror the coming forward whilst keeping her feet on the spot. C tended to grip her partners hands, interlinking her fingers and not able to stay palm to palm, but this stopped after a few weeks. The lads really went for this exercise and continued with the counterbalance version of leaning away from each other by taking hold of the opposite hand and finding ways to stand on one leg. E and his partner Lt took it further into one supporting the other to move around their base – which started to become Contact Improvisation (CI).

New movement qualities developed by introducing ideas like responding to 'gravity'. We locate the centre of gravity in the belly below the navel, and move very slightly from there, feeling the 'small dance' effect in the feet and rest of the body. Then we crossed the floor with balance as a theme – rising and falling, hopping, jumping and bending.

More qualities evolved each week and introduced group focused creativity; the movement sequences lengthened, each person added something new or tried something different. Volunteers led the group warm up and led to a wider sense of spatial orientation – both to the space and to the group, and reinforced key skills like learning to separate right from left, to work on each side, to move backwards or behind as well as in front (as forward is the more general orientation). In witnessing their process, we noticed limitations – such as C who struggled to flex her spine laterally, as if her vestibular (sense of balance) was inhibited. In response, we introduced exercises to instigate and develop a side bend; sat back-to-back with a partner on the floor, they feel their upright position, instigate a side bend and stop when they feel that is far enough and return to sitting upright. This repeats to move progressively towards the floor, which took time as learning to control the descent, to find the path, was challenging.

Six months later, we explored space more, by placing all the different body parts in the space around the body. We moved in all directions, such as lunging forward, back and sideways and the diagonal and took this into travelling in space. Each person's movement had memorable personalised characteristics. L liked skipping right and left rather than forward into space so I partnered him, standing by his side, I introduced skipping forward and backwards, from hopping first. O liked to go forward and back but struggled with balance, so we introduced working with fabrics to stimulate the vestibular system. Holding the fabric vertically like a motile wall, each took turns to roll along the fabric or to roll up inside it. By holding the ends of elastic physiotherapy bands between us in a circle or in small groups, we connected together to explore how if one pulls the others can resist (pull) or reach and we all explored these new patterns. C struggled to move independently with limited mobility and posture. She had some useful sight and worked well with a partner's support; standing behind L with her hands on her hips she followed L's movements. L enjoyed supporting C and eventually they changed roles. We witnessed the group's self-directive confidence grow: E started to do more floor work and learned new steps and qualities like slide and glide, and spatial orientation. He enjoyed partnering and once he questioned sitting back-to-back saying 'That looks like it's for over 18's', which indicated his sensitivity to interpersonal contact or behaviour. We discussed this together – he was shy of partnering and that was respected. Having done the exercise the experience clarified for him that it had a different intention in this context. Lt has found his way into the space and loves finding new full bodied 'moves' – he does martial arts at home and likes wrestling. His energy at times feels excessive but he always moves fully, energetically and safely, with a wide base between his feet, he strides out, never colliding – without vision he loved feeling the freedom of his body in the space, often skipping, turning, putting a hand to the floor and turning around it, folding in his legs to sit down. In the developmental work, we witnessed that everyone struggled to move from sitting onto all fours, and to transition to standing. By regressing through these developmental stages, they all found greater coherence to their movement.

Feedback

O (low vision): I really enjoyed dancing, making my own dances. I love Touchdown.

E (low vision): I am more confident with stuff like this, it has made me more active, it's been a great term! I am a natural, I feel like I am bringing out everything I've got inside me out on the floor. I takes off everything that is dragging me down. When I want to, it gives me energy.

C (less verbal with speech impediment, no vision): I learn new steps, yes more confident with what I can do. (At first, C could barely take a step into the space alone, she was now walking independently around the school, very confidently taking strong strides with the wall side hand just touching now and again as a reference. She really looks independent and confident finding her way.)

Lt (no vision): It takes up where you are, if there is a problem I face in the school environment it helps me release it out. It's boss!

L (no vision): I really didn't know what Touchdown Dance was at the beginning but now I love it. (Having worked with the ball on her own in the session and weeks later getting down on a mat with the ball under her belly): I find it relaxing, feels good, really good. I enjoy it, rolling on my body helped me relax, I feel exhausted, tired. Is it good or bad? I suppose it's good! It's about rolling back and forth, right and left (using her hands to clarify). The body is amazing isn't it! In science we are learning about the blood.

L started off unreceptive to the idea of dance, and unreceptive to participating, but she gradually let go, even to the point of supporting other girls in the class who were more reticent than she was. She voluntarily asked to work with H, 'I will show you what to do, H, look we start like this and then, you follow me OK?' She did so with such clarity and intention we were so impressed. L had learned how to access something new for her body and valued it and demonstrated this in her understanding to want to show and support others. From being self-excluding at the start, she became one of the team which felt like a seal of approval, one that we seek to achieve.

Fabrics and balls

The haptic use of objects is a vital part of the movement partnering and group work for all, particularly those with low or no vision, and those with other impairments. Taking turns for one to lie in the fabric and be rocked, rolled over and dragged across the polished gym floor became a highlight of each session. Some liked it very fluid like being underwater and rolled around as they were dragged, some liked it fast and tumultuous, others to lie rolled up tightly in it, inert and like a cocoon.

The fabric is reminiscent of cradling, of rolling down a hill, letting go and unleashing energy. It brings in receptivity to others through a dialogue directly into the hands and body. This awakens a sense of purpose and clarity of thinking about how to support each person to let go, to release tension, to feel held and supported, without direct touching.

Holding a piece of fabric is a tactile experience until you feel it being pulled on, in your hand, by someone else who is also holding it. This activity awakens the primitive clasping and grasping reflexes of the hand and the desire to hold on tighter or let go. The act of moving the fabric up and down together as a group brings in a sense of the space between the movers moving, the weight and dimensions create a haptic and co-ordinated experience. Two or four people (with one person at each /end or corner) lift the fabric together and feel a whole perceptive field open up before them. Initially tactile and kinaesthetic, the texture and elasticity are sensed in the movement of the fingers and hands, then proprioceptively in the arm and body, in response to the weight and activity of the fabric flowing through the body. This process can be broken down into three types of perceptive response: exteroceptive perception awakens awareness to what is beyond the fabric in my hand; proexteroception enables me to sense the pulling on the fabric from the three different directions; and through exproprioception, I sense the size of the fabric, its weight and quality (Carello and Turvey 2015: n.pag.).[86] These definitions usefully indicate the synchronicity of inner/outer attention and of the stimulus response/sensory-motor cycle of perception and learning (Bainbridge Cohen 2012). As a positive enjoyable experience, the fabric brings people together to dialogue from the tactile and kinaesthetic base line, as equals, attentive to a common shared object and to accommodate each other's needs and limitations. It is a proactive way to introduce a physical membrane through which sensory stimuli flow, are received and responded to, creating a collective sense of cooperation that didn't require intellectual ability or presumed 'knowledge'. The circle of fabric contains us all: we learn to sit, stand and lean into it, or move in response to it, both solo and collectively, learning to change our physical form.

The sensory loading of the body from touch and movement interaction can become 'tiring', as it enables letting go of holding and tension in the body. This manifests as 'tiredness', as a form of self-regulation which we monitor. For some, this regulatory process needs to evolve gently, to avoid compensatory feelings arising too soon – such as fear or overwhelm, or a desire to control others, or to exert too much energy too soon in disregard of others. We separate into smaller groups in this case. The experience of feeling others interacting with the same object is a magical moment. So much happens that doesn't usually manifest for a group of visually impaired children, they learn to sense and feel each other physically and in a spacious way that vision supports more directly in group activity for sighted, for whom whilst moving their peripheral vision encapsulates the space. The fabric substitutes the allocentric orientation of attention to others in space, that for sighted is provided by the visual field, it provides an environment of 'external reference frames' and 'spatial skills' (Heller and Gentaz 2014: 119–20).

Those with some useful vision tend to want to be more active in the fabric and those with no vision tend to slow down and be with the dwelling sensation, particularly when direct contact with others is involved. The sense of 'team' work took on a new complexion, the need to work together to support each other, to change the fabric, to test how much weight it could take, how to explore this realm together. In the fabric, it was as if it absorbed the sensory fatigue from being in school and directed a dynamic relationship in a clear way. It regenerated energy through recuperation and discharged latent unintegrated stimuli.

Interaction with the softly inflated balls provides a sense of agency, of helping oneself feel the body through the ball, to feel it change through movement interaction. It supports spatial orientation as does the fabric. It is a kinaesthetic and tactile tool, a benign instrument to engage with reciprocally, safely– and to modulate and find your own timing and place to be. The ball can be moved around the body, it can be used between the body and the floor, or another body – it supports the widening of the perceptive field, a sense of the kinaesphere around the body and of the three dimensions of space. The ball offers potential support for the body to relate to – to become conscious of itself and its own resources through tactile and kinaesthetic feedback.

Outcomes; social, emotional and developmental

In our annual project report back to BBC Children in Need, we celebrated the unconditional nature of the space, the creativity that we had all experienced and the ability to show care for each other in new ways. The predetermined three differences were – the use of expressive language; to build positive relationships and positive changes in 'self-identity' that became reinforced over time. The key ethic of care – safety and equality – are identifiable in the reported sense of self-discovery, body-oriented learning and physical development, emotional and social engagement, and overall maturation.

1. *Improved sense of positive self-identity*: By using strategies based in creative activities the children built up new or latent positive self-perceptions. We facilitate the body-sense and felt sense as a starting point, for each to attend to their experience and to notice consciously how their mood changes. We found self-touch, description and physical contact worked effectively with movement to experience new self-perceptions. Often excluded from participation in dance sessions, visually impaired children tend to turn away from dance, to say no to it. 'No' becomes a more familiar response, also from others seeking to 'stop' or protect visually

impaired people from harm. Even when we encountered a negative emotion, the 'rule' was to allow this to be expressed/spoken, to be seen, heard and moved. To start with, a negative response meant support was necessary, so it was taken positively; to choose or make a decision to not participate was honoured, to give space for reflection, to weigh up options and give the possibility for an opportunity to re-pattern the apparent negative percept through a positive experience. Children are affected by each other's moods and pick up on each other's behaviours – with visually impaired children, this is less instantaneous as they can't see facial expressions or body language as sighted children do. They can be reticent of showing their feelings in case others don't approve or concur as they can't read what other's think. In our work, we reinforce the positive sense of each person's limitations and contribution, each is *equally valid, equally seen, heard and accepted*. This is an ideal and dependent on the child's ability to communicate her needs; at times, these remain hidden or veiled by resistance and shame, or sense of incompetence – 'I can't see' so 'I can't dance', 'it's not for me'. Having experienced the Touchdown Dance workshop, this shifts to a 'can-do' attitude, a new sense of potency that carries forwards – as we witnessed over the four years of working at the school as maturity, affirmed by the head teacher also as improved autonomy and self-direction.

2. *Increased use of expressive language-how often a person expresses likes and dislikes, and gives more than a one word answer*: We worked with emotional language starting and ending with the Mood Scale,[87] and/or a 'check in'. We experienced different constellations each week and diverse levels of engagement; over-all, there was an increased use of expressive language, and disclosures, and peer pressure was negated.

   Movement language was initially very limited, body parts were unfamiliar, so naming and touching was crucial to affirm this learning. Over time they learnt to describe their own body parts and movements, as a form of audio description. This developed awareness, confidence and positive self-regard reflected in improved posture and sense of importance of the benefit they experienced; see feedback comments above.

   Our key starting point in movement and touch languages is fundamental to their learning – the preliterate 'expression' that manifests through participatory activity, enables each person to establish a connection to self. These fundamental interpersonal connections are dialogic relationships, negotiated by each person within the context of the given environment and purpose – both bonding and defending reflexive responses (Bainbridge Cohen 2012) underlie comfort, safety, emotional attunement and social engagement.

   A form of intimacy, of affection, is apparent; it is not deviance, it is already apparent in the love and affection shown to each person in the ethos of the school. Intimacy in this case is an innate aspect of a primal human 'condition' which nurtures ease, supports agency and so communication. Being able to communicate requires this 'intimacy' – it sets up a base line that manifests as confidence in taking ones place in a group, but is in fact, primarily, a sense of belonging and of complicit agreement. Whilst there is an aspect of collusion and complicity to this 'idea', the overall outcome is the provision of a place to be, to grow, to acquire or 'learn' social skills – and this requires space to practice, process and nurture, through experience, wherein questions and confusions, negative impressions and complex 'engrams' are untangled. This was very contingent on mood, on external factors such as home life and peer pressure. Staff would always share any concerns with us in confidence so that we were aware of those having a difficult time so that it could be part of the session to find a positive way of being heard or supported in order to then express appreciation or change.

3. *More confident in engaging in group activity: The right environment*: In response to the apparent mood shared in the 'scale', we created environments to move in together, from tighter areas with mats close together or the whole gymnasium where we were able to move freely through the vastness of the space, each at their own pace. The receptive mood and co-creative relationship enabled each person to find their place, to attune to their body and to connect to the group in a way that became much more explicit over time.

   Each person contributes to the group learning process. There are many young people struggling with interpersonal relationship and emotional intelligence. Many find it easier to succumb to peer pressure to be included rather than to be different. However, we learnt how well they respond and quickly adapt and change when the right environment, group dynamic and supportive facilitation provide unconditional support unless they put themselves or others at risk of harm.

The use of movement in the realm of mutual play creates a space for personal and community development. The physical expression of mood, feeling the joy of movement and dance to music, and relaxation on mats or with fabrics, brings out many sides of the self that are not stimulated in day-to-day life at home or school. The group starts to find reciprocal inspiration and collaboration builds from the collective sense of the co-created world. We were all amazed at how popular the long strips and the circles of lycra were, in terms of bringing the group into close proximity with each other in trust and collaboration. The fabric was an external stimulus to the imagination creating a membranous environment that held us together, a sense of focus and ease, so everyone felt *safe*.

Physical contact through play with fabrics or each other affirmed strong interpersonal relationships and formed a very good basis for ongoing maturation and identity forming. Social separation was broken down through CI as they learned from each other's limits and gained an ability to do new things. In summary, developmental movement approaches integrate physical, emotional and social 'intelligence'; movement intelligence supports intellectual development; lucidity in movement supports lucidity of verbal communication; movement provides embodied understanding of many aspects of self that remain undiscovered/unconscious and so untapped, till expressed. Used as a framework, developmental movement underpinned emotional and social development, the process of making movement in the intimacy of a peer group nurtured self-affirmation, choice taking and liberation from doubt and uncertainty. For each person, this process was both singular and affective of the whole group and the school community, to the extent that the head teacher stated that pupils should do Touchdown Dance first and then proceed to other activities.

Youth work

Dance your way to wonder.

(Dymoke 2000: 17, the National Federation of Families with Visually Impaired Children)

Our youth work takes place in partnership with organisations such as Henshaw's Society for the Blind, Action for the Blind, LOOK UK and community or educational settings. In all cases, we discuss the general level of ability and mobility. In a one-off workshop, we focus on play and interaction, selecting from the core material and depending on the environment – i.e. a clean floor, ventilation, and how they are dressed. At the start, we undertake a process of 'soft' assessment, for the first half-hour of a two-hour session for example, by doing a warm up and ice breakers that indicate the range of movement response in the group.

In 2000, I wrote an article for their journal called 'Dance your way to wonder'.

> Many of our older adults regret that dance didn't enter their lives when they were young.
>
> There are many gaps in provision, as well as a lack of understanding of the wider benefits of dance in terms of physical and spatial awareness, self-confidence, fitness and pure enjoyment. By creating shapes and sensing physical form, strength and agility, children explore new ways of working together and learn a new 'language'.
>
> (Dymoke 2000a: 17)

LOOK UK

LOOK UK is an initiative of The National federation of families with visually impaired children which have invited us many times since 1999, to deliver workshops for children, young people and young adults up to the age of 29. This included international events as when three of us (Janee Hall, Holly Thomas and I) travelled to Brussels and gave workshops to children and young people. We have also given regional workshops for family leisure days for a mixed group of children and young people whist parents have support sessions. Focused on fun, we adapted the Touchdown Dance core methods in interactive group and partnering exercises. I recall that this involved adapting the 'small dance' practice by standing and moving the torso over the feet and moving from the periphery, then partnering to integrate these such as by connecting hand to hand. These exercises stimulated reflexive responses, supported awareness of gravity, re-orientation to space and balance. Movement interactions in trust type exercises were popular, such as standing back to back and trying to sit down at the same time without holding hands. One or two partners found a way to sit down slowly, together and to stand again, others explored the sensation of falling by letting themselves drop gently onto their bottoms (with mats under them) and gathered themselves back up onto their feet. The thrill of managing to catch oneself from falling lifted the tone of the room and supplied a good dose of autonomy and co-operative dialogue between the participants. Social engagement was not easy for some young people who were used to being sat out during physical exercise, embodying the 'no' that is intended to 'protect' them or others from harm. This cautious but restrictive practice is a form of avoidance that denies the young person an opportunity and choice, an avoidance of working *with* them to find ways to include them (Kuppers 2014).[88]

In 2014, LOOK requested a series of workshops in Manchester to which young people travelled, either independently from places like Birmingham, or together with Vicky Smith, the events co-ordinator at the time. As usual, we discussed at length who the sessions were for and what activity would be appropriate to the group along with all the practicalities. The chosen themes for the day had worked successfully on previous occasions with young people so we felt ready, tried and tested. The focus was familiar, to provide an opportunity to explore movement, spatial orientation and body language, to build self-esteem and provide an incentive to the participants to engage further in physical activity.

The flier had advertised Touchdown Dance, a company of male and female dancers with and without visual impairment, offering 'the opportunity to create amazing shapes, with rhythm and contact through body movement in space' and introduced CI:

> **Contact Improvisation**
> An improvised dance form practiced by two or more people who keep a point of contact while moving freely without music. It is practiced barefoot with loose fitting clothing and has been described as 'art sport'. **Contact Improvisation** keeps the body and brain active, improves strength, flexibility and posture and brings people together through creativity. **It does not rely upon sight in any way and is entirely accessible.** It can benefit those with a visual impairment by
>
> - Improving orientation skills (you in relation to others)
> - Improving co-ordination skills
> - Offering transferable skills for mobility and daily living.

It gave two testimonials from 'people who have attended these workshops before':

> 'It was amazing to feel free to move with the support of a fellow dancer and I enjoyed everything.'
> 'I didn't think of "dance" in this way, it was unexpected and so inspiring.'

Vicky Smith was all too aware of the challenges facing visually impaired youth; their lack of autonomy and dependency on support to travel and attend activities, their lack of access to good quality and appropriate movement education and leisure activity. Having Holly Thomas, a visually impaired instructor, was key to instil a sense of trust and understanding, with the language and guidance being based in her personal experience. We taught the fundamental Touchdown Dance material and added some new group interactive activities with props, like 1 cm thick wooden doweling canes, 1.5 meters long, and strong ten-centimetre wide elastic physiotherapy bands, usually used to rehabilitate muscles after an injury. These were intuitively considered to bring the group together as a community with some indirect contact and reciprocal engagement in perceptive and communicative activity.

The two day-long workshops ended up being an adventure, and the final one, at the Lowry Theatre in Salford, was a celebration for LOOK UK and Vicky's leaving party. The first of the two workshops had up to eight young adults, all of them registered blind, with Vicky and myself as the only fully sighted participants. Interestingly most of them were male, and none had any former 'dance' experience.

Day 1. What we did

Morning: We congregated in the space and made a circle formation. We then spread out to have more arm room. We had a few verbal exchanges about the plan of the day and then went into movement.

Icebreaker: Each person in the circle says their name and does a chosen movement that they describe as they repeat it. We all said their name and did the movement as we perceive it.

Warm up: We used the ideas developed in school settings, starting by moving to music, moving the body segments and then the body as a whole in all directions starting from the feet. The warm up included stretches and indicated locations on the body in the usual way. We did leading and following in twos to navigate the space and to bring in ways to cue change in direction, level, etc. The tactile impulse clearly registered with two of the participants and proved to support them in other 'guide' situations, likewise to inform others. The work was enriching in terms of what we set out to do, providing improved confidence, body awareness and spatial orientation.

Each person did a movement idea that they liked – we create an open space for anything, like 'freestyle'. Having given examples like jump, wriggle or shake, they understood that the bar was not too high. They were all reticent to begin with, shy and clearly not used to having this type of experience.

Introducing CI – using the arms: We create a horizontal ring with both arms in front of the body. With each breath we 'inflate' the ring, so the arms are rounded from the shoulders to the fingertips, almost touching at the breast bone. Then standing back to back with a partner, with the arms still rounded, we turn to be side by side, then we roll around the circle of the arms to face each other. This spatial orientation was a movement puzzle that had to be done in stages till they rolled all the way around and arrived back to back. The purpose was to create a structure that serviced bodily movement and the partnership with another, but moving together around the structure each person explored turning, navigating and orienting towards and away from a focal point, whilst adjusting their own posture to be at the same level as their partner and to remain upright, not stooped.

Leading and following to explore the edges of the space and cross the space, walking and then inviting embellishments like jogging or going up onto tip toes, or backwards, etc. After lunch and after some reflection on this first half, we worked with objects, the physio-bands and the sticks which I come back to.

The evaluation comments that follow were gathered by Vicky on large-print feedback forms that we had co-written and the completed ones were sent to us. The comments reveal the significance of this experience – the travelling in itself was a challenge to half the group, to arrive and meet at Piccadilly Station and then travel by tram to the studio. The feedback on the day and the workshop reveals that they had fears prior to coming, as to what they would have to do and were nervous about having to 'show' their dance moves. Although Vicky had briefed them well, half of the group were nervous at least until after the warm up and having experienced the teaching/facilitating, they were put at ease. I use their initials.

How would you rate your movement skills before and after the workshop?

S: I thought my movement was good before, but I soon realised it could be better. By the end of the workshop, I felt stretched and freer.

D: I felt a greater awareness and understanding of moving around a space and I found at this moment created in me a trance like state which I feel was very much triggered by the body movements.

E: Before: I never really thought about movement before the workshop. After: I'm more aware of movement after the workshop.

Ml: I rate my movement skills to be good, but by the time I finished the workshop, I felt excellent with my movement skills.

M: I definitely feel that my spatial awareness has improved since taking part in this activity.

How much do you feel the movement workshop has impacted on the movement you make in every-day life, (please compare) before and after

S: Before, perhaps when moving around without the use of a mobility aid my movements would be quite tense. Now I'm more aware of my hesitations and I can work on improving my fluidity.

D: I can't think of a specific example of how it has impacted on me when I'm moving around but I am doing some of the workshop exercises when I do fitness stuff like press ups, sit ups and stretches. Before this workshop, I never really considered the space my body took up other than when I tried yoga. I will try some of the techniques when I am walking around an open space next time.

E: Before the workshop, I wasn't really aware of movement in everyday life, but after the workshop, I've noticed my movement is more controlled and confident.

Ml: I feel it has had a big impact on my movement and it has made me feel that I can do things that I didn't think I could do before the event.

Please provide any examples of orientation skills, co-ordination skills developed or improved

S: I am going to work on not dragging my feet across the floor.

D: The body contact exercises have allowed me to appreciate the subtleties that can be conveyed through guiding someone who is visually impaired and I feel I may have improved perceiving possible new feedback I receive when being guided.

E: I think my co-ordination skills have improved after doing the work in pairs in the workshop. E.g. if I'm being guided I'm more aware of how to respond to my guide's body language.

Ml: It helped to get me motivated and realise I can do dance.

Day 1 afternoon

I recall the stick work which we did in twos – it was an exercise I had learned during a theatre training with Philippe Gaulier in Paris, to embody the sense of timing and space in a movement exchange. Initially we did it simply, with one person at each end of the stick, facing each other, with the stick end resting on the palm of the hand with enough pressure to stop the stick from falling. They found a way into moving together from one initiating and the other following till the impulse ended and then responded in return. This dialogue is a physical metaphor for giving and taking, initiating and self-directing, or following and responding. The exercise created a reciprocal relationship in movement that took both into a range of movement beyond the usual day to day, forward and backwards and then around the space. The movement opened up as they became more confident – it was all sensed and felt and fun. Sticks did get dropped and retrieved and needed containing to prevent collision but this was minimal as each couple tended to explore back and forth till they got the hang of it. We then joined two couples together with a third stick so they could be in a circle or in a line, moving together and then tried connecting with the whole group. This brought much amusement as the sticks got entwined and we literally got tied up close together, and some found themselves in close proximity with others they didn't know so well, but with the ease of group non-hierarchical activity. Even the guide dog got excited at one point as we got boisterous and carried away with the improvisation.

Once more touch and movement became a base line for an embodied experience of self-movement that liberated them from the usual and limited levels of movement awareness and informed their mobility skills with subtleties and so more 'control'. The level of creativity and sociability reflects in the positivity expressed and this shows the ongoing need to support emotional wellbeing as well as physical and social independence.

The challenge of presentation

The 'movement piece' referred to below was a movement that each made to introduce themselves at the start, which I described to the others for them to find and 'copy' without seeing it. At first, many felt 'put on the spot' and want to think what to do and get stuck with 'I don't know'. This was very revealing for us all. Some could do a movement that they felt like doing 'now' spontaneously without thinking about it, others showed this was not familiar and a lack of movement vocabulary and/or self-confidence that informs us how much reassurance and support is needed for the rest of the day. Being the centre of attention and not being able to see people's responses is a major factor for self-conscious young people in particular.

How confident did you feel before and after presenting your movement piece to others?

S: Before others started, I felt a surge of panic sweep through me, but by the time it came to my turn I was confident.

H: I was a little anxious about presenting movement pieces to others, especially at the start when we introduced ourselves with a movement. As the day went on, I felt much more confident.

D: At the beginning, I was slightly aware of doing some of the movements wrong but the workshop was very open to our own interpretations and the leader was very helpful with describing the movements.

E: Before: not confident, after: confident.

Ml: I felt a little nervous at first, but I shared my ideas as well.

M: I'm not very confident at presenting, but I felt more confident as everyone was doing the same.

How confident did you feel before and after you created your own movement piece?

S: Nervous at first, completely confident by the end.

H: I was confident both before and after creating my movement piece – it was presenting it that I wasn't so confident at.

D: I can't really answer this exactly but I thought the workshop was a positive experience overall and that helped me feel happier generally as I'm going through quite a stressful patch in my life.

E: Before: not confident, after: confident.

Ml: I felt confident about sharing ideas, but this improved as the event went on.

M: The same.

Please add anything more you feel appropriate

S: It was a great day with lots of fun had by all, which was telling in the large amount of laughter and happy chattering.

E: The workshop was a really great way to meet new people whilst exercising and having fun at the same time. I would definitely like to attend more similar workshops.

This selection of feedback gives a real flavour of the experience and its value. It openly expresses some reticence from those who didn't respond to all of the points and indicates not knowing or feeling ready to respond verbally. This is delicate and further indication of the need for more opportunities – we don't expect everyone to feel responsive or to necessarily wish to continue, but in most instances, everyone who participates finds something useful to take away and continues to use the information and learning. It is heartening to hear that some of them are taking on board their own discoveries whether that is to practice their gait cycle (walking) and presentation skills. As each step into space is a 'fall' into gravity but caught by the ground, the natural response is to slide the feet along the ground to follow the surface and avoid stepping into a void or tripping on a raised part. A tendency to walk flat footed causes stiffness in the ankles and a rocking motion from side to side rather than propelling themselves forward, so lifting the feet may be a challenge and requires confidence. Walking with a cane helps with this, as does the work we do – evidently once the terrain is familiar the reflexive responses can soften and integrate.

Overall we felt the mood lift through the day and progressed with ease as the workshop experience became like a playground, a space to take a step at a time into a new adventure. Their postures improved with the shared sense of vitality and confidence, though for some, we observed insecurity and reticence, others had a greater sense of achievement. Small steps were acknowledged and valued and there was a desire to return for more. The teaching methods, including description, were successful in supporting each person to be with their movement, to engage with their own exploration. This experience would remain in body memory, re-accessible, and potentially adaptable to new or familiar day to day situations – with a sense of bodily knowing an a new attitude towards movement, expressed frequently by the word 'confident'. It was an affective experience over all, a physical activity that stimulated their bodily sense, where psychological and emotional regulation took place, as with D, and the motivation to do more such workshops in future was a decisive outcome.

Adults

There has been a lot of attention paid to the workshop experience for adults in Chapters 1 and 2, here I focus on a particular workshop mainly from the point of view of the participants. Having read the sections on the work with children and young adults, the root of the issues faced by the adults are put in context. The research undertaken by Paxton and Kilcoyne (1993) indicated that the need for access to movement education and practical experience is no different for visually impaired adults as for similarly sedentary sighted people. However, to do this effectively, the teaching methods had to adapt to accommodate lack of sight, whilst sustaining social engagement and interpersonal interaction: everyone stood to benefit from learning more about their moving body.

Touchdown Dance has continued to serve the learning needs of the participants, and, in particular, to be inclusive of people with visual impairment. This has demystified the stigma attached to the 'impairment' construct and questions the justification for exclusion caused by the polarity and the shame attached to mental distress. The diversity of participant experience was not a hindrance, but mediated by the commonalities found in the realm of movement, of being conversant through touch, which opened up an effective relational realm. The schism that previously divided and separated, widened as reciprocity occupied that space: participants' movement became more fluid and mutable.

In the early days, Paxton and Kilcoyne identified movement as the kinaesthetic sense, and rightly so, to emphasise the sensory aspect of movement, the simultaneous sensory-motor counter-current that characterises the innate reciprocity of the body with itself and the world. This embodied experience of movement is the bodily response to incoming sensory stimuli, the quality and tone of the response towards or away from it, so that in effect, both the stimuli (other) and the body are changed. We may only be aware of our own bodily sense being affected, but through CI we notice how, even at the level of the kinaesphere, any change in our bodily form, or internal landscape, changes the external landscape. Moreover, mutual recognition of this is apparent only if the reciprocal membrane is operative. If the facilitation style enables this perceptive state, the membrane metaphor becomes tangible as a result, but also elusive.

Multiple sensory modalities

From a developmental perspective, *movement* is not reducible to the *kinaesthetic sense* alone as all the senses initiate a motor response. In effect, movement responses arise from the intermodal functioning of the senses and the stimulus–response cycle underlies perception and so learning. Bainbridge Cohen (2012, 2018) outlines the developmental sequencing of the senses in human infant development to indicate their layering and intermodal function in perception and learning. For those without vision the spatio-temporal aspect of sight is missing, a difference that sighted people can experience by closing the eyes. As we experience in our work, other senses such as hearing contribute to an optical sense of space and to orientation in space, so that visually impaired people benefit from knowing the location of the source of sound and echo location (Heller and Gentax 2014). Neuroscientist Vittorio Gallese researched into *The Multimodal Nature of Visual Perception* (2016) and found that sight provides perspective and fast object perception, facial recognition and social engagement.[89] In our experience, when sight is privileged over the other senses, the focus of the intention to move is different and generally faster. This helps to explain the differences we experience in the movement responses of participants, but also in the similarities such as the general slowing down of the movement explorations when touch and hearing are the inter-modalities for visually impaired and sighted dyads.

In the workshops, we see how both visually impaired and sighted participants stand to benefit from movement education in the same purposeful ways, through the same holistic pathways, but with different degrees of touch based illustration. Touch can be supported by visual demonstration for the sighted who then receive further insight from their visually impaired partner, and sight can serve verbal description and spatial orientation. In this integrated context we see how touch provides a precise and embodied clarifying process for all, and this new attitude arises for those who didn't consider touch as necessary or useful.

Practical implications – Access and availability

A workshop for visually impaired and sighted people does not happen 'just like that', it takes a lot of time to reach and communicate with visually impaired people, to provide information in the right format, to gather information on access needs, to ensure that access at the venue accommodates these needs, to provide workshop content information and not to assume that the methods are 'understood'. It is necessary to lay out clearly for all that this is not a traditional dance class where you learn steps, or choreography, but a movement class which will serve to fill in gaps, enhance movement awareness and bring people together in that experience who would otherwise not meet. Not everyone wants to move in this way or to share a creative space with such a diverse group. This is not only about delivering dance – this is about changing the cultural landscape, a job that requires 'intercultural sensitivity' and a carefully considered developmental strategy in order to bring participants along with us as co-leaders of the process (Connerley and Pederson 2005: 39).

The effective impact of the movement methods manifests in the documentation of participant responses, such as the comments, letters and the poem by Jane Emerson below. Each workshop evolved with the participants, in the context in which it took place, with the 'intercultural sensitivity' of being inclusive. With this awareness, we create the foundational sense of 'shared values and expectations' and learn that this 'common ground – may be expressed differently in each cultural context', i.e. the location, the people and the intention of the work. The experiential process enables the work to grow with roots in the 'common ground', and for skills to then be determined – both subjectively and shared in the collective context (Connerley and Pederson 2005: 50). Each workshop is held within the Touchdown Dance framework, implementing knowledge and skills but within a space defined by sensitivity to difference and division. To effectively manage the workshop experience requires awareness and sensitivity prior, during and afterwards. These key considerations are reflected in the value given to discussion and dialogue within the workshop time frame and show how to provide the safe container, the ways and means for the project to happen safely – as the following example illustrates.

Move-In 1999

This project illustrates how Touchdown Dance workshops took different formats over the years. The residential workshops stopped in lieu of localised studio sessions which were planned and delivered by the company in London, Bristol/Bath, Glasgow, Manchester and elsewhere. We focused on working with children and young people to ensure the benefit was given as early as possible, whilst adult workshops continued to take place within the CI community such as the London CI classes and national jams. These one-off events tended to focus on leisure type activity in which the social aspects such as making friends and creating a sense of community were important, whilst other more durational projects could take a more incubational approach. Of the many we have done, I have chosen a workshop series that took place in Manchester as an example. *Move-In* was undertaken in collaboration with Dance Initiative Greater Manchester (DiGM) in 1999. Funding was raised for this from the National Lottery community fund, a fund that embraced projects that enrich communities by benefitting those most excluded from the arts and creativity, the least visible in society. Touchdown Dance could claim to have 'expertise', as required by the funding agencies and the ability to project manage and organise, but in practice this authority was set aside, to allow the group to create the content and co-direct the creative process. The workshop facilitators (Patrick Beelaert and myself) facilitated the movement experience, whilst each participant immersed themselves, self-led and/or co-led in dialogue with each other. This gave us the freedom to allow the creative process to unfold and be co-facilitated, and for the participants to adjust and accommodate their differences, generate activity and contribute to the learning we shared. The outcome was a performance installation called *2-Touch* which took place in Manchester in collaboration with Karen Nelson of Image Lab (USA) and co-founder of Dance-ability. Karen and Lisa Nelson (colleagues of Paxton) delivered a two-week intensive workshop on touch and vision for an integrated group[90] at Yorkshire Dance Centre in Leeds.

Prior to the workshop series, we risk assessed the venue, at Z-arts in Hulme, Manchester, to verify the studio was clean and ventilated, in line with recommendations from Dance UK healthier dancer campaign. We checked for any trip hazards, such as electrical leads, and requested the removal or safe storage of unnecessary furniture. Disability facilities and access were also verified as physically disabled people attended as sighted participants in our workshops. At Z-Arts, there was a lift down to the studio and ramps as well as stairs. To avoid access issues, we had to limit the number of participants according to evacuation procedures. Staff were briefed and were enabling and supportive. In the studio, any tables were kept in one area and objects on the tables moved back from the edges as the edge of the table could be used to navigate along. Chairs were available for personal belongings and shoes/canes and bags were placed underneath, and at times two chairs were available per person. Fire exits and toilets and other house-keeping information (such as fire alarm sounds) were identified during the first gathering to enable each person to be as independent as possible.

Prior to the workshop, we posted information for the blind participants in Braille (if requested), as we had a Braille embosser in the office as well as access to services in Manchester libraries, but nowadays digital technologies have pretty much substituted Braille and audio tapes. The purpose of the workshops was explained; we intended to make a performance together, to be performed to an invited audience. The nature of this performance was open to the group to define. The only predetermined factors were that it would be improvised within a framework of scores (frames of reference for each section) and movement based – with the potential for voice, sound, props and so on. The mixed ability group required a mixed ability approach that meant having time to question and discuss as each idea took shape; as to how to use the space, adapt the timing, and for each to decide, at the rehearsal stage, what they wished to commit to.

On the first day, everyone was met in the foyer and escorted to the studio.

The introductory circle was pragmatic and voluntary – we invited each participant to give their name, impairment or injury, or other limitation that other's might 'need to know', as well as where they had come from. On this occasion, the participants included one sighted man with Parkinson's, a woman who used a wheelchair, two who were registered blind, and three sighted, non-disabled participants and us, the two 'leaders'. Introductions and 'house-keeping' were essential to ascertain or verify the type of support each person needed. Traditionally we offered a 'buddy' system so that by the end of the circle time each visually impaired person had an identified person to check-in with or request support from – this person was either another participant, or at times a student in placement or a volunteer has been available. This generally worked out well but could be invasive if an over diligent 'helper' appeared patronising – the solution was then to allow the person to decide for themselves when to request assistance, from whom and then 'how'. Mapping of the space in partners reinforces this – and follows the circle.

In navigating the space, the 'ice-breaker' moments continued as the partners talked as they walked together, with eyes open or closed, around the edges of the space, locating any obstacles like the pillar on the one side. The partners passed each-other on the way round and having changed roles, they returned to the circle where they shared insights and surprises.

What is admitted and not always obvious is that we all benefit from these ice-breakers and introductions; it is not for the benefit of the visually impaired people alone, as the sighted may assume – as social engagement is the first step to breaking down the fears and stereotypical constructs imposed by segregation and exclusion. Everyone discovers something about the space and by touching and mapping it we are then better oriented to it. We start to meet as individuals, equally legitimate as participants, and ready to discover together.

A good hour was taken up with this initial opening but much was gained in terms of narrowing the gaps and dealing with concerns, nerves and discomfort. Anything arising that caused concern about the space was then noted – for example, how to recognise where the emergency exit was and to ensure that no-one was left alone in the space who felt anxious about finding it. By the time we started the workshop content, it was clear that our approach to 'dance' was not going to be a 'see and do' experience, rather a laboratory of movement in which equality of access was embedded in the collective commitment to embrace the conditions that ensure equal status for all. This required that the diverse needs are admitted and adequately provided for.

Session notes day 1

In the Introductory circle, I introduced the planned project stages, gave housekeeping announcements and held the circle for individual introductions. I then instigated navigating the space and allocating partners was undertaken pragmatically by going around the outside of the circle and touching every other person on the back saying the number '1' and all the number '1's partnered the person to their right (number 2). This made the role change easier to manage.

I invited everyone to the space for a guided exploration and 'relaxation' – the purpose was to arrive within oneself and let go of tension in order to be freer to move. Once lying, focus went to breathing, following the breath cycle and sensing the underlying surface or the body in direct contact with the floor, other surfaces in contact with the air and clothes, etc. I invite movement from the periphery, still with a connection to the centre (navel), the sense of radiation from the centre to the periphery and back. I guided with things like – feel the floor rising up 'to meet you', let yourself sink into it, imagine the whole body as one. Movement continued to connect to spaces between the centre and periphery, the flow between these points. Continuing to touch the floor with parts of the body that haven't yet they could follow any inclination/impulse to move, follow the pathway of energy flow.

Touch in twos: one (A) touched the other (B) placing a hand on the surface of B's body – who could move the hand or move away if not comfortable. There was no intended 'agenda' to the touch, other than to bring awareness to that part of the body for B, as the floor had 'done'. But as the partner is a sensate human being, a verbal exchange supports the experience and the inquiry. I facilitated this with guidance such as:

> A, ask your partner where they would like a hand. Place it there and after a moment check if your partner is OK with the touch… do they want more or less pressure? B, you can move the hand to the right place. Wait there with the touch, what is it like? ...Does it change?

After some moments, A changed to put their hand on another requested location which can repeat a few times. After a pause, sitting down again, they fed-back to each other and B talked first. Then they changed roles and we have a group circle to ask how does it feel now, after the dialogue? There was careful consideration in the responses, a sense of being met, a sense of initial discomfort that resolved over time – not knowing 'what' was expected, the purpose, then through sharing the questions were answered and cleared the doubts.

Touch and movement responses – Saying yes or no

With a partner facing each other, one (A) places a hand out palm down, the other (B) puts the back of their hand curled up inside the partner's hand, then B opens their hand inside A's. We ask: Do you meet resistance? Do you want to go along with it and open your hand? Notice the timing? The verbal exchange was animated for some and more quiet for others – the lack of interpersonal touch in the lives of the visually impaired and physically disabled was evident, as touch was a big event, in the first instance and whether or not the hand was resistant needed more time and focus, but the psychology of being actively responsive to touch, or resistant, was insightful.

Patrick brought in core Touchdown movement material and an exercise introduced by Ray Chung in 1997:

With a partner, both stood or sat opposite each other, close enough for one (A) to hold under the other's (B's) forearms, with elbows flexed. A cradled B's arms by placing their hands underneath B's elbows; B put their forearms (pronated) on A's (supinated) forearms with their hands loosely resting on the inside of A's elbow joint. B relaxes, let's their muscles go, then A 'drops' and quickly catches one arm at a time, or both together. The idea was for B to allow and feel the fall and then the catch, to release and trust A to catch them. After a few minutes, including verbal dialogue, they switch. This was a playful way to bring in awareness, modulation and trust. Swap partners.

Leading and following – side by side, the inside arms placed one on the other, both decided whose arm is underneath first, this was the 'leader' taking the others arm into movement, very small initially and then into walking in the space. The focus was on feeling the quality of the initiation, the pace/resistance in the response, how far to go, the pauses, moving into stillness as well as space, meeting each other's response, listening to the contact, allowing one movement to end and when to start another. As the arms can get tired, they could swap roles as and when.

After feedback on these topics and the overall experience, more movement followed with guidance to move from the arm connection to other parts of the body, shoulder to shoulder, back to back, changing levels and taken into free-flow 'dancing' – contact improvising. As I noted, the wheelchair dancer was in and out of her chair, one sighted dancer was really 'impatient to move', the other two were glad to 'be in touch with the body again', everyone enjoyed 'moving in contact'. This first day set the ground for further exploration and the sessions continued weekly.

Session 2. We started with a 'check in' and then shared the 'warm up'. Each person said they are looking forward to continuing. In the introduction, we decided on the content; to focus on the senses relative to moving on the spatial planes, from rolling on the floor, moving towards the vertical and then into space from the different senses, then combine movements on all planes – rotation, lateral and forward and back. The sighted dancers found their way more speedily into the three-dimensional exploration, and the visually impaired stayed longer in the lower levels, in more simple and distinct movements, hesitant to launch into space or to go backwards on their own.

Partnering continued from leading and following to moving into contact, slowing down and using small movement initiations, then followed through into fuller movement. Where the movement impulse ended a new movement started, modulating and changing the sense of spatial dimension and phrasing (space and time) to the movement dialogue. This depended on each person's' sense of spatial awareness and freedom to move in space, which impacted on their sense of timing, and desire to explore modulation and change.

This exercise shows, on the one hand, the multimodal nature of vision, as the body sense and felt sense is augmented by being able to see the space occupied by the movement and the initiation into movement was relative to the space. On the other hand, the limited spatial orientation for visually impaired people was evident in their more inhibited movement range and modulation in flow/timing.

Patrick introduced the idea that our visual perception takes us out into space, working with eyes open or closed. Without sight, it was as if the space was 'looking at you, then it comes alive' to our perception as we worked individually in a part of the room. The aim was to encounter the space through the senses, through sensing the form, structure, stability and/or motility of the object or the structures we meet, so the texture, movement, sound, symbolic sense of connection, etc.

The next couple of weeks we worked with the theme of relating to each other in the space, against the pillar, the wall, around the piano or floor area. We combined our encounters with the spatial landscape with meeting and working with another person. For example, against the pillar, we could wrap ourselves around it and connect to each other, or form patterns with each other against the curved surface, using different levels.

The final stage (three more sessions) involved preparing for the performance. We had a discussion about what to share to an invited audience of peer dance artists and arts professionals, family and friends. We decided to define it as a work in process and an example of ensemble and 'integrated' arts practice – a research project. From this group conversation, I noted some key comments:

J (who had Parkinson's) had not experienced CI before, nor contact with the floor as we had explored. He found 'great joy in floating and suspending my mind', and 'to receive as much as give' (kinaesthetic coherence and affect regulation).

R (sighted) said she felt this was an opportunity for her to explore her movement in ways she felt safe and that she felt more adept now at letting herself move, to relate the self to the environment and other people. She felt that the 'check in' was important as a way to start and end the session. 'It is an opportunity to hear from others', content that each participant 'had bought into what we are doing with no resistance'.

E (physically disabled): It was important that we didn't have to come armed with a diploma. It has been like going into an amazing cave and things were lit.

J (VI): I liked the sound of the wall, the wobble of the panels. I have an image of swimming, of being in water and the metal and pillars can be like the water. I often have the image of filling space with water, moving in it and then draining the water out and still moving.

P (sighted): I got in touch with sensation; you can't make a mistake with the architecture, it is tolerant. I noticed the contrast of the pillar to the heat of his feet. The wall was a delight. Like Jack Frost. It was electric and magic to me, very clear cut, the *immediacy **of sensation** and the movement that has come out of that **without thinking***.

Other comments taken from the list of many:

Under the piano it felt child-like, like hide and seek.

It is a friendly space; I wanted to open my arms around it.

An idea for the performance could be to invite the audience in, in the dark.

I liked the verticality of the wall, it freed up my movement as I could depend upon it, as upon the floor, even the smallest point of contact could be very liberating, supporting me.

> We have just begun to work with objects, I am beginning to get a feel for it. I notice how the pillar supports the room and it is also an obstacle. *But I experienced them in a sensory way, not as an object.*

We worked in groups of four using inflatable football sized soft balls, to lead into contact work. We set up the task of using contact for the visually impaired members to locate the others. After moving each group fed back, some comments were:

> J (VI): It was intimate, like the head to head dance, and I was surprised by the contact with body parts that I couldn't see.
>
> A (VI): I got stuck between two people and felt we must look like a monster thing! It was enjoyable!
>
> H: With eyes shut it's interesting, as you don't know whom you are dancing with, what gender, you just respond to what is given, to the weight.
>
> R: It was a lovely feeling of liberty. The experience of encountering another mover and then as I did it, it wasn't detrimental to my sense of the moment. I could stay or move away. It was a very freeing period of dance.
>
> P: I felt total trust, at each move something was there to move with. With the eyes open you tend to differentiate according to preferences and calculate what to do next. I found points of precarious balance that would otherwise have been avoided.
>
> J: I felt movements happen that I couldn't recreate.

From working with the softness and support of the ball the movement continued with CI. We then gave ten minutes to each group to move whilst the others observed, giving simultaneous audio description to the visually impaired – describing who was where and what they were doing. Whilst observing each other it was noticed how the quality and directionality of the dance changed when the mover encountered another person. It was as if the contact was an answer for the 'search' that preceded it.

As we got near the performance, we reflected on the material in a new way. J brought in a Nelson Mandela text, to add philosophical interest, we read it aloud and then moved in response. We did a check in and discussed what to keep from this material and what we would like more of. We decided to work with the pillar as a focal point in one score and with the soft (dimmed) light state. Some of these ideas became starting points for the final 'scores'.

> H: An image helps me. When we started we had 'building blocks' leading into Contact, more of that?
>
> J (VI): I like text, like the Nelson Mandela text, as a starting point.
>
> J: I would like to work in a 'Touchdown Dance' way with two blind and two sighted movers.
>
> K: We could start in the centre, in stillness, to create a picture as the audience arrives. Each person chooses a part of the building and stays there. As the people arrive we look like part of the building and reveal it. We keep open to it as long as we want to, if we are happy there, we can stay there. Seeing Jane and the pillar is as important as seeing the dance. We can play with the contrast, let the body meet that part of the space?

We decided on a score; to move in and out of contact in groups of four movers, another mover could entered from the outside to replace one of the four. This idea produced the following responses:

> H (sighted): It is important to allow the memory of working in the space to stay in our awareness, to keep my own dynamic, so that if I am cold I put something on. (Concerned that it was not visually interesting if we all drop into the same dynamic.)
>
> J (VI): If I moved with more than one person it felt like a snake melting. It was not as satisfactory as when I moved against an inanimate surface or with one dancer. (Kinaesthetic sense supported by touch and sense of purpose – to explore movement not space, the sensation not what it looked like.)
>
> P (sighted): We could try dancing without contact. To keep a sense of distance, in a situation which is intense but without contact. (Suggesting an ensemble feel and look.)
>
> R (sighted): This is difficult in a situation involving blind dancers where we need a touch based element. (Pragmatic solution, speaking on behalf of the visually impaired but also for inclusivity as a priority and commitment or 'stamp' of the work.)

P: Yes, otherwise it's unequal. (Also pragmatic and acting on the principle – a motor output.)

E: We could use sight as an element of what we are doing, to keep the distinction between the dancers. We could try to keep 2 duets going and not have the 4 merging.

J (VI): We could use sight as one of the senses in the interplay, we don't need to specify.

R: Yes otherwise we are stereotyping. (All are aware of the need to communicate these aspects that are not 'visible' to the visually impaired participants and so best put out in the open/on the table.)

The group participants were not all like-minded people, but had active social and cultural concerns. They all had so much to contribute and jumped at the chance to do so. We all moved on sharing new ideas with ease and complicity and gathered strength as time went on. There was no hierarchy determined by experience or role so each participant was able to explore and define their own experience. We stopped taking in new members half way through which enabled us to really gel as a group and be able to speak frankly.

I noted my concerns that the audience would be unable to read the subtlety of the ideas and not understand the nature of this 'dance' piece. When the work is about sensing others in the space and working off their energy then the presence of the audience could affect the dancers and hinder the dance. This was clearly relative to experience and I was protective of the visually impaired dancers. It always took time to establish an inclusive group. I believed more work had to be done to know more about the perception of dance in the minds of visually impaired people. As the most excluded they were deprived of information, making it more difficult to know how to present and direct the publicity material to reach them. There was a lack of training opportunities, this work was just beginning and the outcomes were more evident in the sessions than in the showing – and visually impaired audiences would require audio description, if they came. This would be next on the company agenda.

Further analysis

With the benefit of hindsight, re-reading these reflections, I perceive the significance of orientation to 'space' and the integration of movement and other sensory modalities, of memory and emotional affect. The modalities of sensory perception are evidently influenced by the way the experience is set up, and by the senses available or arising in direct response to the moment. Orientation to self-movement came in relation to the breath, to the kinaesthetic (proprioceptive) and tactile feedback from the 'earth' (floor) to provide pressure, and more support than cloth and air. Attunement to self through the available senses is evident in the comments above – as is the influence of past experience. As can be noted, both sighted and visually impaired people are able to perceive self in relationship to space through their sensory/motor systems and through autonomous exploration. The spatio-temporal nature of each person's experience involves subjective affinities, personality and the intellect – their past prehensive experience and its influence on the imagination – like the idea of a 'snake' dance that breaks down as the visually impaired dancer cannot attend to so many different foci as a sighted person. The explorative 'mind' of the group (Bainbridge Cohen 2012) was also an influence, as it permits subjectivity, accommodates difference and establishes the 'democratic' process (Paxton and Kilcoyne 1993: 15), this underlies the task of shaping the 'performance'. I consider these aspects separately and then together with reference to phenomenology and BMC® (a phenomenology in itself).

Movement is in essence the animate state of motility and awareness, it establishes the 'animate form' – the body-sense – the 'complete corporeal' state of consciousness of being 'in movement' (Sheets-Johnstone 1990: 336). However, as Bainbridge Cohen (2012) sets out in a developmental context, the senses arise sequentially and those that are available operate together – touch and movement underlie the vestibular system and enable us to orient to the world through the other senses, hearing, taste and smell and finally sight. In their insightful book, *The Psychology of Blindness*, Heller and Gentaz (2014) establish the 'multi-modal' and 'inter-modal' way the senses work for both sighted and visually impaired people, and how lack of sight influences the differences – which further examination of the group experience illustrates.

Jane, who is visually impaired, talks about her experience of the pillar and the metal panels on the wall. She talks about 'an image of swimming, of being in water' and that the panels and pillar can be 'the water'. For the panels to evoke, 'water' is not a common experience and for Jane, it was not the tactile sense of them or visual appearance, rather the sound that came from the panels when touched and brushed – this construction of 'sense' came from the inter-modality of the senses and the activity of being in this open experiential domain. We can follow the process in her description that movement and touch created sound, a vibration and space in which Jane was immersed and her imagination/felt sense unified through ideation. This is significant insight for the sighted dancers who then attune to Jane's experience of the space from a less 'visual' plane, and the group mind and imagination does a 'reconstruct' to accommodate Jane. On another occasion, Jane talks about CI as being 'like swimming on land', so her affinity with water, with swimming, is a pre-sensory motor focus that locates her in this space; she can bring herself fully into her experience as her poems express. The image of 'swimming on land' can be shared and understood by others familiar with being in water and may transfer subjectively as a 'pre-sensory motor focus'; our subjective experience of swimming prehends our interpretation of her experience and may then influence our own. The collective membrane shapes and reforms itself over time and is indicative of the Touchdown methodology in action, as outlined in Chapter 2.

Jane's 'embodied' experience is a form of research. She experiences an affirmative discovery for herself and the group. As I write, I perceive her descriptions in phenomenological terms, as a type of 'hermeneutics of the body', an embodied exploration and a form of research undertaken by each person. The written word, as Jane provides for us in her poem and the documented comments also, provides 'evidence that *shows* embodiment to be an experienced fact' (Sheets-Johnstone 1990: 337).

'Move-In' provided us with an opportunity to investigate our perceptual/ moving beings. Through 'embodiment' processes, the body became a new territory, an inner space for inquiry, where objects were 'sensed' not seen 'as objects'. Habitual 'sense' was suspended, so that different subjectivities could arise to consciousness as 'other' feelings or thoughts. Bainbridge Cohen (2012) and Gallese (2016) appear to agree in 'linking brain, body and language' in the perceptual process (Gallese 2016: 134). The participant statements reveal that, by opening up to the body, many new perspectives arise, more 'voices' seek to be heard. Then, as now, these voices provide an opportunity to explore more fully the vastness and complexity of human experience – the multiple subjectivities and their unpredictable, inter-subjective relationships. The language we each use helps us access each other's experience, pragmatically, and be affected by it. We may draw inferences from it for ourselves, which may be more about what we observe in others (particularly if sighted) rather than our own preoccupations and ideations (particularly if visually impaired) – and once admitted these insights become useful.

Gallese finds that according to studies in neuroscience – the 'findings – the facts – show that the motor system not only controls the execution of movements and actions, but it also plays a major role in perception and cognition' (Gallese 2016: 127). His concern with vision is its dual modality in the perception of space and of objects as they are seen, touched and moved, and he goes further to state that some motor neurons in the brain are 'reused' by different senses:

> Vision is multimodal: it encompasses the activation of motor, somatosensory and emotion-related brain networks. Any intentional relation we might entertain with the external world has an intrinsic pragmatic nature, hence it always bears a motor

> content. Neuroscience shows us that cortical motor neurons also respond to visual, tactile and auditory stimuli.
>
> (Gallese 2016: 128–29)

Evident in the pragmatic comments of the sighted participants above is the wish for coherence and visual interest from the movement and relational interactions. For the visually impaired dancers, the visual impact is not a concern, her transcendent experience is and her movements reveal new perceptions and thoughts that she shares. Relative to the evidence provided by Jane, and others, she illustrates the following point – it's not just about the movement it's about the who and what in the *where and when*:

> We can identify objects, locations in space where sounds may come from with a remarkable precision. All of these perceptual qualities are not the mere outcome of the *impression* exerted by the external world on our perceptual and cognitive systems. Neuroscience tells us a different story. These perceptual qualities depend upon the motor potentialities expressed by our situated body.
>
> (Gallese 2016: 129, original emphasis)

It would then appear that the ensemble intention to move together was possible, not only because we all have movement but also because we all have access to the multi-modality of the senses. We can share our experiences and note the lack of wider spatial orientation for the visually impaired amongst us but also the similar intensity of touch, movement and sound stimuli in relation to more proximal objects/architectural features and each other's bodies. These other senses may communicate relational behaviour but not as mirror neurons do, according to Gallese (2016), 'mirror mechanisms' are 'motor neurons' that respond to movements we make and in the perception of other's, to 'suggest that a direct modality of access to the meaning of others' motor behaviour is available' (Gallese 2016: 131). This aspect is a challenge to visually impaired participants, evidently and particularly to their perception of feelings and emotional states in others, until they are heard/touched/moved and involved in conversation, these factors can remain hidden and repressed within. Our discussions reveal the process of accommodation that took place and the compromises also. We took 'artistic license' with the idea of 'showing' the 'work' which was deeply personal and emergent. This was what was 'seen' and 'heard' by the small invited audience, intimately, with warmth and integrity.

In this context, as the researcher/writer, I have put words upon the perceptual process and interpreted types of phenomena that informed and shaped the experience that I was part of and co-facilitated. This is with the purpose to show the plurality of concepts and discourses that informed our collective creative process that served to generate insights and knowledge about the moving, animate body in human experience. These insights served the creative process, enlightening each person to the value of subjectivity, of the sense of 'liberation'. Further, it shows how this liberty is contingent on how each of us is 'situated' relative to the environment, and socially we become aware of how meaning is inferred and revealed or, more significantly, how some voices are over-written, hidden and used covertly to perpetuate dominant discourses or to manipulate interpersonal relationships. The comments about having 'contact' in the performance to *enable* the visually impaired performers and to avoid 'stereotyping' by not having sight as one of the common senses, perpetuate the stereotypes of 'visual media' as it assumes a priori decisions rather than collective ones. As time goes on, we learn as a collective of visually impaired and sighted dancers, how to define our work in a way that transcends the visual with a multimodal – multi-sensory perspective and reframes the cultural framework. How the audience perceives this is interesting also – how some, as we had imagined in 'Move-In', may not 'get' the idea of what we are doing, may not be familiar with the realm of this experience and may be disconnected from us to the point where 'the idea' is miscommunicated or open to misinterpretation. At times, I felt the tenderness of our experiences, as well as the rigour. I felt the imperative of this creative purpose and anticipated the sense of achievement, whilst also a vulnerability to being 'seen' and potentially judged. This was a learning process, to discover how pluralism can be embodied and the work can be legitimate even if some want it to emulate what they already know and see it only through that lens – we can only be part of a shift, not the shift itself.

The following reflection and poem are by the late Jane Emerson who attended Move-In and other jams for several years.

Why Contact Improvisation?

I went to my first CI workshop in 1996 at Bretton Hall College. Before going to the weekend workshop, I had never heard of CI. I hadn't really been looking for any dance events and I hadn't been to any other types of dance workshop. The advert which I read in Braille magazine just sounded interesting, including the fact that the venue was Bretton Hall, with its open-air sculpture park.

We covered a lot of things during the workshop, both techniques such as rolls and lifting and some improvised dancing. It therefore gave me a really good introduction to CI, and I felt 'hooked' on this surprising new world.

Since then, I have continued to go to workshops and jams whenever I have the opportunity, both with Touchdown Dance and other teachers and groups. CI has become important to me, and some of the factors about it I particularly enjoy are the following:

It is accessible to people of all abilities;

The form seems to have countless possibilities, from playfulness to fun to learning complex movements; from stillness to fast speeds, etc., etc.;

I have found lots of co-operation and willingness for 'trained' and 'community' dancers to work together;

CI seems to have given me a different awareness of physical environments.

I also think that lessons learnt in the dance can be taken into life in general, such as learning to 'be in the moment'.

Jane Emerson, August 2001.

Poem by Jane Emerson, September 2000 (End of the Move-In project)

Dancing with the space,
C.I. moving,
In the landscape of the space
Pillar-dance and floor-dance,
Dance with chord and curtain,
Pillars, like and unlike people,
Unlike- and tree-like.
Soft, wood-fringed metal,
Glidescape and soundscape,
Wall framed.
Then from the edge,
Dancers are space-parts,
Some stay themselves.
And later,
All just floor and people,
And the dance goes on...

6

Performance and Creative Process

The first Touchdown Dance integrated company of sighted and visually impaired members grew from the commitment of workshop participants to performing for an audience. Following the informal performances at Dartington College of Arts to students and guests (such as the one with Gerry Overington, Steve Paxton and Julyen Hamilton mentioned in Chapter 1), the first Touchdown Dance commission *Sixth Sense – Second Sight* and *Moving Prescription*, took to the stage at Dance Umbrella in London in 1992. Steve Paxton and Anne Kilcoyne collaborated initially and then Kilcoyne took over the production due to her interest in the role of light and lighting and worked with an ensemble of two visually impaired dancers (Janee Hall and Bill McKinlay) and two sighted (Caroline Scott and Charlie Morrisey) with Anne Kilcoyne as a 'doctor'. In an interview on VocalEyes,[91] Janee Hall discusses the challenges that creating dance as a visually impaired person raised and which later productions continued to explore. As a trained dancer from an artistic family, Hall had the advantage of being familiar with the arts scene and was able to support the visually impaired dancers who didn't have this background. Hall mentions the practical issues of getting on stage, knowing when to start, how to meet another dancer, getting off stage and so on, and when on tour having to adapt to a new space. These challenges were met creatively in the process of making the work and often with theatrical intent, i.e. through unseen signals or by creating a convention in which visual impairment is not in focus.

As we have seen, by unlocking movement through Contact Improvisation (CI) the workshops unleash creativity: 'Apparently the real juice of the workshops which include the blind is the contact improvisation itself. It is this which produces the feelings of self-empowerment, joy and satisfaction' (Paxton and Kilcoyne 1992: 48). This becomes all the more so in the performance projects – the spontaneity of contact brings up a sense of enactment, of shedding an identity and giving payback to society for the unfair way of treating disabled people. By treating them as equals Touchdown Dance provides the methods and the creative 'juice' to continue from the workshop into a studio and to focus on a creative product – to take this method and its impact on individual lives to a wider audience, where it will speak to others excluded in their own way from participating in society and culture.

Audio description was not available in the early 1990s, nor did visually impaired people attend dance or theatre as often as they have done since the advent of technology and creative ways of onstage description, integral to the performance. Theatre companies like Graeae and later Extant, as well as ourselves, experimented with integrated sign interpreters and descriptions. We added dialogue and multisensory aspects including enlarged simultaneous projection, and describers assigned to visually impaired people in the audience as a way to meet personal needs. Funding support has been essential to provide the technology and admin support to visually impaired people to access the venue, Braille programme notes, entry to the space, and meet the artists on a pre-show touch tour.

Sixth Sense – Second Sight: *Practice-based research – In performance*

In March 1993, Kilcoyne writes in the New Beacon again, but this time about the production *Sixth Sense – Second Sight*, 'an original thought provoking dance event which took place recently as part of London's Dance Umbrella Festival' (Kilcoyne 1993: 117) and announces the seventh annual season of workshops that is due to take place in the summer of the same year. There are photos of the five performers mentioned above and the report gives insight into the content, starting with the first piece *Moving Prescription*: 'The piece was made because of Bill McKinlay's wish to convey his own experiences through performance' (Kilcoyne 1993: 117) and presents his stereotyped identity and the prejudicial attitudes of the sighted world. In an encounter with a 'dehumanising', white-coated physician (Kilcoyne), he is told to 'accept his limitations and adjust to a stagnant and depleted life style' (Kilcoyne 1993: 118). Rather than walk away dejected, McKinlay transforms his cane into a multifarious prop and 'discovers he can abandon the outwards signs of his blindness and enter into an exhilarating high flying field of movement' and does so in partnership with a sighted dancer.

This piece of dance theatre is a satirical portrayal of the all too common experience of visually impaired and disabled people, who face medical intervention and often feel subjected to the authority of the clinician – symbolised here in the scalpel that the doctor brandishes in her hand. Too often the 'surgeon' remains aloof and detached from the patient's experience, from the trauma of the surgery. McKinlay's performance goes further; the subjugation he experienced from the medical intervention is personal to him, to his past experience; it defined him, limited his sense of self-worth and left an old wound held bodily, which arises at this time as a wound to be healed. Within the context of learning to move his body and with equal skill to the sighted, he regains self-worth, trust and a sense of 'interpsychic and interpersonal' support from the team (Martz and Lindy 2010: 28) – he can heal within the protective trauma membrane that Touchdown Dance provides. No longer succumbed to a debilitating life, he is now empowered to stand up against the perpetrator. Not only is this a symbolic enactment in which he shakes off the 'shackles' of a stereotype but he also shows the society what it has not yet seen it is doing to members of its community – the performance formulates new insights for the audience to take away, such as the triumph of the social model of disability over the medical model, and of the physically abled visually impaired man over the assumed insufficiencies of a blind man.

What is missing from this report is a self-report from McKinlay, which has remained in the studio process. We can only imagine the journey that brought him to this moment, one that evolved over time, progressively undoing the self-percept he had whilst a new one emerged, that of a visually impaired performer, one who can move and be part of an amazing experience that had been hitherto unavailable to him. Having known him in the workshop setting I recall how his world changed, how he walked around inside and outside without a cane. He was able to engage bodily in the dance and all the more so for being part of a community of like-thinking people who contributed to his experience. The change in self-percept affirmed his identity beyond being visually impaired and enabled him to take direct intentional action and to be 'heard'. The role taken by McKinlay is one step up from the experience that Gerry had in 1989, who also hoped for the work to have a wider impact on the lives of visually impaired people. The outcome is meaningful for himself and the audience.

Kilcoyne's report displays the empowering role that the Touchdown Dance experience could have on the inner being of individuals, implicitly and over time, enabling access to a personal yet contemporary narrative – to a trauma that would otherwise remain internalised. McKinlay's inner motivation towards change was held by the creative process and shared in an intersubjective relationship with the dancers. His wish to shift the negative perception in society of what it is to be blindly translated into an artistic endeavour, it transformed his own self-perception, and the authenticity of his artistry gave plenty of emotive content for the audience to digest and interpret differently. The degree to which this learning happens cannot be anticipated; however, the experience paid off, and further learning can take place through the dissemination of the report and through reflection upon it, such as attempted here. For Barrett and Bolt (2007), the basis for practice-based research lies in the significance of the level of personal interest invested in it by the artist/researcher:

> The experiential approach (Kolb 1984) starts from one's own lived experience and personal reactions. Learning takes place through action and intentional, explicit reflection on that action. This approach acknowledges that we cannot separate

> knowledge to be learned from situations in which it is used. Thus situated enquiry or learning demonstrates a unity between problem, context and solution. A general feature of practice-based research projects is that personal interest and experience, rather than objective 'disinterestedness' motivates the research process.
>
> (2007: 5)

The second piece, *Sixth Sense – Second Sight*, is 'more abstract' and involved ultraviolet light as a means to enable any visually impaired people in the audience with some useful vision to see the movement work. This piece is more gestural, 'playing on two ways of perceiving the world' (Kilcoyne 1993: 118), one being the long-distance view from stillness, like looking at the moon, the other the view of a cluttered, hyper-mobile consumerist world. The visually impaired and sighted dancers share their perceptual positions through movement.

A short quote from a sighted audience member states how she found the evening 'absolutely beautiful and riveting and has given me a great deal to take home and to think about'. This feedback resonates with the insights encountered in the workshop by sighted people, with the sense of discovering something new leading to a shift in perception, likewise for visually impaired participants and audiences as in the account of Tew. We are all subject to the influence of socially constructed concepts, either as hidden, unconscious 'phenomenal content' (Gallagher 2008: 2), or brought to consciousness by an experience that may disprove them, where they are changed, or a declared conviction; a belief that may be reinforced by experience and made ideologically explicit. In this performance both the tacit and explicit are present – the dance choreography defines the behaviour; a blind man transforms his identity before the eyes of those who defined his 'abnormality' and moves the minds of those watching to perceive a new normal. The boundary of difference between sighted and visually impaired people has been redrawn.

To have two visually impaired people dancing in a dance performance within a major dance festival was a coup. Both dancers had a role that made explicit their views and thoughts on this and more was implicitly evoked in movement. This is a skill that practice-led research foregrounds; Harre and Gillet (1994) clarify the contingency of thought upon experience, and how effectively thought can be communicated through representation:

> Thought is not an overt operation on the world and therefore once the terms and elements have been mastered, the contents of thoughts can be hidden from other people. In this sense thoughts are potentially *private*, constructed apart from the dealings we have with the things they are about although dependent on our past dealings with those things. Thoughts, then, are not objects in the mind but the activity and essence of mind. They reside in the uses we are making of public and private

> systems of signs. To be able to think is to be a skilled user of these sign systems, that is, to be capable of managing them correctly.
>
> (1994: 49, original emphasis)

This performance presented an opportunity to deal with such thoughts, to present a discourse that had remained hidden. We see the stereotypical societal oppression in the figure of the doctor, but as they dance in front of us the visually impaired dancers are liberated from their usual exclusion and removed from 'past dealings with those things' in the present moment on a public stage. The sighted audience member also had thoughts hidden from this experience, but since this experience they admit being open to rethink, to consider a new discourse – one that they had been given from blind and sighted people operating in a way that had not been previously possible.

The concept and content of the performance evolved from what all five performers could achieve in an ensemble, with visually impaired dancers at the centre of a work-in-process about having or not having sight. The content became a 'discourse' expressed with 'the communicative use of signs and symbols, pictures and words in use' and became 'the medium in which cognitive activity takes shape' (Harre and Gillet 1994: 57). In this case, 'cognitive activity' refers to the thinking that arose from the staged juxtaposition of usually distinct realms of being – the medical with the social, and the blind with sighted – in a cultural event.

Paxton and Kilcoyne conceived of performance as an extension of the research already undertaken to discover how CI could be of use to visually impaired people. In so doing, CI becomes an emergent model for understanding further the experience had by so many visually impaired people in a sighted world. Barrett and Bolt (2007) propose that 'the emergent aspect of artistic research methodology' is a 'positive feature', it allows for 'the emergence of new analogies, metaphors and models for understanding objects of enquiry. Hence the capacity of artistic research for illuminating subject matter of both the artistic domain as well as that belonging to other domains and disciplines of knowledge' (Barrett and Bolt 2007: 7).

Productions post 1994

It took from 1994 to 1997 for me to establish a group in the north of England ready to show our process to an audience. I raised funds for a group of three sighted (which included a physically disabled person) and three visually impaired participants who sought to train together with invited internationally known CI teachers. The core ethos was to ensure that all the dancers were able to work to their full potential in order for the ensemble to develop artistically. CI and the modalities were key to the process, supporting the company's aspiration to create original work in which everyone contributed and their unique qualities recognised. We raised Arts Council funding for artistic development projects with the view to sharing the work-in-process in small intimate venues, where visually impaired and more general audiences could access the work. We knew we were exploring new territory and we were determined to create inclusive performances – which was still comparatively rare in the arts at that time, although it had a growing enclave.

My experience of making productions in Denmark gave me the confidence to manage production. The group perceived our different abilities and personalities as our potential uniqueness, and the limitations we shared as a source of theatricality. I realised that we would benefit from another creative mind and input that would stretch us artistically, to position our potential in a realm that could be staged. Performance, as I had experienced it, was a spatio-temporal realm, a creative and inspirational space that would be new for the group. On the Nordic tour of the project in Denmark, I had experienced how to support audiences to see with 'different' eyes and felt optimistic that we too would confirm the value of inclusive, experimental creative practice, a purpose that continues today.

Over time, a strong relationship and understanding grew between myself and the dance officer Mileva Drljaca from the Yorkshire Regional Arts Board, who later moved to the NorthWest Arts Board. The Regional Arts Board granted funding for an expert in CI and performance, who was also very interested in touch and perception, to come and work with us. Chung came from the USA for two intensive weeks at Romiley Forum in Stockport. We devised and performed there and at Tameside College in Ashton Under-Lyne and the Liverpool Institute for Performing Arts in Liverpool. Chung wrote about this experience for the *Contact Quarterly* journal, 1999 24(1). I replace names with initials where necessary.

RESIDENCY WITH TOUCHDOWN DANCE by Ray Chung.

From 13–28 May 1998 I had the privilege to participate as artist-in-residence with Touchdown Dance. I was invited by the Executive Director, Katy Dymoke, to work with the organization as a guest teacher and performer. The goals were to present material pertinent to the practice and performance of Contact Improvisation (CI) for sighted and visually impaired dancers in the format of workshops and classes, and to work with the company dancers/performers, Alan Foster, Patrick Beelaert, Janee Hall, MH, visual artist Amanda Latchmore, and Katy.

Some of the topics presented, investigated, and experimented with included:

1. The role of haptic (touch-based) sensing and perception in the use of human movement, navigation, and in the practice of CI.
2. Finding, making, and developing connections between touch and movement.
3. Finding non-visual ways of working in improvisational performance.

What follows is an account of some of the material presented at a rehearsal session and subsequently used in performance.

Reverse sculpture definition. The idea for this exercise came from a discussion about the sculptor Willem Boschoff's work. (A visually impaired dancer) had brought one of his sculptures for us to use in conjunction with movement. The discussion focused on how Boschoff created the sculpture. He would start with a word, in this case beginning with the letter 'C,' and create a sculpture that would express the concept defined by the word. His Intention was to communicate the concept through a touchable object. The suggestion was made to work backwards: first sensing, touching the sculpture, then finding what the texture, shape, weight, etc. of it meant for each of us, and expressing this in movement, then writing and/or drawing about our experience. In this way, we each defined what the sculpture was for us, what concepts, ideas, images the sculpture evoked, conveyed, and brought forth. Definitions obtained this way were highly subjective but the intention was to find the diversity of meanings within the one object, and possibly correlating it with the original concept, definition, that Boshoff had intended.

We adapted this exercise into a performance score by utilizing ordinary objects found around the theater at Romiley Forum. An ash can, chair, cardboard box, bubble wrap, newspaper, and a length of nylon fabric were selected after a process of experimenting and playing with various objects. The performance score had the collection of objects placed in a corner of the space, and at some point we allowed ourselves to begin finding, touching, responding to them. The intention was to allow the texture, shape, elasticity, weight, and other properties evoke and elicit movement in our own bodies during the process of sensing the object. Once some, or any, movement was found, we would then leave the object and continue with the movements that were found. We let these movements develop within our bodies and with each other until the movements decayed, or no longer held interest for us. We could then find another object and repeat the process. After a time, the objects were gradually removed to a place off the stage, and our memories of the movements and the movements themselves were allowed to continue, develop, interact with each other's movements. This was an excellent way to find movements that we did not ordinarily perform and to find them from our sense of touch. We often would have our eyes closed during this process, but not the whole time.

For me, the main theme that emerged during the residency with Touchdown Dance was how does touch inform the other senses? How does the cross modality of touch and vision inform not just each other but other senses and sensibilities? The application of the information and experience obtained during the research and rehearsal process to an artistic performance was a very efficient and effective way of embodying the knowledge.

Much research, practice, and experimentation has taken place in the field of touch and perception since my residency with Touchdown Dance. Particularly relevant is dynamic touch (Carello and Turvey, 2015) where the sense of touch, as part of the haptic perceptual system (Gibson, 1966) is the primary one for interacting directly with one's environment.

My time and experience working with Touchdown Dance was rewarding, challenging, productive, and educational for me as a teacher, performer, and researcher. The insights I gained have informed my work over the years. The work of Touchdown Dance has been profound from the beginning.

At the first show at Romiley Forum, the performance hand-out introduced the novelty of what we intended to achieve:

Touchdown Dance with Ray Chung. 28th May 1998 Romiley Forum

The Kind of dance you will see tonight may well be the first for you as an audience, in terms of your dance viewing. We work with senses other than the visual sense as there are blind and visually impaired dancers in the group. This means that at times the emphasis is taken away from the visual impact of the dance we are creating and put on the tactile sense, or what the movement feels like in the moment. We work with dance and more gymnastic vocabulary and sometimes the dance is purely "movement", an idea or purely an image expressed in the moment. This way of working is based in improvisation and where physical contact is used it is called 'contact improvisation'. When we work towards a performance we create a framework for our dancing, with specific themes and time frame, called a score.

We integrate dance with other art forms, such as sculpture and visual arts. Tonight Amanda Latchmore will draw or paint as we dance and illustrate her interpretation of our movements. We will also find inspiration from objects in the space which we perceive through touch, with eyes closed, and take this information into a dance.

Part One. Three performance scores.

1. Accumulation.
2. Objects and images.
3. Brushing.

5 minute interval

Part Two. Performance scores with Ray Chung, Patrick Beelaert and Katy Dymoke.

We will be around for a short period after the show if you would like to talk to us after the last dance. With thanks to the Arts Council of England, North West Arts and Yorkshire Humberside Arts and the Ernest Cook Foundation.

I have selected some of the archived audience feedback from both Romiley Forum and LIPA in Liverpool with the performance:

1. Excellent, amazing how it all flowed, starting on the floor on day one and building up with contact made it. The fascinating aspect of the performance (apart from the obvious beauty of many of the moments) was its structure. How much was structure, how much was improvisation within its form – most of it – yet it appears as choreographed.
2. Beautiful, 'touching' moving relaxing to watch, the sound was great.
3. Lovely piece of work, or should I say 'creation', too relaxed to go out now! Thank you!
4. It was so good, so important to see this type of work happening.
5. Loved the show, amazing to see the blind dancer moving.

Sourcing the haptic sense

As Chung mentions, concurrent to the residency and performances, we were involved in a project with the Royal National Institute for the Blind's touring exhibition of wooden sculptures by Willem Boschoff, called *Blind Alphabet C.* Boschoff brought a collection of his sculptures over from South Africa, which were then stored locally at the National Library in Stockport. The sculptures were all made of wood, representing a word that has disappeared from the dictionary under the letter C. The sculptures were in boxes and sighted visitors had to experience them 'blind' with eyes closed. Each viewer took the sculpture out of the box and explored it through touch and manipulation. Some sculptures were kinetic, others contained something, or were solid. There was a Braille description on the top of each box, and a blind or visually impaired 'translator' read out the word and its meaning that the sculpture represented. Boshoff talks about the importance of touch on his website:

> The *BLIND ALPHABET PROJECT* re-establishes the integrity of touch as a socially viable catalyst for interactive discourse. It sets up touch in favour of sight by enabling, if not ennobling, the state of blindness and by disabling the sense of sight.
>
> (Boshoff 2019: n.pag., https://www.willemboshoff.com/copy-of-blind-alphabet-feature-1)

We had the pleasure of giving a movement workshop alongside the exhibition, exploring shaping and moving together using CI, and the connections were evident and inspiring.

I-radiate – 1999–2000

The project with Chung set the ground for the following production in 2000 called *I-radiate*. A series of workshops brought new dancers to the fold and then to the stage with us. We continued to work with improvised scores and further developed the idea of using CI as a source for choreography. The score provided a spatial framework in which creative ideas are defined and sequenced. With cues from the pre-recorded sound score, we engaged with each other through sensing, feeling, and moving together, and connected with a mood or movement quality. The score was ideal as a means to resolve practical issues such as who enters first and from where, what they do practically and thematically, for how long before another dancer enters, etc. The ensemble process of the improvised scores was enriched by a multi-sensory intention, with the aim to give access to visually impaired audiences by the full-screen projection of pen and ink drawings made from photographs of us dancing CI, drawn by Nick Fry, and by atmospheric sound scores created by Scott Smith.

The dancers involved in the research and development process were Patrick Beelaert, myself, Alan Foster, John Hurley and A, a disabled dancer who chose to remain anonymous. Funding was given by Dance NorthWest for artistic development, to create and tour the production as a community dance piece. *I-radiate* was selected and performed at the 'Resolution' festival for new work at The Place, London, in January 2000. With Scott Smith as musician and mentor and Ray Chung's material alongside the Touchdown Dance fundamentals, we took the opportunity to experiment, trusting that movement ideas would emerge from the ensemble mindset. *Resolutions* was a perfect platform. We did not expect to make work that would 'fit' into the mainstream dance world, indeed improvisation was current to the independent dance sector, and like others there, we were on track with something new. We recognised that to stand alone, the work needed to be protected from judgements that would arise from comparing our work that was so different to most dance contemporaries. Being seen on stage and being considered to be a legitimate artist were two very different aims in an otherwise conservative cultural landscape, so we found allies in the experimental and disability arts fields. We realised, like the other disability-led or integrated touring performing arts companies, that to find an audience we would need to be clear about what was on offer. We did not seek to sensationalise or to be more than who we were – a group of individuals within an experimental creative medium and believed this uniqueness would resonate with others. We realised that moments of magic would emerge from moments of more habitual 'traces' and we had to learn to trust in the nurturance of a collaborative relationship from which the magic moments arose.

In our ensemble, diversity and originality were our strengths. The word 'radiate' embodied the feeling that emanated in the work, the collective calm energetic quality that each person contributed to and embellished. The scores were partially choreographed to create a space–time framework, beyond this structure however, fixed choreography became a challenge for the group. Synchronicity is predominantly visual and a form of restraint for visually impaired people, reducing the creative potential of each dancer and the ensemble as a whole. We realised that choreographed dance was not appropriate and that there was substance to the simplicity of the structure and creativity in what we brought to it. After all, improvisation requires performance technique and improvisation becomes spontaneous choreography when the dancers have attuned to each other and found a common identity and purpose. We realised early in the work that the score provided a framework that liberated us creatively whilst containing the ensemble. We realised too that some choreographic devices could be possible and purposeful, without imposing conditions or unrealisable goals.

The movement vocabulary for each score evolved from spending a long time exploring what each dancer offered, both solo and with a partner, building on their ability and capacity to move. Through practice and dialogue with each other, we used each other's strengths and invited dancers with extensive CI and dance improvisation experience, yoga and somatic movement training, to extend our limits and expectations. Scott Smith, who had worked with Paxton and Image Lab in the United States, and Patrick Beelaert, who was working with Julyen Hamilton and others, both supported the process. We found ways to enhance the movement range of the ensemble and to enable each individual's potential to be foregrounded.

For the opening score, A and I did a lot of work without the wheelchair, exploring CI to move from being seated on the floor and levering into each other's centre to move over each other and fluidly away from the floor. In the performance we started separately, sat or almost lying in darkness then light on each of us came up we moved towards each other, rolling and sliding till we met in a combined pool of light. A used my body surface as a landscape to move over and to counter-support her in precarious balance, rising up even onto her feet and lowering back down to the floor. There was no clear indication of either of us being disabled. It was a surprise in a later score to see her locomoting across the space in her wheelchair.

As a blind person who has never seen dance, Alan came to join Touchdown Dance out of curiosity, to have a go and to meet people. Performing was a very different challenge for him than for a partially sighted or sighted person who had danced for many years. These differences entered the space with us, visibly and tangibly, and we spoke them, owned them and then understood how to include this informed movement awareness and relationship with each other in the process. Rather than disguise disability we explored ways to foreground ability, so, where possible, autonomy was facilitated. For example, to enter and exit the stage, Alan was guided to the edge of the stage, a meter or so in front of the back wall, once in position, facing across it he entered walking, counting his steps to arrive where to start the trio score. Alan became confident enough to enter the stage alone, after a few steps he talked as he walked, then stopped to take a position that he described; 'I lift my arms up above my head'. He continued to walk, stopped again and described the next movement; 'I put my head down towards my knees'. This was the simple start to a score that developed in complexity. It showed that Alan could move, interpret and perform. At the third movement, Alan was about half-way round his circle and Patrick entered taking the same path and doing his version of what Alan described. Patrick did not look at Alan, or mirror or mimic him, he moved pragmatically responding to what Alan said and so differently, as intended. The interest and the task was to see the resonant relationship between these two movers. Alan was aware that Patrick entered and that Patrick would, at some point, take over so that Alan then did his version of what Patrick described and we see Alan's movement response. Alan could not see but he can perceive where Patrick is from the sound of his movement and then by his voice, and when Patrick was close by, he sensed the movement in the air around. Alan could also perceive the tone or quality of the movement by the effort in Patrick's voice.

Patrick took his turn, saying, 'I am standing on my head' as he did a headstand. Alan placed his hands and the top of his head on the floor from a squat position – he did not do a head stand. Alan's task was not to copy, but to follow the instruction as he felt fit, to be attentive to Patrick's movement, and to have the freedom to interpret. Alan enjoyed this aspect, at first he responded with day-to-day, idiosyncratic movement, and then took more risks. Patrick, as a dance artist and yoga practitioner, was lithe, flexible and agile, but remained playful and co-creative as Alan worked with his body, his limitations, his patterns and found new, often precarious movements. Their complicit relationship was the world we witnessed or heard described; a dialogue of different equals, of mutually supportive interplay – a meaningful metaphor for inclusion, an example of equality of access, of empowering participation.

In the third section, John also entered from the edge of the space describing what he saw as Patrick and Alan continued to move and started to connect in CI. From this distance, John established another more objective viewpoint that contained them in the space. John placed himself relative to them and told the story he imagined himself witnessing, such as 'I see two people in a wood …' and the audience then share his imagined world. John's voice provided spatial references for Alan and for the audience and the aim was not for Patrick and Alan to change or do as John said.

This score acknowledged the validity of variance and the use of description became an ongoing aspect of all our performance and film work. We discovered that it was not necessary to over-compensate with description but also that there are limitations as to how much information is usefully transmitted and that choreography can be substituted with other performative concepts. When it came to the theatrical device in the idea that Alan and Patrick don't do the same movement, this was described to visually impaired audiences when they attended, in the script or by a person sitting next to them, but they were few at this time despite listing the work in the talking newspapers.

The third score, *the Sweep*, involved standing side by side across one side of the stage, we then 'swept' across the stage, walking slowly side by side, with the back of our hands touching so we retained our pace and line. On the return, one or more dancers could drop back from the advancing line, into the space and improvise to the soundtrack, as the remaining line continued to the other side of the stage, at the same pace. A practical issue arose if Alan was left in the line without contact with anyone to orient him, and this meant adjusting the space between the dancers in the line. On the next 'sweep' the dancers could re-join the line or continue improvising and potentially be joined by another dancer and continue to dance until the 'line' swept them up. This structure worked really well; the dynamic adapted easily and duets or trios could evolve out of solos. It was a wonderful ensemble piece that taught us many practical solutions; if Alan wanted to re-join the sweep his partner ensured they faced the right direction or he stood still till one of us in the line connected with him. The piece ended with one sweep including everyone in the line.

The programme notes stated:

> 'I-radiate' establishes a common ground where dancers connect and disconnect working with touch, energy, sensitivity and fluidity. The dancers connect and disconnect with touch, sound and object to keep visual cues to a minimum. Blind and sighted dancers compose improvised scores with a quality to be met to be known.

This was the first Touchdown Dance production that toured outside of the north-west region, featuring workshop participants who, like Gerry, wished to perform. Knowing that our work was new, it paid to inform the audience about what we were doing and showing. At Romiley Forum, we started with an optional audience participation exercise involving all the dancers on stage prior to the piece. The intention was to give a brief introduction to the CI experience, inviting anyone interested in the performance space for a 15-minute CI class. We had over twenty takers including children, who joined in, with eyes open or closed, to do some brief leading and following in pairs. There was a fun atmosphere and it was effective as Alan demonstrated the tasks. After the show ended we invited the audience to stay and chat with us, those who did were really appreciative. Some talked with us in the space and others later in the bar. The feedback was that the pre-show experience was really helpful and an insightful reference to our touch-based practice before watching us perform. Many could appreciate the nature of the dance relationship and how much more we were doing – how much further we were taking CI and working it into a creative context.

With the view to testing the pedagogic process with younger people, we sold the show with a youth participant project called *impro-motion*, to the Contact Theatre in Manchester. We delivered a series of sessions to an integrated group of disabled and non-disabled young people. This was the first time we did a residency with disadvantaged young people and they chose to give a preshow performance before ours. We explored identity and had fun with hats and props and they really wanted to perform. This memorable opportunity was in partnership with the youth theatre team and support from the Contact Theatre, Manchester.

To get to perform at *Resolution*, we received an Export Bursary and Pivot Commission from Dance NorthWest. *I-radiate* was produced in collaboration with Contact Theatre Manchester which offered rehearsal and technical facilities to support our artistic development. Two audience responses we noted at *Resolution*:

> 'Sublime' (Arts Council Officer).
> 'I have to learn to see dance differently, I have to learn how to watch this, it was different' (Ballet dancer in the audience).

These early works were 'game changing' and revolutionary, in terms of 'breaking new ground'. It became obvious, as we made funding applications and marketed the work that Touchdown Dance was still unique in providing opportunities for visually impaired people to be included in dance training and performance. It was evident also from the challenges that appeared to be common to visually impaired people, that inclusion required access to movement skills from a young age. This observation informed the future partnerships that we made in education and community settings and the catalyst came from the opportunity to promote our performance work as a form of repertoire for workshops and this format could make it accessible to 'new' audiences. As dance is a visual medium that visually impaired people could not see, then participating and doing the movement appeared to be a potential form of inclusion – this we learned as we found a language that described and named what we were doing and enabled us to create new work.

SENSE-8 *2000–01*

SENSE-8 is a ten-minute dance film (Figures 6.1 and 6.2).

Initially, this film was a Dance on Camera Commission, but the panel stepped back following the research phase when it materialised that Steve Paxton could not

FIGURE 6.1: Six dancers improvise CI in the filming process for the *SENSE-8* film.

FIGURE 6.2: Two duets in warm up, Holly Thomas and Scott Smith forehead to forehead, and Alan Foster connecting his arm to Rick Nodine's chest.

FIGURE 6.3: *SENSE-8* filming the Round Robin score.

commit to it. Katrina McPherson and I applied for an Arts Council Capture Commission as recommended to us, and were successful. The dancers were all known in the dance world apart from Alan Foster and Holly Thomas (who had only recently joined the company and was a performing arts teacher at RNIB Redhill College). As the film was about CI we collaborated with Mary Prestige, Jo Blowers, Jovair Longo, Rick Nodine, and Scott Smith, I was the eighth. Katrina and I met to share ideas; the movement ideas and concepts for the camera came from myself and were developed further by Katrina with her experience of different camera work, from close circuit TV (CCTV) to hand-held cameras, steady cams and a 'dolly'. Neville Kid (camera operator) had worked with Katrina before as had Simon Fildes (editor).

This film was audio described for visually impaired audiences as well as sighted. I recall sitting on my lounge floor with paper and pens, drawing concepts and spatial relationships, seeking ideas that would be accessible to visually impaired audiences.

The following were the nascent ideas that became filmed scores:

Waves: We all crossed the space in 'wave' form, rising and falling as a group, lifting and lowering each other, sliding and rolling into the floor. This was a dynamic piece that would be captured from a distance on the dolly (camera on rails) and potentially with hand-held camera work.

Shaft of light: I explored the idea of seeing colours against black/dark backgrounds with visually impaired people who had this type of vision. I considered the possibility of moving through a narrow shaft of light and capturing what came 'to view' rather than needing to 'see' the whole picture. The sound of the movement would be crucial to this.

Pool of light: I had the idea of a fountain, a duet that stayed within a pool of light and moved constantly over and under each other within the light and the camera capturing the flow rather than the whole body-scape of the dancers. These felt accessible to the visually impaired dancers and audiences would perceive the flow of the CI dyad within an aesthetic context.

These two sections were filmed in one day having blacked out the whole studio and using steam to minimise the distortion of the image by the beams of light coming from lanterns fixed on the ceiling above.

The Round Robin: In a community context CI traditionally takes place in a circle, with two dancers entering to connect and dance together in the space whilst the circle witnesses. At some point, a third mover enters to connect with one of the duets and a trio continues for a time before the person released from the dyad finds their way back to the circle. The round robin continues as another third person enters seamlessly to join the dance. The circle can be standing or sitting and the dancer can enter walking, rolling, running, sliding – however feels right. In our circle, the six sighted dancers held the space for the two visually impaired dancers who couldn't see but could ask questions about who was dancing, or what was happening. Anyone could respond and use audio descriptive techniques to speak about what they witnessed. On the day we filmed, it took several takes to capture the dynamic of this form on different cameras and to record the voices describing movement to use as audio descriptions of one or a few words, dropped in for visually impaired viewers. The filming of the round robin required an 'onion ring' format; filming from around the outside on the dolly or steady cam, to capture the whole circle, then from behind the dancers, and then from the inside with hand-held cameras.

Duets: The eight of us formed four duets, Rick Nodine and Jovair Longo, Mary Prestige and Holly Thomas, Alan Foster and myself, and Scott Smith and Jo Blowers. We were filmed dancing CI from mid distance to capture the uniqueness of each duet and shots of two duets taking place simultaneously. The performative quality of the more experienced dancers contrasted with the sensitive and more introverted quality of those new to dancing together, which gives a sense of the diversity of CI and its accessibility.

We filmed in the top studio at the Yorkshire Dance Centre in Leeds. The pre-production tasks on the first day were to set up for filming, to move together and progressively work through the scores and to ensure what we had was good enough quality. We had five days of filming to get enough footage to edit down to a ten-minute final cut. Katrina directed the camera and technical side, and my sister, Vicky Dymoke, who had produced the Touchdown Dance CDROM and website, was our production assistant. Jovair and Rick had danced and performed CI a lot together; Scott, Mary, Jo and I were also familiar with each other and had participated in the CI scene for many years but had rarely performed together. This made the *duets* and the *pool of light* relatively easy to capture. Scott and Jovair did the *shaft of light* score and all of us did the *round robin* and *waves*. Sadly the *waves* section had a lighting issue and the rushes were not clean or good enough for the film. However, the CCTV camera caught moments of transition and breaks that added a pragmatic and insightful 'fly on the wall' perspective.

Simon and Katrina went away and spent months on the editing, sending their ideas back and forth bearing in mind the intention to create a film in which seeing the dance was a multisensory and implicit endeavour – evocative of movement rather than choreography. The edit was the choreographic modality. Fildes decided on a looping technique, repeating a short sequence to show the improvised dance evolving from a moment of connection into a longer sequence made perceptible in close-up or through repetition. Another technique of blurring the image takes the viewer into a vortex to give a 'sense' of how it is to be on the 'inside' of a movement moment when we no longer 'see' clearly but perceive continuous movement. The final cut was always going to be a compromise but the final version is a rich rendition of the dynamics of CI and its accessibility. Adding the words such as 'the touch of a foot' or 'who is it?' brings the viewer more consciously into the image and personalises the dancers for all watching.

Multiple layers of perception emerge through the moving cameras, the looping of footage in the editing and the soundtrack, all creating a multi-sensory and accumulative experience of CI. The concept of observation is important as this happens between the dancers and the cameras and amongst the dancers themselves, not only through sight but also through hearing, with sound by John Cobban, and through kinaesthesia, captured in the energic flow ensured by the editor Simon Fildes in the final cut. If *SENSE-8* is viewed for its 'choreography', then much is missed; the choreography is very much in the edit. The whole series of the Arts Council of England's 'Capture' commissions premiered in London at The Place, in November 2001. *SENSE-8* was distributed by Capture, the UK's national agency for dance and the moving image, and has been screened at many venues around the world, including the Sydney Opera House and the Reel Moves tour of Australia and New Zealand. *SENSE-8* is viewed in many higher education dance programs, as exemplary of integrated dance on-camera work that is accessible to visually impaired audiences.[92]

TACT *2002–03*

This performance project with choreographer and contact improviser Julyen Hamilton was initially devised in a week, from 12 Monday to 18 Friday August 2002, as part of the 'Free Summer on the South Bank' season at Queen Elizabeth Hall, London. As noted in Chapter 1, Julyen Hamilton had performed with Steve Paxton and Gerry Overington in Dartington in 1989, wearing a blindfold. However, this was the first time he had worked with an ensemble including three visually impaired dancers, Janee Hall, Alan Foster and Holly Thomas, and three sighted, Patrick Beelaert, Jovair Longo and myself. There were also three musicians, Scott Smith, Barnaby Rourke and Jim Paris who provided a score of music and percussion, mostly improvised. They created an atmosphere for the narrative pieces that evolved, providing a spatial and temporal structure for the dance.

We spent the five days on the Ball Room floor, or downstairs in rehearsal rooms, devising the performance pieces in duets, trios and ensemble scores. Hamilton created a rich spectacle involving strong individual and personal performances, along with group scores that were intimate and persuasive. The Ball Room was an open space, with a café/bar and shop area, with people coming and going. We delivered public workshops in between rehearsals, or these continued downstairs whilst two of us entertained participants. Together we strove to challenge the art form, to stretch perceptions and to question held assumptions. In this short time, we devised a piece together that raised many questions and started to define what we stood for, and we grew from this achievement.

The process

Hamilton took us through a process that tested us and himself, with the notion of being attentive to the precise intention of the movement as we made it, as we 'see' it at the moment prior to making it, which was a challenge to the inclusion of the visually impaired dancers. It was a concept that required a sense of the body schema or image, and so a refined sense of movement. It meant not faltering in the moment of deciding where to go and what to do. It meant being able to 'see' where to go next, which is a spatial and visual compositional concept (Hamilton 2008). Such was the clarity that Hamilton required of his dancers and his rigorous, uncompromising sense of purpose: each movement had to have meaning, to be pragmatic, deliberate and to 'ring true'. This conscious magnification of movement awareness enlivened the dance and was also demanding and conflictive. It brought to the surface the profound difficulty for blind or partially sighted dancer to relate their movement making to space. The task to master was apparently simple; improvisation was not about letting the body move 'as it wants', to simply 'be' in movement, but instead to move with attention to shape and to the space, to the tone of the space, in response to an impulse to create at the moment. The task was to enter the space and take a position already perceived in the mind's eye, and required an idea of the space available. This initially excluded Foster, but in time he became more familiar with the space, and with moving in it, particularly when accompanied by an audio description.

For Hamilton, the dancer is the Kinesthete, the maker of movement, the engineer of the imagination, responsible to her/himself and to the art of dancing before an audience. Not being able to see the audience was not necessarily an issue, but not knowing what is intended for the audience to see from what is 'shown' was. In the Ballroom our rehearsals were open to the public and we had undertaken to do a work-in-process, to make something 'worth seeing'. This naturally caused unease at first, but being seen made Hamilton's task all the more real and a challenge we all had to face daily. Foster had performed several times and yet in this context, the expectation of seeing him dance in Hamilton's style, wasn't legitimate, it wasn't in and of itself, a meaningful experience for him. The idea of 'performance' remained a challenge and Hamilton accepted that in the short time together, we would each absorb and shape our contribution in concert with the whole.

This was where we brought in a personal story to create a theatrical narrative as well as dance content. As the title suggests *TACT* explored different interpretations of touch, including the tactile-based discovery of the environment in childhood, the primary relational sense, and the issues at the basis of some cultural differences. The piece took shape by turning a series of lived experiences into a theatrical incident, working with memories told as stories, sketches or monologues, impressions translated into movement composition, with film, light and sound. Two of the visually impaired dancers (Foster and Thomas) and two sighted (Beelaert and Longo) created pieces from a memory that involved touch or tactility in a way that connected them to the environment and the event that took place. This was interesting as it led to thinking about how and why we recall certain experiences and how the commonality of sensory experience (such as bright sunlight) impacts differently visually impaired people compared to sighted people, due to having different sensory thresholds or sensitivities.

With the fusion of elements, dance, movement, sound and narrative, we created an inclusive ensemble piece, all be it a work in process. All these aspects are interwoven to create a theatrical, multisensory and poetic production so that dance was not the most important aspect. Short film clips were produced by Becky Edmunds for projection and voice-over recordings of the stories were used for some pieces and incorporated into sound scores. Scott Smith, on keyboards and guitar, also did voice-over during the movement pieces and transitions, as a form of audio description, and to amplify what the dancers said.

Alan Foster was up first, he told his story using a microphone, of when he was rolled in the snow at school by 'so-called' friends. The musicians added sound effects and Becky Edmunds edited a snow scene she had filmed in Vermont for projection behind him. Four of us then lifted him up and turned him around in the air above our heads, in front of the snow scene. The image of heavy snow gave further sensory colour to the moment as Foster was lifted aloft.

Holly Thomas pre-recorded the story of her childhood memory of playing on the edge of a sandpit, blinded by dazzling sunlight. This was played back as she moved on and around a chair expressing the embodied and visceral memory, and with Barnaby O'Rourke accompanying her on Cello. Thomas was also filmed in close-up as she moved and projected on a large screen to make her more visible and accessible to visually impaired audiences and speaks about this experience in her point of view chapter in Chapter 8.

Jovair Longo pre-recorded his story about his touch-deprived childhood in Brazil. His nostalgia indicated he was missing home, but not the lack of touch. We amplified this by using physical contact as we moved with him in a spatial score. Jovair has since moved back to Brazil.

Patrick Beelaert's childhood story was of when he had soldiers on the sand and the drama of them getting lost and buried, we enacted this in movement as the musicians played loud percussive sounds to evoke a war scene.

Janee Hall and I didn't have stories but performed in all the other pieces. This would not be the last time we worked with Julyen Hamilton. We continued to work with his teaching and the material and performed *TACT* again in November 2002 as part of the Changing Perspectives platform at Lilian Bayliss Theatre (programmed by the Xposure Disability Arts Festival) and had questions from the audience afterwards. We also performed at the Greenroom (Manchester) and travelled to the school for visually impaired children, The Royal Blind School, Marburg, Germany, where we confronted demands for more traditional audio descriptions.

Josephine Leaske (2002) wrote a review for the Dance Insider online magazine:

> The bright filmic images together with the music and dancers' monologues, in which they describe colors, sunlight, smells and sound create a rich palette of shades and textures that doesn't need to be appreciated by sight alone.
>
> (Leaske 2002: n.pag.)

Transition

After the film and touring of *TACT* the company members changed. Janee Hall and Holly Thomas remained but Alan Foster, who had been with us from the first performance in 1998, stayed with us for these five years. He stepped back to join a rambling club in the Peak District. I often reflect on the challenges of dance, of the entrenched 'view' of dance as a visual medium. I note the wonderful times we spent exploring levels and pathways that had been neglected in his life and which were reflected in his performances.

Stuart Jackson joined Touchdown Dance in 2004. He had initially attended a workshop in 2001 set up by Christopher Thomson of the education department at The Place (the London School of Contemporary Dance) with the view of finding visually impaired participants for a dance company from Japan. I was asked to deliver a series of preliminary workshops for visually impaired and sighted dancers. Stuart came from the RNIB Redhill College (run by the Royal National Institute for the Blind) with Holly Thomas who was teaching performing arts there. I will never forget the first dance class with Jackson and Thomas as the only visually impaired people in a group of over twenty sighted dancers. (Thomas also writes about this in Chapter 8.) Jackson seemed fearless, walking through the space with me spotting for him, but he soon took off in the space and the sighted dancers moved out of his way. After four years of travelling back and forth to Japan, and touring as a dancer with Saburo Teshigawara around the world's main stages, he was dropped from the company. His time to dance seemed up, but when I got a call from the college I welcomed him to come and dance with us, to give us a try, and he is still dancing with us today.

CLOSER. *Created 2005–08 reworked as* APPARENTLY NORMAL *2010–12 (Figures 6.4, 6.5, 6.6)*

CLOSER was funded by the Arts Council of England to tour venues and agencies and with different touring packages on offer. The most popular was a weekly residential including workshops in the community and one or two public performances, which we specified should take place in more intimate small to middle-scale venues. We sought to maintain a sense of intimacy with the audience, using the sound of the dance, atmospheres, light and audio description from the stage. *CLOSER* was a natural continuation of *TACT*. We worked with spatial relationships and with the underlying themes of proximity and distance to others. We explored ways to sustain awareness of each other, without sight as the predominant connection. We explored relationships in solo, duets, trios and full company dances and wore differently textured and tactile costumes so the visually impaired dancers could recognise who they partnered with in the improvised scores. In the scores, we chose layers of different movement qualities and atmospheres, supported by sound, video projection and lighting to define the multi-sensory aspects and to transition from one score to the next.

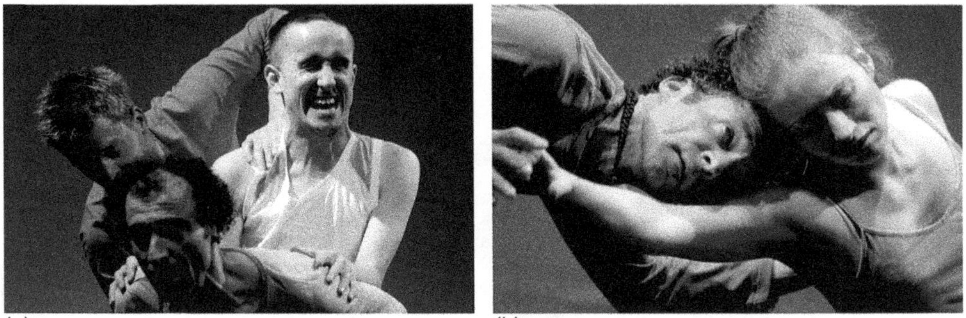

(a) (b)

FIGURE 6.4: *CLOSER* performance at Waterside Arts Centre Sale, 2006. (a) Upper body shots of Patrick Beelaert, Stuart Jackson and Robert Anderson – enjoying their repartee in the 'boys trio' score 'Where are you?' 'I am here!'. (b) Upper body shot, a moment of contact, standing in a side bend, Holly with her arm extended, and the underside of her head resting on Patrick's head as he offers support, and cradles her arm in his. Photographer Brian Slater.

FIGURE 6.5: *CLOSER* photo shoot, Robert Anderson standing with Holly Thomas across his shoulders in a high lift. Photographer Eric Richmond.

FIGURE 6.6: Katy reaches to the left over a flexed left leg, supporting Janee from the side to reach out into space. Photographer Eric Richmond.

We decided to work with Julyen Hamilton again; this time at Waterside Arts in Sale, with video artist Mark Minard to film our workshop process and discussions. We had an intense seven days, with Hamilton determined to engage with Jackson and impressed by his professional training with KARAS. We explored scores in which three moved and one stayed on the edge describing what the others were doing. We all took a turn, including the visually impaired dancers. This brought another layer of awareness as we heard about things going on we were not perceiving, and it gave a narrative that supported visually impaired audiences and the audio describer role that focused more on an overview rather than specific movements. We explored ensemble improvising and took time to discuss the experience. Jackson needed lots of clarification as he was not used to improvising having had 'set pieces' to rehearse with KARAS. He had learnt to point his toes and shared other warm-up exercises and sequenced movements with the breath, moving forwards, backwards and in circles back to face front. He had also learnt to run full circle in the space on his own and developed a great joy for dance and an eagerness to learn and perform.

Hamilton had researched in two areas throughout his career as an improviser, teacher and internationally acclaimed performer; 'one of COMPOSITION – how things are made how they might go together and one of TECHNIQUE – abilities needed to have a close relationship between the creative self and the mind/body' (Hamilton, workshop description 2008). Hamilton has defined his interest in kinaesthetics, in the way the body moves or is moved, and the interplay between technique and composition in the moment of performing – the role of perception. He expressed two aims in a workshop description written for his work elsewhere in 2008, which I recall as a thread in the week we had together.

> 1. to be able to have an awareness and a refinement of intention when deciding spontaneously in the act of improvising.
> 2. to have some references as tools of simple analysis to look in depth at what is made, both by ourselves and the others when working in a group configuration.
>
> Without this ability – and the tools for it – we are often at the mercy of our subjectivity and seldom have the complementary power of our objectivity, a state which can reduce the nature of our 'compositions' to the merely personal and lacking in the 'universal' power.
>
> (Hamilton, workshop description 2008)

Hamilton taught us further improvisation skills, fundamentally about presence, awareness and focus on what appears bodily and kinetically, to make the dance concrete, clear, fully conscious and uncompromised – which appeared to be attuned to a visual aesthetic. He would guide the dancing, moment by moment, often moving with us to support Stuart, Janee and Holly; facilitating by being in the space and talking directly to the dancer in voice and/or movement. We all shared a belief that this process of rigour and awareness had to be clearly understood bodily, for us to work as an ensemble in which each mover was the author of their individual movement, but within a shared sense of purpose that evolved in each score over time. The visual aesthetic proved to be more multi-modal; fundamentally kinaesthetic, it was about moulding and locating, about situating the movement in space, as if the body was clay that we each morphed into shape without splashes – the focus was on the duality of temporality (such as phrasing) and spatial orientation. Hamilton's role was to demand of us, to demand attention to the subtlest of changes so that as we moved we did not drop into vagaries; every movement meant something and spoke something that was minded. I recall how our movement was enlivened in this process; challenged to attend to the bodily sense that operates in dialogue with the mind, we were more thoughtful. Or, having trained in CI and with a refined sense of movement, the body moves with this wisdom, responding to and serving the purpose of the moment – to be and show the correspondence of the body–mind.

Having worked with Hamilton for a week we then worked with Angus Balbernie for a further week, in which we refined our ensemble skills, more specifically those to do with spatiality and timing, and thematic concepts. The scores are shaped up around these principles, sustaining improvisation as the key performance modality and freedom for each dancer to interpret the score in response to each other in the space. The challenge was to sustain the sense of solo within an ensemble – to sustain the overall picture without too much of the same happening. Angus reiterated out the uniqueness of being an ensemble of six very different individuals and the pros and cons that this presented to us. For sighted audiences there was richness, and for visually impaired audiences we had to use the other senses to translate our individualities and our interactivity in each moment – our sensitivity to co-creativity and trust in that being interesting 'enough'.

We were in the unknown, neither Hamilton nor Balbernie could really know how this performance would work out. The abstract nature of improvisation was conceptualised by questions that were inherent to a group of sighted and visually impaired dancers, such as: how do I know when I am close to someone when I cannot see them? How do I move close to someone and not be inhibited, or afraid of collision? How do I move into a relationship with someone in the space who I can't see? How do I get to feel close to someone when I feel distant from them even when connected? How do I move freely in the space and create an atmosphere when I cannot see what the others are doing? How do I enter and leave the space coherently when I cannot see the edge of the space? How can we know that we understand the concept in the same way? These questions were the starting points for hours of exploration, to hone the themes, the ideas on spatial formation, and choreographic elements that each dancer improvised with in each score. Often seeking clarity and sense led to a loss of vibrancy, of the essence of the score, so we had to always return to the collective idea of the group. Being strongly individual we also learnt to respond to each fellow dancer in an intrinsically sensitive and focused way. Rather than dropping into more of a murky sensory state, we arrived at each moment in movement emphasising the performative. The timing of each encounter could not be as direct as when all sighted dancers were moving together, so the scores encompassed these questions, incorporating these challenges into the frame itself.

Balbernie introduced concepts such as sculpting movement in space, paying attention to slow sustained improvised movement on the spot, or fast sustained movement, and moving into space from that. Another aspect was the idea of moving in synchrony, with the sense of moving as an amorphous group, with a shared focus, rather than precise timing. Taking inspiration from flocking birds, we referred to their navigation system, how they calibrate and sustain the space between them with sensors that operate similarly to echolocation and by riding air currents. We used this as a metaphor for moving together without vision, to find a way to traverse the space on a diagonal (the longest distance available on stage) whilst keeping to a cluster – which became the name of the score.

Having completed this research and development period we undertook a two-week residency in Brighton, with Scott Smith in attendance offering sound, and Becky Edmunds who made a short film in which each dancer introduced themselves in a voice over as we moved together in the space. This would be screened before the show as the audience settled in their seats. We had a warehouse-type venue, adapted into a studio with a wooden dance floor. We had time to revise the work and to really connect as a group; we walked on the promenade, swam in the sea and sat on the pebble beach. Stuart explored walking independently with his cane – striding out confidently and those coming towards him moving out of the way. This time nurtured our relationships and camaraderie. Stuart was away from the RNIB residential home, one of us enjoyed his enhanced independence.

The six scores

1. Solos

One enters the space to solo for 30 seconds or so, a second enters in counterpoint or in attunement with the first, and then after two or so minutes others enter, dancing their own solo. This score was influenced by Hamilton's improvisation skills and awareness, as well as Balbernie's concepts of working with spatial frames and dynamics. One clear concept was the idea of having an anchor – one person standing or holding a space whilst the others moved, which avoided having a group of dancers all moving in space with no sense of the whole landscape. This was supportive for the ensemble, particularly for the visually impaired dancers to sense when the dynamic and space had changed from the lowering or rising of the amount of energy and movement. The sighted dancers held the space overall and maintained the timing if needed.

Balbernie gave us the following options for our solo, and the instruction was to enter in counterpoint/contrast to those already in the space moving and transitioning between the options. It was a discipline to keep the movement going and to attend to the group.

1. Fast Sculpt (find a place and stay on the spot).
2. Slow Sculpt (a slow version of 1).
3. Slow Solo (moving around in the space).

This ensemble score continued for eight minutes with a soundscape behind it compiled by Balbernie from sampling sounds in different landscapes.

On a sound or light cue, we moved to the upstage left corner (top right from the audience point of view) leading and following as we do in CI, and congregated in a group with some standing and some lower, with the person in front ready to start the next score.

2. Cluster

We congregate in the corner in contact with each other facing the opposite diagonal corner close to the audience. The group cluster score was based on the concept of flocking birds that sustain pace and direction, each is part of the whole formation; maintaining constant velocity and distance from the centre of the cluster. The motion of each within the whole is influenced by the nearest person and distance between others, with the option to solo away and to return to a collective CI dance across the stage. Closeness comes when we connect with rhythm and movement, not just contact, and it resides in our commitment to the theme of complicity and ensemble, a metaphor for inclusion.

We cluster in the corner and start to move forward on the diagonal. We stay clustered and physically connected at least with one other person keeping to the diagonal. The one at the front provides the direction for the others who lean into each other, sharing weight, or reach and pull, drop or lift, the ones trailing behind can create a CI duet, stretch the group back along the diagonal line and then resume the forward direction to rejoin the cluster.

This score used a different dynamic of slow, sustained, interconnected movement with some 'options' that supported the audio description. The dancer in front started slowly and we all followed this initiation, sensing through touch and responding, using CI to move over, under or around each other. After rehearsing this many times we decided on the options: one may break away from the pack to solo and then rejoin the line as migrating birds do, or instigate a brief duet with a clear commitment to an ending and then both return to the cluster, or one can return to the corner we came from to stretch the line back, creating a metaphor and sustaining the scenographic image. All these moments are also audio described.

Another aspect of the cluster was a sense of 'loading' the space in stillness, for one or more of us to 'anchor' the image in space, to foreground the physicality of the movement taking place. The conscious positioning of the extremities, such as the hands, feet or other contact points with the floor, created the connection and support for the rest of the body to organise around. The anchor gave clarity to the moment, rooted in gravity and to the earth, a sense of place and position, in time and space, and a structure for weight bearing – it created an environment that worked visually and kinaesthetically for the dancers and audiences.

Our notes emphasised the need for clarity in the relationships of each of us when in solo/duet/cluster. This relied on each of us keeping our own clarity, in terms of the bodily sense in the present moment and the score, as we didn't have choreographic phrasing to rely on. We sought clarity in each touch connection, in taking a decision to break away or return, and a narrative that has an ending we agree on – a duet is left to complete while the others have gone.

To finish we gather in a cluster close to the audience downstage right, except Holly and Robert who continue a fluid duet as the rest leave. Holly was initially challenged to trust being lifted off the floor and high onto Robert's shoulders but learned to be carried aloft and to slide feet first back down Robs back to the floor. She sits ready to reach and be pulled back up again as Rob scoops her up. This dialogue continues centre stage for a few minutes as Stuart prepares himself to enter from one side when Robert and Holly leave an empty stage.

3. 'Boys' trio (see Figure 6.4)

Stuart enters and paces a number of steps that should bring him stage centre. A light comes up and he is sitting on the floor, reaching into space, brushing the floor as if looking for something. He stands and tells a story in movement, reaching up to the sky, reaching forwards and running forwards then backwards as if pulled and pushed. Lights come up fully as Patrick and Robert enter, not necessarily together and at times via the audience, and start to approach Stuart. As Stuart cannot engage visually with them there is a sense that he is doing his solo in his own world until one of them speaks to him, or he notices them. Stuart connects with Robert or Patrick in response to whoever interacts with him physically or to their vocal cues that locate them in space. They explore ways for Stuart to know who was where; e.g. Rob or Patrick say 'where are you Stuart?' playing with the idea that they cannot see him when they clearly can. Stuart then knows where they are and can respond either by moving towards them or playing the same game saying 'I am here!' The play continues with them all interchanging 'where are you?' and 'I am here', playing with distance and proximity, different levels and even all three trying to hide behind each other. Fun is had by all. Stuart rises to the game and starts to move into space with Robert or Patrick. Being blind Stuart takes advantage of feeling free to respond and move, to play and entertain the audience by stepping up to the game. The trio support or counter each other and continue for five minutes. At times Robert and Patrick move out into the audience or the edges of the space and converse from there. The dynamic is left open to the dancers to build so they can even play off the audience, breaking down the traditional 'fourth wall'. This is a raucous and comedic score in contrast to the previous ones and always gets the audience vocal in amusement.

4. 'Girls' trio

Holly, Janee and Katy enter into the space from the edges. The six of us are on stage momentarily, like a cross-fade, to give time for Patrick or Robert to connect with Stuart again and exit without directing him. Hamilton's compositional improvisation influenced this piece. There were three large rectangles of light on the black floor, one for each of us, and on the back wall, three adjacent frames projecting different videos of waves rolling up a beach. Each of us improvizes a solo within or on the edge of our frame of light, inspired by an imaginary location or memory. Mine is inspired by childhood memories of being on a beach on the Isle of Wight and the sea view. After three minutes the frames dissolve in more light and we move as a trio, or as a duet and a solo. This is an abstract score and proved hard to audio describe.

In our rehearsals, we worked in separate spaces; first alone and then with moments of space/time connection, with the limited sight and kinaesthetic attunement we strove for clarity of connection rather than unison.

5. Atmosphere duets

In devising this score, we explored movement qualities that Balbernie gave us and others that we found and named individually and then shared with each other. We created a list of 'atmospheres' and chose four each: two inner states and two external states, and decided the order we did them in. The four inner states that we kept were: *anti-gravity*, *silver balls*, *drunken joints* and *bubbles*, which gave the sense of upward flow, mobility, loose joints and lightness, but were open to interpretation. The two external states were *forest* and *beach*, which gave the option to be in a landscape in which to move together, whilst a third *waiting*, was a state, literally the sense of pausing, of time passing, so different to the anchor.

We explored the qualities in duets and changed the choice/options each time to then decide which four we would use and the sequence for the improvisation. The final score began with two duets, myself with Patrick (both sighted), and Janee and Stuart (both visually impaired), and halfway through Patrick and Janee changed sides. This meant that two sighted and two visually impaired dancers of opposite gender partnered with each other and then changed to two integrated duets, one of two men and the other of two women.

This score contrasted with the previous ones as it was uniquely for duets. Both duets started in the space, in a separate spot, but side by side. Patrick and I moved with the atmospheres of *forest*, *reduced gravity*, and *beach*. One of us rising up and being held back by the other, it was physically spacious. Janee and Stuart started by moving more in the space, and then more sedentary, with three states: *silver balls*, *bubbles* and *forest*. After swapping partners, Katy and Janee did *waiting*, *reduced gravity*, *forest* and *beach* with a physical dynamic involving contact and weight bearing. Stuart and Patrick did *reduced gravity*, *silver balls* and *beach*, working with unison, support and lightness. We chose a music score that was atmospheric and referenced the states and a forest 'gobo' and green lighting state created the impression of light coming through a canopy of leaves. In this environment the duets contrasted vibrantly and dynamically and the interpretations were evocative and meaningful.

6. *Stone garden*

The score was immersed in a soft lighting state and enhanced by a large screen image of stones and seaweed across the back wall, again by Becky Edmunds. The concept was minimalist, evocative of time and space again, from stillness unfolding into movement in the whole space. We took positions on the same diagonal line as Cluster, each person progressively closer to the floor, from Rob standing at upstage left, to Patrick lying downstage right. We remained in stillness for some moments at the start and began moving internally and slowly with gentle movements of the torso, barely perceptible, starting sculpturally and repeating two or three movement options, on the spot.

In this formation, we had to use our peripheral vision/kinaesthetic sense of each other's movement and audio cues to indicate key timings for the start of each section and when to transition to the next. In this score we worked with an idea of unison of intention and theme, without unison of timing being necessary. The scores imposed limitations such as a minimal use of space and time, which enabled each individual dancer to comply and mutually complement each other; the sense of ensemble evoked the 'closeness' of our relationship.

Balbernie gave us the score concept and structure; we remain on the spot, from stillness, we make almost imperceptible '*micro*' movements, limited to the torso, initiating from the spine, ribs or pelvis with the intention of 'I feel'. We each make a small brief movement that we feel and end in stillness. After transitioning to two further movements, we repeat this sequence. The next phase, '*Meso*' involved making movement that expressed the intention 'I am', by opening up to the horizontal dimension with sculptural shaping, communicating presence within a self-membrane. We then started to open up from the situated sculptural mode to more peripheral shaping, adding new movements and extending into a wider frame of the kinesphere – as if acknowledging or being receptive to other, 'you are'.

Finally, in '*Macro*', the intention was indicative; we opened to the 'beyond' into the wider collective membrane, as if saying 'there is': each dancer opened to the space on the sagittal plane, gradually leaving our position and travelling towards, away and around each other.

Our energy states increased as we opened to the space and generated a magnetic intent in our movement. This was a new theme and exciting to feel attracted towards or repulsed from each other. It meant that the sighted dancers changed the space and the visually impaired dancers responded or initiated a change. We had the option to form a duet, trio, quartet or quintet. We explored using the voice and description but dropped this on stage. The quiet aspect of this piece was intentional, inviting intimacy with the audience – it would require clear audio description and we also decided that visually impaired audiences should sit close to the stage.

Stuart solo – a post-script

Stuart entered from stage left having counted the number of steps previously, he walks purposefully to the centre of the stage, where a frame of light came up on the floor to contain him. His choreography was set – from sitting crossed-legged, with his head in his hands he found his way to standing, he ran forwards reaching with both arms, fingers extended, then, as if pulled backwards he ran backwards to his spot, arms lowered and then reaching back behind him, still in the frame of light. This reach and pull continued till he stopped, reached up and on the out-breath he walked in a circle back to his spot and repeated to the other side. He spans on the spot, immaculately without any disorientation. He stopped and as the light faded he returned to the floor to sit. This movement narrative ends with him sitting down again. The lights faded out completely and we all come on for the applause.

CLOSER was created from a sense of conviction in the ensemble process, with the guidance of Balbernie. He read from various philosophical sources and from his own reflections, introducing themes to support the innate characteristics of each dancer and also of our ensemble. Our identity was based on the idea that we focus on our unique qualities, that differ from already existing preconceptions of dance. Angus was influenced by Howard Gardner's theory of multiple intelligences, and Clive Bell's book, *Civilisation*, which speaks to the inherent nature of art as a means to retaliate against civil society. Our creative process was a form of protest – a means to claim a place in a cultural landscape that was otherwise closed territory. We were creating and taking steps to contribute to the cultural landscape, and despite disinterest from the funding bodies, we chose to buy into the necessity to achieve creatively together, and have faith in each other.

In this piece, we learned that connection does not have to be visual, if it's physical or internally felt it manifests kinaesthetically as a quality. We could feel the elasticity of the connection with a partner, through the trust and sense of fluidity. We learned we could be physically close to another even if we did not physically touch, and that we could also be touching someone and not feel physically connected if our attention was elsewhere. Robert commented that *CLOSER* was a piece about physical, emotional, perceptual and intellectual closeness, in relationship to each other and the audience. This also resonated for Stuart, with the fact that the negative – not having closeness – is the usual experience for many visually impaired or disabled people, most of the time. Patrick commented on the fact that when we improvise we challenge the space, ourselves and time, and create space for change by holding the attention of the audience on what our ensemble shows to be possible (see Chapter 8 for further reflections on this process).

Touring

Once completed with lighting designed and set up by Alan Jackson, and a sound score recorded by Scott Smith and Angus Balbernie, *CLOSER* toured around England following a premiere at the Nuffield Theatre Lancaster in November 2005. The highlight of the tour was our participatory performance at Sadler's Wells in the Lilian Baylis Studio, with a matinee for a group of blind children transported from a local school. They had a Touchdown Dance workshop in the morning, partnering and working from the floor to standing. We introduced the key movements involved in three of the scores, the idea of sculpting, leading and following, and the cluster, which we did with the sound score. Then in the afternoon, we performed the piece with them on stage with us, fully costumed, with lights and sound, we performed and they moved around close by and with us. The response of the staff was quite emotional, and they said the children had got so much out of this experience of moving together, as they did also. This was impactful and determined our desire to reach out even more to educational settings and to children and young people themselves.

Tour to Scotland

Karen Wood of Dance House in Glasgow and City Moves in Aberdeen booked us for workshops and performances. She asked us to deliver a continued professional development workshop for seven professional dancers which would also involve a performance alongside our own. In Glasgow we did community workshops at

Touch Base, the centre for SENSE-Scotland, for adults with complex needs, learning disabilities and sensory impairment. We then had a week of training with seven dancers from all over the United Kingdom, based at the Centre for Contemporary Arts (CCA) in the city, we taught improvisation, core Touchdown Dance fundamentals and worked the dancers into the first three scores of the performance. This meant we had thirteen dancers on stage for the first half of the show.

Another highlight was our collaboration with the Glasgow Improvisation Orchestra – they had booked the CCA for the same dates we had hoped for, and Karen arranged for us to discuss the possibility of sharing the evening. This led to the idea that after some of their own sections we would improvise with them for five 5-minute sections, and then, following an interval, we danced *CLOSER*. The first half had thirteen musicians (two drum sets, two pianos, several strings and wind players) and thirteen dancers potentially on stage together. As it worked out, it was extraordinary and we all really enjoyed ourselves.

Touch tour and audio description

We also had really good press coverage from the talking newspaper journalist and had several visually impaired audience members. The CCA was very efficient for these audience members, meeting them and accompanying them up to the studio where we had a 'touch tour' with the dancers and talked them through the programme and had a brief demonstration of CI. They sat at the front and clarified this was a 'new experience' for them. The cacophony and poetic first half with the orchestra was very much a visual experience and impossible to audio describe. Our performance was more inclusive, with simultaneous description from a describer sitting next to them and from the stage.

The programme and publicity stated:

> Using senses other than eyes, six characters take flight aiming to bring the separate closer. Liberated from taboos of touch, they share a world where contact is not cautioned, but is the starting point of freedom and trust. In this dynamic new show of structured improvisations, the unique company of visually impaired and sighted dancers combine voice, video, live music and a wide range of dance styles to break the rigid regimentations of traditional choreography.

At Saddlers Wells in London, the audio describer communicated through digital headsets. The script introduced each piece, stating who was dancing and what the key aspects of light, image and content involved. The prestige of the venue, their administrative support and provision of audio description meant we had more visually impaired audiences than expected. They were given clear directions and met at the door. We collected audience feedback which was varied and very insightful – some considered that the description and vocal interaction from the dancers were more useful, others enjoyed the mix, and some felt it could be more descriptive of the actual movements that the dancers made. Some people found the facilities and venue difficult to navigate and even to find the entrance. Interestingly, the director of Sadler's Wells at the time was visually impaired and his own research had found that visually impaired people were the most excluded from dance. We were unique in the way we invited the audience into our experience by using a multi-sensory and intimate format. There was a lot of interest in the fact that we were an integrated company working inclusively in both community and education settings. There was also interest in the nature of our improvised scores, light, sound and projection: it was a different type of audience experience to a contemporary dance company in the repertoire.

Follow the frame

By 2016, Angus, Stuart and I had evolved a strong working relationship by making impromptu performances for events over previous years, such as the *Rabbit* improvisation weekends at the Northern School for Contemporary Dance in Leeds. *Follow the Frame* was one such duet that Stuart and I made with Angus accompanying us on guitar. In this piece we started holding our thumbs and index fingers in a triangle in front of our foreheads to mark the spot where we then put forehead to forehead and started to do CI. We then held these triangles in front of our faces, as if a frame that we looked through. I did so but Stuart mimed as if he was following it. Stuart enacted a solo piece about a man who went to the outer atmosphere in a balloon and then parachuted back to earth – a story he had heard on the radio and it ignited his imagination. Stuart rose onto his toes as if being pulled up by a balloon that he mimed holding in his right hand. Whilst I waited with my 'parachute' to encounter him and then connect in CI joined by four students from the dance school.

The parachute man came from a conversation between Stuart and me.

Stuart tells the story of when he is 10 years old, his mum and dad bought him a parachute man from a shop;

> St.: I used to have one and used to play with it in the house, I would throw it up and it would come floating slowly down (I was about 7). It's nice to have a memory of different things.
>
> Kt.: How did you feel when you got the toy?

St.: I felt good really good about it. I remember the good things I used to go off doing my own thing. When I came downstairs I threw it...

Kt.: Are you used to doing that as I child?

St.: Yes, I used to have a baseball and I used to throw it up the stairs and it used to come bouncing back down to me, down the stairs right to me, I used to catch it.

I could hear where it was coming from I guess.

Sometimes I used to take it up the stairs and roll it down the stairs. It sounds like someone walking down the stairs, that's what it sounds like...Cos it goes down on its own.

Kt.: I wonder why you wanted to throw it up the stairs?

St.: I wanted to know if it would come back again, sometimes it stayed up there of course.

Kt.: So if it stayed up there what did you do?

St.: I'd go up and get it down again. I always used to like that when things came back to me by themselves. That was *clever*.

343 m/s – The speed of sound (Figures 6.7 and 6.8)

This production by Angus Balbernie was commissioned by the Buzz Cut Festival in Glasgow, initially in April 2017. Angus chose to make a duet with Stuart Jackson and asked me to assist with it, to support Stuart in the process. Angus chose to make 343 m/s with Stuart as a more theatrical piece involving story, dreams, conversation, improvised dance and movement. This piece grew into a trio with myself performing and contributing my dream stories by invitation from Karen Wood to share the stage with I-Dance in Kilmarnock in October 2017, where it triumphed! (see Figures 6.7 and 6.8).

We performed again in April 2018 at Experimentica. About Time festival at Chapter in Cardiff and Dartington for an artist-led initiative to re-enliven the dance studios in September 2018. Finally, we performed at the Voarte Disability Arts Festival in Lisbon, in February 2019 when Robert Anderson joined us to support Stuart and then joined us in the performance. This is an account of that open process.

FIGURE 6.7: 343 m/s performance Angus Balbernie and Katy Dymoke sat on chairs watching Stuart Jackson jump and spin in the air. Photographer Mike Scott.

FIGURE 6.8: 343 m/s performance duet with Katy and Stuart as Angus watches. Stuart moves to his right, Katy folds her arms over her head Angus blows feathers over the floor. Photographer Mike Scott.

343 m/s Lisbon

In the round, we sit on chairs as the audience enters. Angus advises the audience to leave their phones on, and that he will say why. Usually, this is not allowed in performances space. Angus plays a track by The Who, 'I Can See for Miles and Miles', and leads Stuart around the audience to say hello and introduce himself.

Angus talks about Owls and then introduces us all. Stuart is last, and he says he is a professional dancer with Touchdown Dance. Angus plays a Beatles track 'Come Together Right Now', and asks Stuart to 'prove it'; he jumps up, improvising a solo, Katy joins him, then Robert.

At the end Angus steps forwards to ask Stuart – 'So Stuart, do you dream?' and he states: 'Yes, I do. I have had scary dreams.' Angus asks him further questions: 'How do you dream?' 'I feel movement.' 'What is your dream like?' Stuart shares a dream with the audience: He is much younger, staying at his parent's house. Suddenly the house starts to move and he has to hold onto something to hold himself up. He goes downstairs and it's like going into a large dark hole. He has the sense of 'falling and turning'. Angus responds: 'That's quite a dream!' Katy responds, 'More of a nightmare than a dream?' 'A dream nightmare', says Stuart and asks 'How do you dream Katy?'

Katy has her turn to tell a dream from childhood and recalls it in black and white.

> I am walking on the curb of a road I see it continuing ahead, I know that cars are coming up behind me, on the road, I feel it, and I fear that when they come past I feel they will suck me into the road. I hear a car coming but wake up before it passes me.

Stuart states 'A dream nightmare.'

With both hands, Rob, Stuart and I carry threads of fairy lights clustered together to make a 'nest' as we walk ceremoniously centre stage towards the audience. This is a quiet and meditative moment, in an acoustic space. The lights are almost in a blackout. Angus asks people to close their eyes, to consider 'how do you hear?' He moves as he talks about the conical eyes of owls, their ultraviolet vision – about difference and disability. Once done he puts on 'Shoot Out the Lights' by Richard Thompson, on queue Stuart does another solo gesturing to the lyrics and kicking, pushing his hands strongly out in front as if blocking something. It is a very real rendition of the words, visually powerful and evocative of his agency and personality.

Over a recording of an owl hooting Angus gives owl facts; an owl has one ear bigger than the other, different types of feathers, acoustic hearing and their eyes don't move – because the ocular nerve comes right to the surface – so an owl can see 100 times better than humans. He adds facts about other birds, e.g. a Peregrine Falcon flies at 180 km an hour. Angus ends with a poetic moment about silent feathers; owls fly silently with a layer of feathers down underneath. As the lights lower, he spreads duck feathers on the floor around the space to a soundtrack of an owl flying, and Stuart does a slow-motion roll in the feathers, assisted by Katy, diagonally across the stage.

The lights come up and Angus addresses the audience to play their favourite track on their phones as we walk around to listen and converse; we hear all sorts like Amalia Rodriguez. To end Angus plays 'I'd Rather Go Blind' by Etta James and invites the audience on stage to dance with us. We continue talking and socialising together in the space.

Ana Barata of Voarte wrote 'your performance as a very intimate and warm hearted moment'. This performance is still in exploration and returns, in different iterations, at different festivals with different guest artists.

Two further performances, *APPARENTLY NORMAL* and *Attention* are written about in Chapter 8 by Jamus Wood and Robert Anderson.

7

Final Words

The paradigm shift – Towards the individual and collective – Embracing the membrane of inclusion

The initial ambition that inspired Paxton to collaborate with Kilcoyne was serendipitous, and their initiative was very timely, as embodiment practices and practice-led research were making waves. They evolved an inclusive practice in which the body was both the medium for learning and the research instrument. The underlying philosophy and ethos were one in which interaction and understanding, body and mind, practice and research are considered porous, not separate. Within its self-defining two-way membrane, Contact Improvisation (CI) was a container for self-realisation. As I discovered, this corporeal territory was the locus of learning that I now perceive as a body–mind continuum, from the cellular to the societal, a membranous space for growth.

From the opening of this story, I have been drawn to consider the historical context and ethnographic influences surrounding the emergence of CI as a movement practice. I have considered the battles of the community dance and disability arts sectors against the 'citadels', challenging perceptions, introducing a paradigm shift in which practice and experience, inter-disciplinarity and subjectivity are foregrounded. The intention to explore, to establish a dance practice that was not conceived by a single source, choreographer or school, founded this democratic form.

The ethos and moral sentiment that shaped the ambition of the research undertaken by Paxton and Kilcoyne came contemporaneously to other integrated arts projects, disability-led projects and social justice campaigns for change. This has become a continuum, a process in space–time, of practising, questioning, reflecting, and repeating that allows for the potential breadth of contexts and variable applications from which new meaning and learning arises. In CI, the phenomena of touch and movement are subjectively experienced but always in relationship to something other – another aspect of self, of another person, or the environment, another entity. Consequently, the outcome is indicative to those experiencing this relationship, both subjectively as 'I' have the sensory information which informs 'my' response, and inter-subjectively, as together with 'you', we learn to accommodate our differences as valuable information. A new attitude, a change in the state of awareness, and availability to share with others enriched the practice and opened up the potential for change, for elucidation and liberation from pre-existing perceptions or beliefs.

This transitional state is innate to human existence unless anchored and/or veiled by societal diktat regulation, or by oppression and control by others with authority given or self-ordained. This loss or lack is the given status for individuals lacking autonomy (Dymoke 2017) and for subjects such as touch that have lost their essential role due to misapprehension (e.g. fear) and lack of open judgement (due to taboo). A pedagogy such as CI, founded on the body, is inherently non-judgemental, as no singular body is the same, no singular touch experience repeatable in exactly the same way – without the displacement of an emergent sense by a new altered one – a different hue or shard. The value is from the present moment – each time as if for the first time, and the past unfolds into the future.

Practice-led research is essentially about being in the world and the study of the phenomena present, to identify gaps that emerge and contribute to new ideas, methods and approaches (Smith and Dean 2009). This applies to all, and variably, so that difference is demystified. Once normalised the participants invite this change inwardly, as well as outwardly in the social body, having thrived within this membrane they go forth with this support. Ethnographically, CI offers a territory where the body is the locus of the lived experience and shapes the understanding that evolves. It can be considered a form of popular culture that seeks the re-appropriation of lost territory, by re-assembling aspects such as touch and movement into an 'assemblage' of modalities, to reclaim the body, as I do here, joyfully, if discursively (Deleuze and Gautarri 1972).

The research accomplishments and the return of non-touch

In the bodily encounter, the membrane between otherwise distanced entities, whether blind and sighted, erotic and non-erotic touch, begins to exist and its potential to be experienced. The physical and mechanical aspects of CI render bodies to be able to move with ease, and minds to open to new ideas and to recognise the importance of dialogue, interpersonal exchange and the role of touch. The recent pandemic has ironically reinforced this – as the regulation and prohibition of physical contact has re-associated touch with fear, fear of a threat to life, and touch as a conduit for this. We have become accustomed to not touching and have to adjust to the possibility of touching – to be 'in-person' rather than 'on-screen'. The vulnerable children have been deprived of the liberty to be together with their peers, and they have been told not to touch each other and to not let others touch them – removing the membrane that enables relationship and substituting it with fear. So once more touch is not the problem, but in the act of touching the virus is transmitted, and the solution is to touch safely. Once eradicated the potential for touch is to be relearned – and this has become possible using props, and interpersonal contact with those in a 'bubble' – the membrane of the bubble is literally the new relational sphere, controlled and subjected to isolation by those entrusted with the authority to do so.

We have been told 'don't touch me' by some of the children. This instruction was respected, and in time trust returned as did touch, within the safe bodily realm of the session, where we cohabit the space, provide comfort, ease and support, and the freedom to move. Touch returned, fear was displaced, and fun, laughter and liberation resumed through the different self-selected modalities of being sat back to back, or side by side, dragged on sheets of lycra or rolled on the mats. Our feet touched, 'like wires connecting' and then we melted like icebergs under fabrics, or huddled side by side, 'snishing' together under fabrics as we talked about climate change (as they called it). This is the continuum of change, the place of transition, where the triad of membranes instils safety, trust and human-to-human co-visioning, co-creating and collective cooperative exchange. In this sensory realm of being, adjustments come fluidly, as needs arise, or voices speak for them. They resolve in action, in a sense of alignment and organisation of the body – a coherence of self that then grounds the mind.

Simply put, a sighted partner accommodates no sight and a visually impaired partner accommodates sight, and both discover that through touch these senses are far more interconnected and present in movement as in life. Whilst Touchdown Dance has demonstrated the simplicity of this, it has also demystified the apparent complexity of touch by evidencing the joy of its re-discovery as a right.

Creativity requires a motile and mutable membrane; it is there to be discovered, honoured and respected – a principle of non-control, a principle of the right to be human. As a self-regulatory construct inherent to the work the membrane underlies the methods of communication and relationship, as a result, the work has been experienced as inherently therapeutic, life-changing or life-enhancing, for some. Over time CI takes its place within contexts and communities in need, with health and development of the self as its purpose – with the view that the primary senses of movement and touch are foundational to that purpose.

Capturing the experience – The multiple membranes

As the participant accounts in the previous chapters document, the territory of CI is one of transcendence and transformation, experientially it brought up the sense of shifting boundaries, shifting ground, and the removal of divisions by altering the underlying perceptions and beliefs upon which such divisions depend. In the context of the estrangement of visually impaired people and sighted, the positive experience of learning through the body negated the pre-expectation of difficulty. Where incompatibility was concerned, any uncertainty dissipated as the dialogue became mutually informative and beneficial. As the body felt comfortable the values that created discomfort were then undermined.

As a rule, as the basic principles identified, we are used to examining our experiences to deduce meaning and establish relationships or boundaries. As adults, we are used to re-consider and re-configure the perceptions that enforce the necessary boundaries and shared beliefs as pre-judgements. We may not always choose the time or take it, but this self-re-evaluation was what happened here with the experience of touch and movement methods, and often in unexpectedly fluid ways – as the body reconfigured with the cooperation of a partner, so did the mind in relation to the same. Both partners would need to accept the movement experience as the mutual place of learning, it was the sum of what both could bring to it, and so both were equal partners in its making, co-responsible for how much the whole experience meant overall.

The practice-centred focus of Touchdown Dance is at the juncture of a meeting place between blind and sighted people on the one hand, as well as with each person and their experience on the other. Whether blind or sighted each person learns from their movement experience and is changed by it, by extending self-movement potential, the sense of 'core self' extends with the accumulation of experience and shapes a new self-percept. The change takes place at the meeting point of the subjective experience of the mover with their body, of body to another body, each body moving in the way it knows, finding its way to explore – and embody itself. The membrane of each person shifts as they move and again in relationship with a partner, which combined will shift the membrane that once separated and stigmatised the minority community of visually impaired people from the dominant sighted one in which they live. By inhabiting the body, embracing it as the site for learning, the language of the body becomes central to the endeavour to write about and discuss the practice. The process of growth and transformation, of motility and mutability takes place within a space (the body of the dancer), within another space (the sphere around the dancer – the kinesphere), inside a wider space (the studio), and each space operates within its own 'membrane'. We have, in effect, a multiplicity of membranes; we could say around each person, each dyad, each group, each studio, each community, the wider society and the world.

As a kind of counter-movement to the mainstream dance scene, CI was nonetheless movement, advocating self-movement as a form of self-making, a form of democratisation of the self-body relationship and that of the self-social body relationship. By establishing equality of access, CI shames inequality of access; by accepting difference CI shames stigmatisation of difference; by creating the freedom to move to include those whose needs are different, CI shames the oppression and suppression caused by exclusion.

The three continua – A new three-dimensional, embodied perspective

In approaching the methods by which CI practices work, it has become apparent to me how inter-systemic the world of CI has become. It is not easy to separate out one method or 'system', one discourse or principle, without detracting from the rich inter-disciplinary nature of the work. As a result, there is cross-over and repetition between chapters, as the weaving of the writing unfolds and the 'fabric' of the work, as Paxton perceived it, takes form, multifariously and multi-subjectively and as a reminder that CI 'forms' us as we are forming it (Paxton 2008). We are all present within the expanding framework that is motile and mutable, fluid and strong, a three-dimensional continuum – like fascia.

In my view, on the one hand, CI is the core research practice, and on the other, the concept of CI as a research 'framework' has become incompatible, too rigid for the complexity and vitality of creative and alive practice-led or practice-based research. As the ethnographer and author, I have established three types of continua, each a temporal–spatial concept that contains and communicates with the others and represents the inseparable nature of the interrelationships that I perceive enfolded within the cultural landscape, and intrinsic to the research endeavour undertaken. These continua can be seen as separate 'dimensions' that provide valuable insights into their respective realms and assist me to reflect upon my experience of the praxis.

From my 30 years of experience as a research participant and practitioner, my embodied understanding has construed a corporeal environment in which the research takes place, an 'animate form' from which the methods emerge. The practice of CI is perceived diversely as a dynamic three-dimensional landscape, an ethnography of different cultures, a social situation in which habitual or customary behaviours are explored relationally, as continua rather than static polarities

The first continuum expands on the first basic principle; the subjective perspective of the participant who relates inwardly to the deep bodily sense whilst moving, shifts outwardly with an expanded sense of interrelationship with the external landscape and the infinite possibilities this offers. I took inspiration from Simon Jones, in his chapter on the complex and often paradoxical self-positioning of the artist research subject in relation to their research endeavour. Jones refers to a continuum between 'the interior void of the soul and the exterior void of the absolute possibility' (Jones 2009: 25). As an artist-researcher and BMC® teacher, the 'void' is a familiar percept, in the constant meeting with the unfamiliar aspects of the bodily sense. CI is a place of potential that is nurtured through relationship, and it demystifies and animates both the interior and exterior voids.

The second continuum encompasses the exchange between the self as participant, learning CI, in polarity with other types of learning that society dictates as 'valid'. Polarities arise in the learning process when the self-position of the participant experience conflicts with existing norms. Inclusive research requires working with participant experience, involving participants in decisions that can affect them directly, and brings into question the pre-established rules and beliefs that justified the exclusive system in place. The political ramifications characteristic of Emancipatory Research (Nind 2008; McClimens and Allmark 2011) arise when service inadequacies are assessed by users, and as a result, changes are made that impact positively on the quality of life of those involved (Gilbert 2004; Walmsley 2001, 2004; Redmond 2005; Williams et al. 2008). In their experience of CI, visually impaired participants negate the socially constructed stereotypes of dependency, in an environment that makes such definitions 'void', imposing a rethink and ultimately changing people's lives.

The participant experience is key to practice-led research and their role is to contribute to the 'findings' by sharing the nature of this experience; how it has changed their bodily sense and sense of self (the first principle and continuum 1) and what type of learning that has occurred. As the experience depends on working and communicating with others, with participants from very different backgrounds and lived experiences, the process requires debate and, considered, reciprocal exchange (the second basic principle and continuum). With this degree of participant involvement in the research, there is frequent reference to a sense of liberation, of self re-discovery – implicit of an emancipatory release from societal repression and control, and a shift in consciousness towards participation and direct action – which indicates a new 'attitude towards learning'. The second, I suggest, is therefore grounded in the third continuum.

The third continuum encompasses the interrelationship of body and mind – the literal and metaphorical sense of what it is to have and be a body, and what it is to have and be a mind/minded body – an animate form. The realms of embodied practice and reflexive critique are expressed in the languages of the body and verbal description. This third 'relational aspect' is a process that brings experience to conscious 'know-how' through dialogue and debate – an aspect upon which the liberty of those excluded from participation in such learning depends. It is the means by which to challenge and change existing belief systems – and will be as variable and as subjective as the participants themselves. In finding the unitary sense of the bodily self the discussion references the intricate inter-subjectivity of emotion and perception, of the multiple selves there-in.

I do not perceive these continua as linear or one-dimensional, rather as fields, characterised by the qualities of the different spatial dimensions – which also co-exist in three-dimensional space–time; the first appears to be more horizontal as the soul radiates towards the possibility of finding what is 'out there' in the field of practice; the second is more vertical, with the subjective perceptual mind state, separate and self-contained in face-to-face relationship with the rhetorical social diktats; the third is moving through the three dimensions of the body to a more directional (sagittal) and intentional interrelationship with the body–mind of others. I notice these relational aspects within myself, conversant and mutually informative, and I notice them differently in others.

Bridging the interior and exterior void with CI – The 3D realm

The brief vignettes and first-person descriptive accounts reveal how a movement experience can be brought to conscious awareness without losing a sense of subjectivity. Self-observation adds layers of understanding and opens doors further into the exterior realm of 'absolute possibility' (Jones 2009: 25) within an emergent embodied, practice-led research process (continuum 1). The implicit becomes explicit, a permeable process that brings into question existing values or 'diktat' – the emergent process of continuum 2 arises within continuum 1 and evokes continuum 3.

The first-person perspective introduces an introspective process (Varela and Shear 1999) in which the singularity of a participant's experience reveals aspects that can be shared with the reader or questioned. Commonalities are also found that can be further explored through interpretative methods, for others to learn from (Butler 2005; Barrett and Bolt 2007). The vignette is a pragmatic method, purposefully drawn from primary research material. These accounts were made of an experience that was meaningful at the moment in time, without any idea of how this may contribute to knowledge beyond the personal sense of the first continuum, of the inner soul meeting itself in the realm of practice. However, each account gives voice to, and affirms in some way, the value of embodied perception and gives credence to the type of knowledge that arises from the inductive and inter-subjective reality of the learning experience. This includes the previously undervalued interrelationship of perception, emotion and consciousness (Butler 2005; Gallagher 2008; Bainbridge Cohen 2012; Totton 2014) and the philosophical and metaphysical aspects of CI.

These phenomenological accounts provide a lens through which CI practice is seen to reference theories of perception and consciousness (Varela and Shear

1999; Gallgaher 2008) This dimension is one in which introspection is valued for the nature of the contribution that it makes to the inter-subjective reflective dialogue that captures and translates the practice into a form of documentation that then transmits the experience for academic and other research arenas. In this process, the three continua are integrated into a praxis, called CI, within a culture of practice-led research, an animate form defined by a mutable and permeable membrane, rather than a framework. The three continua interlink to provide space and time dimensions that contextualise the narrative and different participant voices. Finally, the mutable membrane concept serves as a meta-metaphor and lens through which to see these continua in operation, with reference to a plurality of influences and experiences, with the discussion of their contents in the time–space of the present moment.

Over time CI proves to be a form that provides a space–time for the evolution of a new non-autocratic approach to teaching – one in which the mover defers to their movement experience, to their body as the locus for learning new, corporeal, 'know-how' in 'equal' dialogue with diverse others, each with socially conditioned beliefs. How this happened was through trial and error due to the apparent complexity of social diversity, but simplified by reducing interactions to dyadic relationships in which differences revealed the learning taking place, bodily and then mindfully through touch – the fourth, integrative continuum.

8

Three Touchdown Dance Artists' Points of View

Introduction

In these point of view chapters, Holly Thomas, Robert Anderson and Jamus Wood give a personal account of their journey with Touchdown Dance, sharing key aspects of their experience and the impact of the work on their daily and professional lives. They describe exercises already mentioned but offer subjective value and this repetition, like in learning Contact Improvisation (CI), deepens understanding, reveals the development of personal and interpersonal relationships and sustains professional practice. I decided to keep them as discrete chapters so that we hear the voice of each person telling their story.

Holly Thomas is an independent dance artist and collaborator who is exploring inclusive choreography and creative audio description in performance. Holly is visually impaired and collaborates on inclusive dance projects. The Touchdown Dance methods have nurtured her towards her independent artist and workshop facilitator status.

Robert Anderson is working in health and social care and has taught dance, CI and embodied approaches to movement in higher education settings for many years. An independent dance artist, he is involved in many collaborations and as a Touchdown Dance artist Robert reiterates the significance of the Touchdown Dance ethos already contextualised but from a personal, self-directed research perspective.

Jamus Wood is visually impaired and shares his personal journey into dance as he confronted his eye condition and the challenges this presented to his ability to access the dance sector. Touchdown Dance provides a place not only for learning but for healing – as we all discover, but more particularly for Jamus, it supports him to overcome the impact of sight loss and the trauma of medical intervention. His journey with Touchdown Dance took him into performance work and eventually to Sri Lanka.

Holly Thomas – Dancer and facilitator

First encounters

I first met Katy while I was teaching performing arts at R.N.I.B. Redhill College and I was participating in an education project with The Place in London.

Katy led the first movement session with a second teacher from the education project. There were two moments from the class that I particularly remember. The first was an exploration in trios, in which three dancers stood side-by-side. The two dancers on either side of the central dancer offered their forearms as ledges, on which the central dancer could rest their arms, establishing contact from elbow to fingertips. The central dancer had the role of traveller and the others acted as supporters. The traveller was reminded to relax their shoulders and to find support from their travelling companions. The supporters were also reminded to soften their shoulders and to find support from the ground. We began to move together as a trio, navigating our way across the space, exploring shifts and exchanges of weight, support and resistance, whilst finding momentum, acceleration and elevation. We discovered an increasingly greater range of possibilities throughout our journey, and I distinctly remember a sense of excitement as our trio gathered momentum. As the dancer in the centre, I was able to 'push off' from the springboard-like ledges provided by my fellow dancers and discover an unexpected upward trajectory as my feet came away from the ground. I was assured by the physical support from my fellow dancers as my feet returned to the ground. This was the first time I began to understand the concepts of 'finding the floor' through the soles of the feet and of maintaining a vertical line through the centre of my body, ensuring the weight of my head remained balanced, all of which enabled a soft, safe and balanced return to the ground.

The second task, led by another teacher, was to learn a simple movement sequence, which was taught as a described movement journey. We were asked to imagine that we were walking around the edge of a lake, then to find a stopping point, move towards the ground and dip our fingertips in the water. The direction and timings of each movement were then layered into the description. Through this process we were able to commit the phrase to memory, capture the quality of each movement and move together, without having to rely on eyesight to do so. This was my first experience of creating a movement phrase from a sensory description and something that continues to be part of my own dance practice.

Joining the company

When the education project drew to an end, Katy invited me to spend time with Touchdown Dance in Manchester. At that time the company were about to begin rehearsals for *SENSE-8*, a collaborative project between Touchdown and film-maker Katrina McPherson as part of an Arts Council of England Capture commission. This was when I was first introduced to Contact Improvisation (CI), a dance form that was completely new to me. This initial period led to spending time in the studio and participating in weekend workshops with Touchdown company members over the following years, through which I explored and learnt many of the principles and techniques of this dance form.

Early days explorations; Stepping in

What I most remember from those early days in the 'studio' (be that a church hall, dance studio or community centre) was the sense of stepping into a creative process that was both familiar and unknown. The familiarity came from being in a space with other visually impaired (VI) people and where the teaching was designed to be inclusive. This was similar to my approach when teaching performing arts. However, what was new to me was the practice of CI itself and an approach to teaching that was less formally directed, allowing an extended time for exploration and discovery as VI and sighted people moved together. This exploration of movement through a deep sense of embodiment and through both contacts with, and relationship to, another person, the ground and the space around me was something that I thoroughly enjoyed and brought with it a new understanding of dance and of physical and spatial awareness and connection.

Opening circle

We began our practice with an opening circle, in which we introduced ourselves to each other and said our names. This enabled all of us, whether VI or sighted, to know who was in the room and to become familiar with each other's names and voices. It also meant that each person had the opportunity to ask questions and talk about what they required to work together safely. For me, the opening circle provides an opportunity to let my fellow dancers know that I am visually impaired and to say something about what this might mean when we dance and share space together. I have found the opening circle to be a fundamental tool in establishing an inclusive space in CI classes and jams.

Navigating the spaces

After our opening circle we worked in pairs, a VI dancer paired with a sighted dancer, exploring the edges of the space together. The sighted dancer guided a VI partner around the perimeter of the space before the roles were reversed, with the

VI dancer leading a sighted partner, who worked with their eyes closed. The guide was simply invited to give their partner time to explore their environment. I soon became aware that this exercise was not about one person 'guiding' another, but rather it was a practice of two people sharing a journey where the 'guide' became the 'listener', observing and supporting their partner's exploration of the nuanced tactile, auditory and sensory environment around them.

I remember a number of things from my discovery of the room on that first day. There was a moment when my hand was tracing a path along one of the double doors that marked the entrance to the room. A column of cold air touched the side of my face. This stream of air was passing through a visually imperceptible gap where the double doors met. This awoke a curiosity within me to explore the space through the surface of my skin and rediscover the sensory properties of the room. I became more aware of the soles of my feet and the surfaces underfoot. It reminded me of the sensation of stepping into a shallow river, where the tactility of the riverbed and the coldness of the water on my skin triggers an immediate shift in sensory perception. I remembered that the space that we inhabit is only partially revealed to us through any one of our senses and that the process of tuning into our environment takes time and practice.

During that first day and in the research period that followed, I experienced and learnt many of Touchdown's basic principles and the developmental movement practices drawn from Katy's knowledge of Body-Mind Centering®. Below I will offer a response to those that have most stayed with me.

Finding pathways: Tracing the lines

In this exercise, we worked in pairs with one dancer, the 'receiver', standing with their arms open in front of them, as if holding a large beach ball ('aikido arms'). The second dancer, the 'tracer', stood directly behind the receiver with sufficient space to allow them to make contact with the receiver's back and upper arms. Using their fingertips, the tracer drew a 'V' shape on the receiver's back, beginning at the sacrum and travelling up and across the shoulder blades then along the under surface of the arms. This movement was repeated several times, allowing the receiver to gain a sense of connection within their own body along this pathway.

When I first encountered this exercise as the receiver, I experienced a sense of softening in the muscles and joints in my arms, and a heightened awareness of the support available from my internal architecture. I noticed weight and weightlessness. I became aware of an expansive quality within the palms of my hands, of the connection from my little finger to thumb, and the potential to reach through my hands and fingertips into the space surrounding them. With this, I began to sense the energetic pathway along the under surface of my arms, along the underside of my palm, through my little finger and beyond. For me, this was the beginning of a new understanding of spatial depth, one that is not available through my visual perception.

Over time, working with Katy and the core Touchdown team, my own felt sense of connection from centre to periphery has developed and with this the possibility to reach beyond my kinesphere, extending my reach and movement beyond the edges of my visual perception.

Tracing lines and pathways across the body is an approach that we often use to un-puzzle many of the pathways used in CI, pathways that in regular classes I have sometimes found to be taught from a visual starting point. For example, when teaching a lower-back lift, the Touchdown approach might be for the under-dancer to trace a line with the edge of their hand along the outer edge of a partner's leg, thus enabling the under-dancer to position themselves in relationship to their partner through touch-based contact and offer their lower back as a surface for their partner to take a ride. This contrasts with the more commonly taught method of initiating this lift by stepping towards a partner or into the space directly in front of them. In this approach the under-dancer is required to use sight rather than touch to locate and position themselves in relation to their partner.

Finding the floor

For me, the material that we explored through floorwork practices became the foundation from which I was able to find freedom in movement. The material that we covered included developmental movement patterns, flexion and extension, rolling, travelling, spiralling, reaching, pulling, pushing, yielding and softening. The time given to fully experience each element allowed the body the necessary time to absorb and assimilate new information and to try out new movement possibilities. With the floor as my partner, I was able to find a softness in my body and an awareness of other people moving in the space. There was therefore no requirement for additional physical support to enable freedom of movement. Having the capacity to soften into the floor and to feel the floor meeting you was then, and remains now, an intensely reassuring sensation when working without visual depth perception.

Back-to-back

We spent time working in pairs sitting back-to-back, finding rotation in our spines, sharing weight, and finding extension and connection from centre to periphery and from the sacrum through the arms and into the fingertips. We found shared pathways into the ground, surfing along the floor and then sequentially reversing this pathway, making our way back to sitting. This is an exercise that I have used in many settings, as it provides an equality of challenge for VI and sighted dance partners; these pathways do not rely upon sight and can only be achieved when both participants find the pathway together. It also requires negotiation based on the range of movement and comfort of both dancers, therefore both are required to listen and move within an agreed shared movement range.

Eyes open, eyes closed

For many months I danced, for the most part, with my eyes closed. I needed this time to let go of using the sight that I have. I became aware that using my sight whilst dancing CI diluted my awareness of my own body and of my partner. Spiralling and turning with even a soft gaze would trigger a lateral eye wobble (nystagmus) that was disorientating and tiring, and I found myself trying to see the room, which was frustrating and of little useful purpose. However, there came a point in my practice where I became more able to integrate my visual perception with my moving body. This seemed to coincide with allowing my centre to engage with the sensation of reaching, primarily through my arms and beyond my kinesphere. As I allowed myself to more fully commit to this reach, to take my body off balance and travel through space, I found that my eyes were inclined to open. I was, in some way, more able to let go of trying to see, and I no longer needed to close my eyes in order to do this. Throughout this process, I worked with company members: dancers with skill and experience, and whom I trusted to support this exploration. Sometime later I began to notice a sense of reaching through my visual perception so as to move in the direction of my gaze. This was not about trying to see. It was more about movement intention through an imagined sight line.

Sharing practice

Touchdown has allowed me the space to learn from and share practice with other members of the company over an extended period of time. I have personally benefited from receiving teaching from accomplished dancers in an inclusive environment. One example would be the practice of working with 'the bones'. During one session we brought our awareness to our own skeleton and the 'liveness' of our bones. I imagined my own skeleton transformed from a heavy-yet-brittle, hard, dry and impermeable framework into a living and responsive structure. A structure that provided support, shape and form to the body and that enabled me to receive and support the weight of another.

We worked with a sense of softness in the bones and fluidity in our joints, allowing an ease of movement. We found support from our sitting bones when sharing weight and working with counterbalance. We worked with the lightness of our skeleton and the weight of our skulls, finding rolling points of contact and momentum. We allowed the weight of our head to lead our bodies into movement, finding pathways towards and away from the ground.

There are many more examples of the movement explorations, methods and practices that I learnt with Touchdown, all of which continue to inform my dance practice.

Approaches to inclusive teaching and facilitation

Throughout my work with Touchdown, I have co-facilitated classes and workshops with VI and sighted participants, dancers and teachers.

I participated in Touchdown Dance weekend workshops with mixed groups of participants, dancers and teachers. We would usually come together for two days to learn and work with the company's core movement material and inclusive approach to practice.

The workshops focused on how to bring VI and sighted people together in such a way as to establish trust and create an inclusive space in which to explore and share CI. My learning from these weekend workshops formed the basis of the Touchdown workshops that I went on to co-teach with other members of the company. It has also influenced and informed much of my own work in other contexts.

Workshop for VI young people

Together Katy and I have delivered weekend workshops for VI young adults. Participants ranged from school-age children to those in their early twenties and included students attending college and university.

We would often work with the 'Fundamentals' material which can be found in the Touchdown Education pack.

We would always begin with an opening circle (as described earlier), where we all said our names and introduced ourselves before moving on to a series of ice-breakers and physical warm-up games. Participants would then work in pairs to explore the perimeter of the space. In this exploration, one of the participants would guide a partner around the edge of the room using the sighted guiding technique, and the person being guided, traditionally the VI person, would take the elbow of their partner who guides them as they travel together.

Beginning at the studio door, the person being guided uses their free hand to make contact with the surface of the wall and together they take a journey around the space: a journey in which the person being guided leads the exploration, taking the time they need to explore the edges of the space through all of the senses available to them. The roles are then reversed with the 'guide' then becoming the journey taker.

In the young people's workshops, all participants were visually impaired, therefore those participants who were confident to navigate would initially take the role of the guide. As facilitators, myself and Katy offered additional support and guidance when required, particularly to younger members of the groups, to enable them to maintain focus and direction whilst navigating the space. I noticed how young people would often approach this exploration with lightness and humour. There was a sense of ease in navigating the edge of the space through tactile and sensory pathways.

We often worked with 'back-to-back' exercises. Beginning from the ground working with 'Finding Pathways' into and away from the floor. When working with younger blind participants Katy might suggest they use a yoga mat, which enabled them to maintain a clearly defined and self-contained working space.

In the latter part of sessions, having moved through a range of guided paired work and observing how participants were responding to and experiencing the work, we might begin to work in trios and travel through the space.

Beginning in stillness, the trios would establish the forearm on forearm connection (described earlier in the chapter) before finding a moment when they could begin travelling across the room together. Participants would often find moments of stillness and movement. We would encourage them to move between different levels and negotiate shifts and changes in direction. With support from myself and Katy, some participants might begin to work with momentum and even move into supported jumps and safe landings. As facilitators, we were mindful that we did not offer the more dynamic elements of this exploration until participants were ready and had thoroughly grasped the skills and alignment required to enable them to explore and experience this safely.

As with all Touchdown workshops, the young people's sessions were about responding to the interests and capacities of the group members. At the end of the sessions some participants expressed an interest in participating in other workshops and in sharing their own artistic practices. All in all, these were very creative exchanges.

Workshops with VI children

Whilst all of our workshop sessions are rooted in the 'Touchdown Fundamentals', the capacities, age, abilities and circumstances of the participants we work with can differ greatly. It was our responsibility to be aware of and responsive to the strengths, needs and requirements of everyone participating in the work. Therefore some of the sessions I facilitated with Katy would draw upon a range of approaches.

When leading sessions for VI children we worked with small groups, which sometimes included children with additional needs. Our approach was to create a session that allowed the time and space for the children to engage and participate in a way that was meaningful and accessible to them. Therefore each proposed activity, task or exercise, whilst having a specific focus or intention, could be explored or experienced in a way that had meaning or interest for the individual taking part.

The opening circle

Beginning with an opening circle, we would all introduce ourselves by saying our name and/or making a movement of our own choosing, each movement was described either by the participant or by Katy so that we could all then make our own version of that movement together.

The shape of the circles we created was both approximate and fluid. Those who 'drifted' from the circle were invited to reorient themselves. Within the circle, we would move through physical warm-up games such as 'pass the pulse', 'body parts' and 'rainforest', establishing a sense of fun, togetherness and connection with our own bodies and with others in the space. From here, and when the group was ready, we would transition from working in a circle to moving through the space.

Both Katy and I would gently invite and encourage those on the periphery to join the main activity, whilst allowing each child the time and space they required to feel safe and comfortable in the space.

As facilitators our invitations to join or 'be with' included coming away from the circle to move with participants on the periphery and working alongside participants whilst respecting the space they had chosen to work in. We might join in with movements or vocalisations, all whilst being attentive to non-verbal cues and communication.

Over the course of a session and as trust was established, children came into and moved away from the main activity as they wished to.

Moving from circle to space

Having established a sense of togetherness and warmed up our bodies in the circle, we would introduce a movement game such as 'Conga', that allowed us to travel through the space in connection with each other. Katy and I would assist the children to form two conga lines and, slowly at first, with Katy and I taking the role of leader for each conga line, we would set off on our respective journeys. Time is required to maintain the connection through the conga line and as the 'leader' and VI teacher it was really important for me to remain open in my 'back body' and to 'sense' back through the conga line so as to feel whether or not we were all connected and moving together. I might also call back, using participants' names to check that we were all moving together, encouraging participants to use their voices and call to the conga if they needed to. This allowed us to locate the conga line and regain connection using sound rather than sight and also encouraged a sense of physical independence and agency when reorienting ourselves within the space. We might also guide the children back towards the conga when needed.

Once the line was established, we would introduce music to support exploration and maintain the energy and focus of the exercise. When using music, we are mindful of sound levels, ensuring the comfort of participants who may be particularly sensitive to loud noise, whilst also ensuring that the sound of the music does not block out the ambient sound from the room that VI participants may be using to navigate and orient themselves in the space.

As our conga team became more confident, we began to explore levels and surfaces, changing direction and making our way from the centre to the periphery of the space. With assistance, the children took turns to lead the conga, whilst myself and Katy supported the children to maintain connection and move together.

Working in pairs

During sessions, children might also work in pairs as we guided them through exercises designed to bring awareness to the front body and the space in front of them. One example of this was to introduce the idea of aikido arms through felt/tactile imagery. We might ask them to imagine that they were each holding a giant inflatable beach ball in their arms, which they could inflate and deflate with their breath. When they began to lose the curve of their arms or their awareness of the space in front of them, we would remind them to inflate their beach ball, fill the space with their breath or imagine just how much air is inside the beach ball (see Figure 4.6).

The children would then work in pairs, standing back-to-back, finding a rolling point of contact and pathway across the back and along the curve of the arm and back again, returning back-to-back.

In my own practice, this exploration strengthens my sense of connection from centre to periphery and has, over time, supported me to find a felt sense of spatial depth and physical reach that is not available to me visually, a reach originating at the centre of the body and extending through the arms and into the space, with a counter reach extending through the legs, the soles of the feet and into the earth.

Revisiting the circle

We return to our circle formation regularly throughout sessions to regain focus and foster a sense of togetherness, before returning to working in pairs. Over the course of a relatively short workshop session, the circle becomes the space where everyone feels comfortable to participate together and by the end of the session, children who may have begun by participating from the periphery, often find their way to the closing circle.

Workshop reflections

When reflecting on these workshops, one of the things that stood out for me and that I learnt was an appreciation of the many ways through which we can experience movement practice and the fundamental movement principles that underpin Touchdown's work. Whilst one person may engage on a linear path moving through exercises and tasks from beginning to end, someone else may take time to arrive and observe, to engage in moments of tasks and explorations and bring what they have learned to the group in a new or unexpected way. With each exchange, there is something to be found, often a new understanding of ourselves, each other and of the movement material that we are making.

As a VI facilitator my preference, when working with VI participants, is to either co-teach with a sighted facilitator or to have an access worker with experience of inclusive and somatic dance practice in the space with me. This affords me the freedom to teach and fully engage in the moment, whilst staying present in my own body.

Performance work

SENSE-8: 2000–01

The first performance work that I was a part of was *SENSE-8*. *SENSE-8* is a dance film for VI audiences that explores observing dance from multiple perspectives. The process of making the film included explorations of movement scores in duet and ensemble, as well as layering audio description into the improvised movement practice. This later became part of the sound-score for the film. The 'round robin' was one of these scores: two dancers duet in the centre of a circle of VI and sighted 'observers'. At any time any one of the observers could describe to the circle aspects or elements of the dance that they noticed. Equally, VI dancers could ask a sighted observer about the improvisation, to elicit information that they were curious to understand or know more of. This question could be as simple as 'What's happening now?' or 'Who is dancing?' Some questions were more specific such as 'How is Jovair moving?' or 'Did Katy lift him fully off the ground?'

Another feature of the round-robin format is that an observer can at any time enter the dance space to create a momentary trio. They connect and improvise with one of the dancers from the original duo, which signals the other to find their way back to the observer's circle. To make this process more inclusive, sighted observers would say who was entering the improvisation and who was returning to the circle.

As the scores developed, moments of description were collected and woven together to create a descriptive sound score that sat alongside the visuals of the film. There were also moments of close-ups of parts of the dancers' moving bodies, giving a sense of the movement dynamic without needing to visually capture the entirety of the dance, creating an interesting exploration into how we see and perceive dance.

Live performances – From the circle to the stage

Between 2000 and 2009, I performed and toured in *TACT* and *CLOSER*. Under the direction of Katy Dymoke, Angus Balbernie and Julyen Hamilton, both pieces used a series of focused scores to hold improvisations.

TACT

Drawing on personal experiences of visual memory, perception and touch, we worked together to find movement scores through which to express our felt responses to these themes. For example, having recalled and retold a memory from my early childhood in which I was sitting at the edge of a sandpit feeling the coarse grains of sand between my thumb and fingertips, whilst bright shards of sunlight danced on my eyes, I worked with Katy to find a movement response to this sensory experience which became a solo and later a duet with cellist Barnaby O'Rourke.

Working with Julyen Hamilton during the research and development phase of *TACT*, I gained a greater understanding of the tools of improvisation. An exercise as simple as 'saying yes' to every moment of improvised movement, whether beautiful, expected, surprising or imperfect, offered the possibility to fully commit to each moment. This practice was, at times, a challenge, having worked for so long with my own movement exploration I realised that improvisation within the context of performance required different attention. I became more aware of working with time: of making clear decisions at any given moment.

We worked with the practice of finding clarity in each moment, whether that be a decision to 'jump forward 30 seconds' or to 'really' rather than 'nearly' touch when in contact.

As an ensemble, we explored and practised: holding and sculpting space, finding and using phrasing, and identifying the dynamic quality of each improvised movement phrase. All of these brought a clarity to our movement composition in relation to each other, the performance space and our audience.

CLOSER

When working on the development of *CLOSER*, we spent time working with different 'states' or 'atmospheres' within each score: deepening our awareness of the space we were moving through and our movement quality within each 'atmosphere'. We considered the visual composition of our scores, and how we might communicate each atmosphere to our audience.

In one particular score, entitled 'stone garden', the dancers began, on the diagonal, holding the space between us and occupying different levels, creating sculpted tableaux. With direction from Angus Balbernie, we structured the score so as to move from stillness into 'micro', 'meso' and 'macro' movements. Our starting point was to find barely visible core proximal movement punctuated with pauses, before finding subtle shifts of weight, as our 'stone bodies' began to find movement and levity. We transitioned into the 'meso' phase of the score, becoming more expansive, working within our own kinespheres. Our movement developed as we began to reach beyond the kinesphere into and through the space or 'garden'. We worked with the idea of attraction and repulsion, finding pathways towards and away from each other.

What I noticed in working in this section of the score was the process of both 'signalling' and 'listening' from the centre of the body, creating a felt connection through the space, enabling us to move with momentum towards and away from each other. I was also aware of the fully sighted dancers' capacity to visually 'spot' the VI dancers, finding moments to connect with us or signal to us through the space. All of us had to make ourselves available to be located and seek out those opportunities to connect and move away. When working on this section of the score we often referred to this as 'magnets', which describes the quality of the relationship that was at the heart of this part of the score.

Inclusive design and audio description

In *CLOSER* and *TACT*, the company used a range of approaches to create an inclusive design for the performances. Both performances included pre-recorded and live spoken autobiographical moments. Alongside this, the performers' movements were enlarged and captured in shadow or projected on the back wall of the performance space, providing some partially sighted audience members with enlarged images of moments of the live performance.

We worked with an audio describer who delivered live descriptions of the set and improvised dance scores for VI audience members, via headsets. We designed pre-show workshops where VI audience members could meet the dancers prior to the show and through a combination of hands-on demonstration, contact work and spoken description, we demonstrated key moments and movement qualities from the material within the performance. We also linked these to the accompanying moments in the sound scores. These then became a series of audible reference points within the live work.

Future imaginings

In providing access to live improvised and scored dance performances and in working with small focus groups of VI audience members to hone our methods, we began to challenge the notion that dance needed to be witnessed through a purely visual medium.

Whilst performing with Touchdown, I became interested in how we witness dance and the extent to which the somatic practice of de-privileging sight might inform how we receive and experience live performance. Is it possible to make choreography that is not predicated on sight as the primary sense through which the audience receives the work? Moreover, what if the movement vocabulary and navigational skills and techniques practised by blind and partially sighted people become the starting point for making choreography?

It is these questions (born out of the rich teaching that I have received and the explorations that I have been a part of as a company member of Touchdown Dance) that continue to influence and inform my own practice going forward.

Making space in the dance sector for VI inclusive and VI-led dance is essential. My hope is that the work, methods and practice developed by Touchdown continue to influence current practice, raising expectations and providing opportunities for visually impaired people to participate in, and take a lead in making, creating and enjoying the dance.

Robert Anderson – Dancer and facilitator

Introduction

I've been a company member of Touchdown Dance since 2002. My contribution to the book focuses on the particular influence of the dance form on CI in Touchdown's work. I will focus in-depth on two examples: a weekend community dance workshop with VI people and a weeklong training with a professional dance company in Sri Lanka.

Personal journey

I became interested in movement and the body through participation in weight training, judo, yoga, tai chi and massage training during my mid twenties. Leaving the United Kingdom at the age of 25 to travel through Asia and live in Australia for 3 years opened new possibilities for me to explore the world of movement and somatic practice. I took my first dance classes whilst living in Sydney from 1991 to 1994 and was further intrigued and inspired by the creative potential of dance by watching Sydney Dance Company in performance. During my return journey to the United Kingdom, I spent six months studying with a classical Javanese dancer, Mas Mbesur on marble floors of Kraton (royal palaces) in Surakarta, Central Java. He gave instructions via imagery to elicit some of the movement qualities of Javanese dance. This was my introduction to dance improvisation where I began to learn to move from impulse, memory and imagination within my body.

I continued to study contemporary dance back in the United Kingdom at evening classes at The Place (London Contemporary Dance School). Although I enjoyed the challenge of learning contemporary dance through the conventional method of copying the movements of a teacher, I also struggled with this traditional approach. However, I immediately connected with the somatic and interactive nature of CI. I discovered a weekly class held at Bedford House in Holborn delivered by a pool of London-based teachers. It provided a form of movement that I could learn rapidly with my earlier grounding in bodily practices. Learning to move from the felt experience of the body in relation to the floor, the space and others made much more sense to me than working with imitation (of a teacher) and a mirror. It gave me a framework for developing my dancing towards performance and teaching.

Development of a practice

My enthusiasm for CI gave me the confidence to begin a BA programme of dance studies at the University of Surrey in 1997, approximately a year after I began practicing. Dancing CI broadened my understanding of movement analysis and choreography. Training in CI also gave me a sense of trust in my body and confidence in interactions with others on the dance floor and in life in general. The CI teachers I worked with in London in the early years of my practice were remarkably generous and supportive in the way they delivered the work. There was a culture of inclusiveness and encouragement for people of diverse ages, abilities and backgrounds to come together, learn and practice. This was combined with a social element where we would commonly go to a cafe or pub after class and continue the 'exchange' of ideas.

As my skills and confidence progressed, I went to larger jams and more advanced workshops with international teachers. I also travelled to CI workshops, meetings and festivals abroad. After completing my BA, I spent six months in San Francisco studying CI, modern dance and composition with internationally recognised teachers, including Scott Wells, Kathleen Hermesdorf and Kim Epifano. This experience of immersion in the rich dancing culture of the Bay Area in California provided the inspiration to create the organisation London CI upon my return to the United Kingdom.

In my final year of my BA Dance and Culture programme at the University of Surrey, I wrote my dissertation on the theme of the therapeutic potential of CI. I was interested to research and consider how CI could be of benefit to people in a variety of ways. A specific area of interest was in the inclusive nature of CI that enables people with disabilities to participate, study and perform dance. As part of this research, I attended a Touchdown Dance workshop at Romiley Forum in July 1999.

Workshop at Romiley Forum

This first experience of a Touchdown Dance workshop facilitated by Katy Dymoke introduced me to their unhurried yet focused approach to dance education. Participants worked sensitively with safety and a perceptible appreciation of the fundamentals of the CI form. Usually, a sighted person partnered with a VI person. Exercises were simultaneously demonstrated and verbally described. On occasions, those without sight were invited to touch the bodies of the people demonstrating to clarify the positions being sought. Commonly the sighted partner was the active party first, which seemed the most time-efficient way of learning. However, this order was sometimes reversed when a non-sighted dancer more familiar with the work was partnered with a less experienced sighted participant. Either way, the roles were reversed so that both partners experienced giving and receiving. There was a sense of equality amongst the participants regardless of experience and physical or sensory ability.

Through sensory observation of these passive exercises, I became even more conscious of the weight of my body, to my connection with the floor and to the quality of touch of my partner. I was able to gain a sense of release in my joints difficult to achieve if I had been warming up alone. This seemed to result from a relinquishing of physical weight and control to my partner. The degree to which I could 'let go' or 'soften', in the hip joint, for instance, related to my sense of confidence in my partner's touch and trust in their ability to work with sensitivity. In the reciprocal 'hands-on' role, I was able to practice giving my attention to the range of motion, the weight and the level of muscular tension in my partner's body.

The approach of working simply and gently with the body, established in the workshop, enabled me to let go of anxieties about doing the exercises well or being a 'good dancer'. Thereby, I was able to focus fully on the research in a manner that felt new and rewarding. Comparing my performance with others in the workshop seemed unnecessary. Rather, I could focus on the physical interaction at hand with a sense of expanding attention and an abundance of time available to explore the exercises. This in turn created a mood of relaxation, which supported the learning process through enjoyment and curiosity in meeting and dancing with fellow participants.

In the next stage of the workshop, both partners became active in exploring the contact between them and the possibilities of moving together. Whilst exploring the range of movement, such techniques accustom dancers to new ways of using the body and moving that are helpful within an improvised duet. These included working within spherical space, using all the body's surfaces for sharing weight, and controlling movement through rolling, extending and rocking (Paxton and Kilcoyne 1993). These structures develop further sensitivity to movement and nurture an ability to apprehend and remain in contact with a partner's pathway of momentum.

These exercises provided 'starting points' (Paxton and Kilcoyne 1993: 31) from which improvised dancing could begin. Here, the teaching structure is left and participants move together using their familiar vocabulary. The new patterns that had been studied were practised and integrated 'organically' into the physical memory of the dancers. The movement itself becomes the teacher as the dancers learn from their experiences. There is no set formula for a successful contact dance and each new partnership offers different possibilities for creative exploration, self-expression and learning that are not dependent on how many skills have been acquired. It is the dance itself, according to Paxton and Kilcoyne (1993: 50), 'which produces the feelings of self-empowerment, joy and satisfaction'.

New perceptions and empowerment

By observing the way in which I was able to learn during the workshop, some of the benefits of Touchdown's approach to promoting mutual learning between VI people and trained dancers became apparent. I experienced certain changes in perception and attitude during the Romiley workshop. I realised that the work had elicited a sense of newness and discovery in CI for me. There was a pleasure in working slowly and thoroughly with the fundamentals of the form. It enabled me to experience the quality of contact between bodies and to improvise with enhanced awareness and clarity. There was a sense of relief in letting go of expectations of achieving 'high flying' virtuosic contact work. This was replaced by a curiosity to explore new possibilities and an enjoyment of the mutual generosity in the partner's work.

During the Touchdown workshop, I experienced a new level of sensitivity to my partners' bodies and movements. This made me feel very 'present' in the work and my sense of time passing diminished. I also realised that I'd come with an expectation that the work would be of a basic level and therefore unchallenging. However, I discovered instead that the work had a lot to teach me. My habitual patterns of dancing were put on hold and I found a new openness and receptivity to the ongoing physical dialogue. In so doing, I could perceive new pathways and principles of movement with a level of detail that previously would not have been available.

Rachel Freeman, director of Blue Eyed Soul, an integrated dance company based in Shropshire, told me, in dancing with people with different abilities nothing can be taken for granted. Dymoke commented that dancing with the VI makes sighted dancers more attentive to the sense of touch and kinaesthetic awareness. This results, she believes, in an enhanced sense of presence in the body and thus, dancing embodied with awareness.

For dancers with visual impairment, the significance of working with touch and the kinaesthetic sense is amplified. The sensory information that an individual receives in contact work provides an opportunity for reassessing the body in action and for extending the ability to focus on inner sensations (Paxton and Kilcoyne 1993). Learning to move and walk with confidence instead of a sense of fear is an outcome of the work identified by Jane Emerson, a VI participant of Touchdown's work (Deitch-Dey 1998).

Touchdown Dance collaboration with nATANDA Dance Theatre of Sri Lanka[93]

The work of Touchdown contains both educational and artistic motivations. There has been a performance orientation within the work of Touchdown Dance since 1989 when Gerry Overington, a non-sighted dancer, performed with Paxton and Julyen Hamilton in a piece called *Blind Faith 1*. I became a company member of Touchdown Dance in 2002 and participated in many educational and performance projects over the following years. I performed in three major works: *TACT* (2002–03), *CLOSER* (2005–10) and *APPARENTLY NORMAL* (2010). These works were supported by direction from Julyen Hamilton and Angus Balbernie and toured at theatres around the United Kingdom.

The value of Touchdown's inclusive approach to dance training was evident to me again in December 2013 when Katy Dymoke and I travelled to Colombo in Sri Lanka to work with nATANDA. We had been invited by choreographer

Kapila Palihawadana to share some of Touchdown's working methods with the company in support of their new work entitled *Touch*. The dancers of nATANDA are trained in modern and classical forms including traditional Kandyan dance, western ballet and modern dance, yoga and Bharatha Natyam. Our aim was to support the dancers in their solo, partnering and ensemble work and in how to support the inclusion of three deaf-mute students in the rehearsal process.

We started our workshops with simple exercises to encourage the dancers to listen to their bodies and take time to be able to follow and respond to sensation and impulses. The deaf students were able to follow a combination of visual and tactile demonstrations by lip-reading instructions passed on by the dancers (Figures 8.1 and 8.2).

FIGURE 8.1: Workshops for a performance project *TOUCH*. Four dancers practice high bridge work as described in Chapter 4. At the Goethe Institute Columbo, 2013. Photographer courtesy of nATANDA Dance Theatre of Sri Lanka.

FIGURE 8.2: Two pairs sitting back to back, the 4 dancers are beginning the back to back exercises. At the Nelung Arts Centre studio in Columbo, 2013. Photograph courtesy of nATANDA Dance Theatre of Sri Lanka.

We presented techniques, new to the dancers that encouraged the release of muscle tone in the body to allow movement through the joints and to facilitate ease of travelling on the floor. Finding efficient healthy ways to allow weight and forces to travel through the moving body is one of the elements of CI that is beneficial to dancers working with solo phrases of movement. The use of appropriate muscular tone and alignment of skeletal structure becomes instinctual when the senses attune to listening. The dancers of nATANDA were remarkably strong and flexible but seemed vulnerable to injury because of being accustomed to moving with a high muscular tone.

In sharing our understanding of CI principles with the nATANDA dancers, we stressed the value of attending to physical sensations of weight and touch. An internal focus enables the dancer to integrate perceptions and impulses into their dancing. This attention to the felt sense of the body in CI has been present since its inception with principles transposed from the martial arts and yoga. During the consequent development of CI over the past 40 years, somatic practices such as Body Mind Centring, Authentic Movement and Feldenkrais Method have influenced the teaching of CI and enhanced this kinaesthetic approach to dance training.

Susan Foster (1992: 482) claims that dance training 'creates two bodies: one, perceived and tangible; the other, aesthetically ideal'. This 'ideal body' provides an elusive blueprint of perfection towards which dancers labour and map their deficiencies. Peter Ryan (Ryan cited in Novack 1990: 181) argues that CI highlights the 'difference between your real body and your idealized body' by asking what your body can do, not what you can make it do. He points to mental attitudes of 'let, allow, release' as opposed to 'get, achieve, take, put' (Ryan cited in Novack 1990: 182) which allow dancers to use their kinaesthetic awareness and impulses to guide their movements and protect themselves from injury.

We introduced elements of partnering and contact in the workshops to help the dancers become comfortable with touch-based ways of working. The use of touch with trained dancers helps to establish listening skills, clarity of intention, direction and response in partnering. Spiralling or arcing patterns are utilised to optimise smooth and efficient flow of movement within spherical pathways. The dancers were quick to integrate the new principles and could progress to more advanced techniques working with weight and momentum. We offered sequences of lifts, balances, jumps and catches for the dancers to practice, working with extended follow-through and momentum.

We presented scores and games to bring in playfulness to the dancing. We witnessed the dancers enjoying the creative and light-hearted nature of the work. Our working methods offered a new or unusual experience for the dancers to own and create their dancing in the present moment. They were starting to take more initiatives in the improvisations and to trust in their choices.

The contact work evidently began to support compositional elements within the ensemble scores and improvisations. Instead of always being guided the deaf students started to actively contribute to the group and were fully accepted as equal participants in the work by the company. There was a sense of collective responsibility amongst the company where the dancers were finely tuned to one another responding and developing material at the moment. In working with CI the immediacy and the sense of connection with others gained from being in a contact engagement remains when not in a direct touch-based dance. So, in a way, the dancers are always 'in contact' with each other.

The work was able to cross boundaries of language and ability by a process of collective problem solving and shared responsibility. We observed the deaf students to visibly gain in confidence during the course of our week together. By the last couple of sessions, they were picking up the movement material from witnessing demonstrations very quickly. They were also communicating with each other and the rest of the company much more both during and after rehearsals. They appeared to be enthusiastic and energised by the work. Their support worker from the Sri Lanka Central Federation of the Deaf also noticed the positive change when she came to witness the work in progress. She observed that the female dancer was less held in her muscle tone and had found a more fluid movement quality. She also noticed the speed at which the two male students were picking up the new material being presented and the openness of the communication between all the dancers.

A collective or non-hierarchical approach to dance-making has been a characteristic of CI since its inception. Part of Paxton's and others' inspiration for developing CI was a desire to move away from the traditional hierarchy of a choreographer setting movements on 'his' dancers. He argued that this structure was stifling and resulted in dancers appearing as 'watered-down versions' (Paxton in Novack 1990: 54) of their teachers. CI has inspired many companies working in dance and theatre to adopt more collaborative and democratic working methods.

Inclusion and accessibility for disabled dancers

The description of Touchdown's work in the two contexts, the Romiley Forum workshop and the training with nATANDA, illustrates how the company utilise the principles of CI to make participation accessible and inclusive for people with visual impairment and other sensory impairments. The work creates a forum for communication and the sharing of knowledge based on principles of equality and respect for different abilities. VI dancers have not only attended Touchdown Dance workshops but have also been present in the broader contact community, at national and provincial jams over recent years. The integration of people with disabilities within CI events indicates a willingness to explore new territory and ways of working and to challenge aesthetic codes within the dance world (Albright 1997). As such, it has the potential to demystify and enlighten misconceptions and prejudices that able-bodied people and those with disabilities might have about each other.

The nature of contact improvisation, where individuals collaboratively explore certain principles of movement to create a physical dialogue, means that participation is not limited to any specific body type or ability (MaCabe 1997). It is this inherent respect for the physical possibilities of dancers' bodies that enables people with different types of disabilities to participate in CI. Adam Benjamin, the co-founder of CandoCo comments that 'the exploratory and pioneering work of Steve Paxton [...] has directly and indirectly, led to the studio doors of our schools and colleges opening to many who could never have dreamed of gaining access' (Benjamin 1995: 47).

Until recent years, disabled people have been excluded from dance. Albright points out that historically the role of the professional dancer has been reserved for the 'glorification of an ideal body' (1997: 57). Measured against the aesthetic values of classical and modern dance, disability she argues represents the cultural antithesis of the fit, healthy body. Disabled people encounter inaccessibility in many areas of life. Deitch-Dey (1998: 89) claims that being disabled in western society can be equated with invisibility or 'living in a culture of silence'. The inclusion of disabled people within CI provides a platform for the 'silent body' to speak and challenge prejudices towards people with disabilities (Albright 1997). In this respect, CI promotes change not only for the individuals involved but also on a socio-political level by promoting acceptance and inclusion of differences.

Returning to the personal journey: Reflections on the development of Touchdown's work

Whilst the influence of CI has been central to my experience with Touchdown Dance over the past eighteen years, the underlying principles have expanded and developed our work in various ways.

Dancing relationships have deepened over that period of time through the shared knowledge and experience acquired during research, rehearsals and performances. Even after periods of time when the company hasn't met, we are able to reconnect and tune into one another promptly because of this depth of understanding acquired over time through our detailed and patient approach to working with the body.

The company members have grown in confidence to bring forward ideas and personal experience to support our work in teaching and performance situations. This reflects the egalitarian and democratic nature of CI where all dancers' contributions are valued regardless of gender, (dis)ability, or experience. Over my time of working with Touchdown, I've seen an organic evolution towards collective methods of researching and devising material where all company members can contribute. We often choose to 'pass the baton' between us in providing starting points in our research and leading sessions in the studio. The VI members of the company (Holly Thomas, Jamus Wood and Stuart Jackson) have all taken the lead at times in directing company rehearsals or in delivering a workshop.

In devising one of our most recent performances, *Attention*, I shared my experience of working with elements of Stephen Batts's *Poetic Movement* material to create a number of scores. The scores provide frameworks, variously descriptive and prescriptive, that each dancer can fill with their own movement choices. Thereby, there is a creative process of interpreting the guidelines through movement and providing direct feedback to the other dancers. This helps to keep our work fresh as each company member brings their personal experience and interests acquired from their work and life outside the company. The use of scores helps to focus our dancing whilst enabling and promoting playfulness and creativity in our improvisations.

Masterful support has been provided, by guest directors Angus Balbernie and Julyen Hamilton, to refine understandings of clarity of intention in solo dancing and in spatial, temporal composition between dancers inhabiting a performance space. Our grounding in CI and somatic movement techniques has been expanded through the support of invited artists to embody ensemble understandings of relationships in a performance space. The addition of music, visual images and lighting adds to the framing and composition in making our work inviting for an audience.

The new concepts relating to spatial connections brought to us by Angus Balberine, during the development of *CLOSER* (2005) brought challenges to the company in making the principles accessible to the VI dancers and in particular our fully blind colleague, Stuart Jackson. We had to learn to work with all our senses collectively to enable individual expressivity whilst looking for a sense of an ensemble cohesive relationship to space and each other. This process required patience to problem solve and to remain in constant dialogue with one another. Whilst touring the work, each new performance venue and stage setup would provide new challenges for the company of six dancers to navigate. Unfamiliar stages, audio and video projections and lighting equipment provide challenges to VI performers. Our abilities to listen to one another and communicate sensitively were invaluable to responding creatively to these unfamiliar situations.

An important element of our working process is documenting the research and rehearsals as they unfold. We will individually make notes in our journals and someone will make a record of our work as an ensemble. These documents support reflection and the development of new projects. We created a formal document recording our working methods with the publication of the Touchdown Dance Education Pack in 2015. The process of recording our teaching methodology was again a democratic process where different members of the company would initiate a revisiting of the core material with their own individual perspectives on what to deliver and how to make it available to prospective students.

My personal involvement with Touchdown Dance has supported my dance practice in many ways. I have developed and expanded my abilities as an improviser, dancer, performer, teacher and collaborator. To some extent, it has provided a bridge between my starting point as a community-based dance enthusiast and the professional dance world. My experience with Touchdown has supported me to take on teaching roles in higher education and present papers at academic conferences. It has also been an important part of my journey in belonging to a community of people who care about inclusion and collective enterprise for advancing our society. As Rosemary Lee describes dance can be of significant value in people's lives and within communities: 'Dance enables people to explore and find creativity for themselves, to share physical languages, to become artists for a moment or more, and feel a well-being in the whole body' (Lee 2001: 26).

Jamus Wood – Dancer and facilitator

Journeys into embodiment

In being asked to contribute to this book, I thought to consider my journey into discovering movement and somatic practices. This includes the work of Helen Poynor, dancing and teaching CI and the experiential approach to the body systems and human development, Body-Mind Centering® (BMC®). On reflection, I landed on two questions to address in considering how my life has changed.

My Questions:

How somatic dance practice and CI have helped support me in meeting the world with a sight condition and a vulnerable health condition?

How the Touchdown Dance work has inspired me to explore and open the world of space around me?

History – Living with a visual impairment

I have an eye condition called corneal dystrophy. This condition will last my lifetime and has involved my undergoing, to date, six operations, including three corneal transplants. This condition has a number of effects, the main one being changeability. The surface of my eyes is irregular and breaks down frequently, leading to multiple journeys to the eye hospital. One of its challenges is that I am continually losing sight and wanting what I just had back, so there is a continual sense of loss and grieving for what has just been.

So what is the effect of all of this? It leads to a sense of vulnerability in the world, always having to have a sense of the condition of my eyes and being aware of potential hazards; things like dust and sand can wreak havoc and lead to a journey to the hospital. Changeability is another aspect of the condition which can affect committing to work on projects and can make travelling abroad a risky undertaking as my eyes are just not stable. Emotionally there is a resistance to being realistic about my capabilities regarding my eye condition and a sense that my eyes stop me or are 'letting me down'.

One of the challenges is to find self-compassion for my eyes that are doing the best they can, and for myself.

I find myself in a transitional place, not feeling like I fully belong to the VI community but also not being fully sighted. So my journey has been one of finding support and capacity through working with Touchdown Dance, CI and somatic movement methods.

My early somatic journey – The gifts of corneal dystrophy

As a child, I had many visits to the hospital as my eye surfaces had a tendency to break down. I remember trips out, in pain with my eyes tightly shut, highly sensitive to the light and being led by my mother's hand up or down steps along hospital corridors, sitting in a room of waiting for people. I remember a very sensate experience of being led, with qualities of light, sound and strange smells all greeting my young senses as I was guided. Qualities of the voices all around me, some kind and caring, some more frightening all these touched my senses and my body. And then there came the doctor's examination with pain, bright lights, watering eyes and stinging drops. Sometimes there were the interested student doctors who wanted to see this 'rare case'. All this left a visceral imprint on my memory and more deeply in my body. Yet these early-guided journeys I see, in some ways as being very similar to the somatic dance experiences I would have later. Today, I still make very similar journeys, though not with my mother's hand as support, and now those imprints seem quite familiar and there is a sense of familiarity about my relationship with them.

These early experiences of pain, light, sound, smell, touch and movement were all very vivid and dropped me into my body in a way I rarely tasted in the rest of my life. The doors to experiencing embodied experience were well and truly opened.

There is another aspect to what my family called 'Bad Eyes' and that was when I would arrive home lying on my bed. This would be a time when I appreciated being on my own, in the dark and where I would be with the sensation of pain: my mind would turn inward and trigger my imagination. This was a time when I was close to myself and distant from the outside world, but the inside world was most vibrant. In many ways these times were rich.

As I grew up these episodes of 'Bad Eyes' were ever present and would arrive when I became stressed or distressed. In my early twenties, one arrived as I was writing my dissertation for my BA in Photography. Photography seems a very odd direction for someone who has deteriorating sight and looking back it was not the best decision I ever made, yet it is the one I made. My brother, who suffers from corneal dystrophy to a lesser degree, also ended up being a cameraman on television. He recently had a graft to his eye that has worked out well for him and continues to work in television.

In more recent years, and as I come back to explore these early years, Katy, Rob and I made a short trio around this kind of journey, meshing movement, dialogue and our histories into trio score called 'Looking into the light'. These bad eyes were my education into the world of sensation and now-ness.

My father's gifts

My father, a man who has his own vision of how the world is, has constructed his life around this world-view. What is important here though is that my father had a very particular aesthetic as regards to space and structures. He was a designer and a 'sometimes' architect, one of his gifts to me is a sense of appreciation for space and an interest in how I could inhabit space. This is a very useful tool as my precise vision disappears and the environment appears as shapes in space.

Challenges

This all sounds great, almost perfect, but it was not. I suffered as a child; somehow there was just not enough of me to meet the challenges of my childhood world. So I had little self-confidence or belief in my place in things. I was quite depressed as a child and remember going to a psychiatrist. However, none of this really helped. So I emerged out of my teens with a determination to make my way, but with little actual self-confidence. I was forever being upended by the challenges of life.

My twenties were spent on studying photography and then the attempt to be a photographer. At the end of my twenties my eyesight became poor and I was trying to hold together a photography business whilst not seeing.

In a way I grew up very un-embodied. I remember a couple of years into dancing, hearing someone describe themselves as 'a head on a stick' and that felt true for me at the time. The journey into embodiment has been painfully slow at times, yet I have persisted and further gifts have been received.

Journeys into embodiment and the visceral world

During my BA in photography, I started to photograph alternative spiritualities in the United Kingdom, and in so doing I began to interweave this art with dance, through the doorway of Shamanism. I met a lot of interesting people on that journey, but dance was something very real to me. The spoken word has touched me, particularly in the form of storytelling, but movement and dance do something very different for me. It evokes something inside me, it calls qualities in me forth, in movement. During the last year of my degree, I changed direction and started to study spiritual dance rather than an alternative religion. This broadened my perspective of dance.

From photographing dance to dancing

At 25, I discovered CI and a little later five rhythms dance. They did very different things for me.

Both forms touched my life deeply. In five rhythms, I was able to touch on my pent up frustration at being in the world with this eye condition and the feelings of being impotent in the world. I was not VI but I often was: I was between worlds trying to make my way as a sighted person but my vision was limited and my eye condition would alert itself in times of difficulty and become very disabling.

CI was not about that at all; in many ways, it was healing for me, a welcoming of myself into this human body, and learning to move again. This form was transformational and I was a very slow learner.

My first CI teacher was Malcolm Manning someone who was a great exponent of both the Feldenkrais Technique and Body–Mind Centering®.[94] He introduced his students to both forms and I have continued to work with these forms for the last 25 years.

In many ways, I would say both my mind and my body were dense with fearful armouring you could say the rigidity of trauma, so it took a long time for the somatic material to work its way into me. Yet immediately on lying on the ground for twenty minutes, working slowly with awareness in a contact class, my body knew this was good for my nervous system and I felt a slow 'thaw' starting to happen. I have since discovered that this is a very common trait in CI; I have seen how people arrive hard armoured against the world and then six months down the road they are melting nicely into their bodies and they start to radiate warmth to the world around them. For me, it was a slower process and took many years but I persisted.

In the end, what it actually took was to take myself away from my life in the United Kingdom, it meant giving up photography and committing myself to the study of somatic dance in Finland for two years in a little place called Outokumpu. As an old mining town that had become a college town, it was remote, cold and yet a place where I submitted myself to somatic dance practice five days a week. In the mornings, we had a two-hour Feldenkrais class and in the afternoon technique classes based on the developmental movement patterns from BMC®. I was probably one of the worst students yet the work went in, my nervous system rewired and many body seeds were planted in that place, that continued to grow over the years to come. In Finland it was as if I had paused my life; I took a breath and then another. I had made a choice for myself to meet myself – to give time for my body intelligence to rise and for long-forgotten developmental movement patterns to re-emerge.

On eye operations

On returning to the United Kingdom, it became clear that I needed to attend to my eyes. In the summer months, I had a very serious abrasion to my right eye that got infected causing many additional complications to the eye. The doctors used a very expensive and strong antibiotic to treat the eye but it never fully recovered. My vision had deteriorated to such a degree that I realised a corneal transplant was the solution. The operation went well, yet I did not receive the hoped-for vision. This was the first of six operations over the next fifteen years.

The pattern was to have an operation and then the vision would slowly degrade once again, then have another operation. These operations also induced a lot of trauma into my system leaving my body easily braced, especially around my eyes.

Meeting Touchdown

I came late to Touchdown. I would say I have been in or around the company for ten years now. In many ways, it has brought my attention to being sight impaired, more specifically to the slow transition from being sight impaired trying to get by as fully sighted, to embracing myself as sight impaired.

Angus Balbernie has been a constant figure shaping the content of the company's performances during the time I have been there, which is a different experience for some of the other company members. As before I joined, Julyen Hamilton and other teachers had pivotal influences.

There have been many aspects to Touchdown Dance's work but in some ways, the simple things are the most profound which I consider next.

The orientation score of one guiding another; one walker is eyes shut, the other is eyes open guiding them through supporting their arm in a very similar way to how a sighted guide guides a sight impaired person in walking. But here we take this protocol further into movement, moving forward and backwards, side to side, up and down, exploring places and spaces, including pausing. In all these different ways, the walkers open to their senses and experience a sense of time and space. This exploration triggered my early memories of the hospital, being guided by my mother and reawakened those early somatic experiences. I feel these exercises are very potent and healing. I also feel they give a sense of what it is to be led as a sight impaired person for both sighted and sight impaired. There is a power dynamic always at hand as one moves through a world that is always about having a vision.

Inner voices

For the performance *APPARENTLY NORMAL* (2010–14), Angus Balbernie was brought in as director and shaper. Since then Angus has been a continual presence in the work. In a simple way, Angus is an architect of space and context; he loves words and Touchdown's movement scores often have included the spoken word.

One may ask what's all this got to do with my story? Well, Angus interweaves several pieces together for me. As a child, in front of the doctor's lamps, I had no discernible voice and experienced, in general, a high amount of pain. Angus gave me the possibility to touch those memories and have a voice, as we shaped and moved these memories of the past. In the mini performance 'Rain' a trio piece with Robert Anderson, Katy Dymoke and myself, there is a short section called 'look into the light' (mentioned above) which vocalises some of my experience with the hospital and the doctor's lamp, which are so normal to me yet still quite disempowering. Touchdown's work has been deeply empowering for me and many VI movers who have been involved. It gives the body not only movement but voice. In my experience, many sight-impaired people have very rich inner worlds, yet for some like myself, expressing ourselves outwardly and being seen where we cannot see, there can be a sense of separation and isolation. To allow touch to be our eyes, to feel the vibration of the voice, are embodying and locating. We know where we are in space when we know where we are coming from, this gives us many possibilities and a great starting place to create movement together.

Space

Space is another important contribution that Angus brought to Touchdown and for one that touches my father's appreciation of space. The nature of my eye condition is that my vision is soft and that I see shapes and space better than I see detail such as people's expressions and the written word. As such working with space is somehow locating for me. When I am in space and can appreciate being in space, I know where I am. This naturally draws me into a sense of my body, in waking up proprioception (my sense of my body shape and form) and interoception (the sense of my internal visceral space) and together, the general felt a sense of the body as a whole.

Spatial concepts support me when I am moving and dancing. With my history of sight impairment and many traumatic incidents over time, my nervous system has become contracted. The work with space involves mapping different kinds of space, such as kinesthetic space, social space and object space. This is supportive as I know where I am, and as such, creates elasticity and resilience in how I relate to the world around me. One of the most comfortable places for me has always been on my own with no one watching, yet there has always been a hunger to participate and be relational within the environment around me. So understanding space through different lenses has been hugely supportive. I feel this work on space has softened my nervous system and my sensory capacities, which has led me to a place where I can respond to my environment rather than reacting with rigidity and shock. This transformative process is true for many visually impaired people who commit themselves to this kind of work over a number of years, as others in the company have testified.

Becoming a relational being may seem simple and a natural inheritance for many. But for me, it's been a long journey of both coming closer to my own experience of my body, of my feeling state and simultaneously coming out into the space around me. Crafting my body to shape space, in time.

Much of the difficulty has been due to levels of developmental trauma in my body, from the many eye operations and eye poking and prodding from doctors when I was young, which left me dissociated from the experience I was having and took up residence in my thoughts. So I would be thinking about the experience I was having rather than experiencing the experience itself.

It is so easy to live outside of our experiences, the modern world invites it. Imagine that you are trying to see the world like I remember the sense of my mother trying to see: there was no peace at being sight impaired, always the attempt to see. I was brought up to make this attempt, which meant I was always ahead of myself trying to be in control of my experience of the world rather than letting the world come to me, where I could shape and relate to it from the inside out.

Teaching in Sri Lanka

In 2015, I travelled to Sri Lanka with Katy Dymoke. It was touch and go if I could go as the cornea in my left eye was in bad shape, with repeated cornea breakdown and the doctors telling me I would need a cornea transplant. I had a bandage contact lens permanently on it and was due to have a replacement corneal graft two days after returning from Sri Lanka. My vision was poor and I was constantly monitoring my eye.

My experience in Sri Lanka was wonderful. I had never been there before and the experience of dancing with nATANDA Dance Theatre Company of Sri Lanka was widening for me. Katy delivered parts of the Embodied Performing Arts Program that she had developed with Mark Taylor from BMC® material. I had participated in this programme some five years earlier, so it was good to revisit the material in a very different context.

One of my primary understandings from that trip is that we are all very different. The nATANDA dancers' bodies were so different to the physical bodies of the dancers I was used to dancing with. Both are strong, thin and very bendy. To put it in simple terms, they are way more flexible, and as a result, had less capacity to support the weight. And there was a concept that 'being bendy' was good, which is interesting when it comes to CI and weight bearing.

It is interesting that I was used to western culture of somatic dance; it is almost like the water I swim in. I take it for granted, and these principles of moving and ways of perceiving are normal to me. So to be in an environment where this way of working is new fresh and revolutionary and refreshing to me, to understand the value of what this work can do.

To be dancing in a different culture with a different mindset and very different value system is valuable as it showed me the tight frame of my western understanding. To put it simply, it was different, which freed up a lot of space for new movement and expression to emerge.

So what I received from watching, dancing and interacting with the nATANDA dancers was the process of being orientated from the outside, and a balancing quality of internal movements started to change the quality of their movement. Movements softened, a new movement emerged and a sense of containment entered their work. Touch was a significant part of the work and for many of them initially felt taboo, especially between men and women. Their next performance would be called 'Touch'. In many ways, we were involving and inviting nATANDA to inhabit their bodies, to allow perception and internal movements to emerge.

The work we offered was inherently developmental, moving from the soft qualities of cellular awareness, through the watery movements of prespinal life and the qualities of the limbed starfish. Later we embodied qualities of our organs, taking some time with the heart, lungs and kidneys. Only in the later days did we touch on the skeleton. I can only imagine what these teachings were for the nATANDA dancers. I imagine that they were 'out of the box' and radically different to their understanding of dance. To witness and be part of the freshness and the learning space was a real privilege in those days in Sri Lanka.

The work of Touchdown

I wish to speak of my learning experience through the Touchdown work. Why it has been rich for me and what is particular about it for a VI person like myself?

I have been considering what the difference is between CI in general and the Touchdown material. What I am coming to understand is that it is not the *what* but the *how*. The Touchdown environment gives time for development to happen in its own time, whereas in a CI environment we are trying to develop skills ... there is an impetus to fly well if you like. In the Touchdown environment, it is OK to be a beginner, and to be a beginner for a long time. For me this had a massive effect.

I needed more time just to inhabit where I was, I was able to lie on the floor for a very long time. Slowly over time, this softened my body tone and my nervous system. I was receiving somatic and developmental material for a long time. Apparently from the outside not much was happening but on the inside, I was receiving deep nourishment and small things were changing. Giving time for us to develop is so important, it's preparing the ground, I can see it like gardening, I clear the ground, I prepare the soil, so I think that a good half of the somatic work is about preparing our soil, so we can actually take the material in and then suddenly outwardly our movement and presence develop in a spurt when the moment is right. It all comes down to when the time is right we are ready to change.

So when I go back to my early days of CI with Malcom Manning, in his classes, I was changing my base level, my relationships with my body, he was teaching CI where most of the contact skills I was just not available to, but I was learning in a very rich and deep way; there was something about lying on the floor … letting gravity through, and there was development and softening in my reflexes and nervous system from just allowing; it was a kind of deep nourishment. I think it's so common for people with visual impairment issues, that this idea of support is just not there, so actually lying on the floor with the 'permission' to lie on the floor, is a very important piece of the work; and for some people like me, they have to do it for a very long time. Learning to move close to the floor is revolutionary. And then over time, the availability starts to arise and then, after what the teacher might be wanting you to understand, starts to become available to you. Or sometimes our learning goes in a very different direction and some development that is quite 'left-field' can arise.

Coming back to Touchdown and my own evolution as a dancer our work with Angus Balbernie was transformational as he opened us up to a different dialogue with space and time. There were ideas of exploring surfaces, objects (people and environments) and different explorations of space. In many ways, Angus presented interesting, irreverent questions to us through movement and I realise that I had many ideas of how things are locked down and categorised. For example, starting points such as 'find a center in your body, any center and move from there to your surface' were, in my case, revolutionary, having spent so much time trying to find my one true centre. This freed me, not only my mind but also my body. It allowed inquiry to open and arise through my dance. I stopped looking for the answers and became more interested in good questions and exploring them.

Some VI people have a similar sense of looking for an answer. So Angus's questions can help create plasticity in our thoughts and our body. It is amazing what such a little shift of perspective can do.

Reorientation …

Let's now talk about our 'mid-line' and some of the core work of Touchdown. In a way, we are teaching and dancing CI but what if this is just the context? What if this work is actually about growing up, widening ourselves and developing our movement capacity through a deeply nourishing developmental practice? What if we are waking up our capacity to be in a living body here and now and through our perceptions that are rooted in our body? Step one seems to be inviting a sense of the body that is safe and supported, where it is, I mean that it's actually OK to be in your body. When I think for myself and maybe many other VI people, the body was a place of pain and unsafety in the early years. My body was in fact a hostile place.

One powerful example of the work is rolling, I don't know why that is, but there is a quality about rolling and finding your central axis, finding your spine, finding a kind of beginning space in you, to orientate from. This I guess is Steve Paxton's whole work really, to allow the sensory input to orientate around that mid-line.

For myself this was a shift, from being oriented around what I find in my outer environment, to coming home to the very core of my body, and find this as a place to orientate to my world from. Very different. New body. New world.

Onto the stage

Touchdown's studio work has often revolved around creating dance pieces to share. I have received so much from the process of creating and working on material for performance.

In our last piece 'Attention', performed at NATFEST in Sri Lanka in 2017 and in London we worked with several fairly simple principles into a movement based on the work of Steve Batts that Robert shared with us.

The first exploration was into levels; moving from lying to jumping in a series of seven layers of movement, numbered from 0 to 6, that I recall as follows, in essence:

0. Lying into gravity and allowing our cells to fall down into the earth are fairly passive. Perceptual: self-orientation and a quality of being.
1. Moving close to the earth, exploring surfaces, weight and levering. Swimming across the floor. Shifting, sliding, moving through space on the floor. Prevertebral movement patterns. Perceptual: sensation, weight, surface, immediacy, fluid body.
2. Sitting, kneeling and on all fours. Hands are very related to the floor, finding hands and feet. Crawling, creeping and animal quality of movement. Perceptual: orientated down towards the earth. Orientated around physical structure; skeletal, muscular. Hearing very awake. Can easily shift from parasympathetic to sympathetic (autonomic nervous system) activation.

3. Squatting, high kneeling, hands on the ground. Perceptual: wide vision, horizon gaze. Sense of being ready for action, responsive, awake.
4. Standing with bent knees and dropped pelvis (as in martial arts); dropping into the pelvis, sense of kinesphere, travelling through space. Spherical sense of space, hara sense. Perceptual: shape, movement and colour, feeling the space through the ears. Proprioceptive, sensing, moving. Awareness of being seen.
5. Standing, social: Walking, standing, everyday movement. Perceptual: Social engagement, composition, mind, social space and connection, naming, quickness, receiving and responding; action, reaction interaction. Gaze comes back in.
6. Jumping up. Tends to come from 3, 4 and 5. Perceptual: Go somewhere. Sympathetic nervous system. Brings energy.

We learned to inhabit each level, noticing perceptual changes in how we related to the world around us, and the qualities of the movement at each of these levels as an individual mover. For myself I found this rich material to dive into. Noticing that some of the levels were familiar and I had a habit of moving here, whereas other levels I avoided, so developing new ways of moving in the comfortable levels and learning to move in the places that were less familiar. In so doing, my movement vocabulary expanded and changed. As I developed and embodied these movement levels and qualities, as a group we also looked into transitions and how this could create poetic tension. Later we developed the ability to move on one level whilst adopting the perceptual qualities of another.

The second exploration was with space and proximity and the relationships between ourselves and the other dancers at different distances or levels of touch. We worked from distance to proximity and then through layers of touch: skin, muscle and bone, exploring the notion that we can touch from one space through into another space, which was interesting. Once again transitions between layers of space or layers of the body became interesting to explore.

Our third exploration was around attention and how we orientated our attention. This was a kind of classical theatrical performance technique, yet we took it into an embodied exploration around how it shapes my body when my attention is on myself, or when I am addressing an audience member. Very rich material.

On the stage, it's an interesting experience for me. Generally, I can see less than usual as the lights can be blinding and my eyesight intensifies this quality. The audience is generally not there to my vision but I can feel a sense of listening and attention from beyond the stage. Strangely I can feel a little in my own world on stage with less vision, so essentially it is me and the other Touchdown dancers I am aware of. As we come and go from solo to dancing with each other, in many ways it feels like being in the studio together, only I can see less and there is the intensity of 'know we are performing', it matters and there are people watching! So it has a surreal quality of being in the moment and being on my own at the same time.

Somehow I do like performing, I like the way a performance brings together all the threads we have been working on together into a cohesive whole. The threads are never complete yet together with the other threads they create a cohesive weave.

This does create a sense of something is complete, like finishing a song or completing a project.

Afterword

Steve Paxton

Contact Improvisation – A contemporary game

Contact Improvisation is a dance game. Seventeen people began in 1972. Several of us are still researching and enjoying it. *Contact Quarterly*, a dance journal publishes about its progress, with other relevant research. Within twenty years, Contact Improvisation was practised in various cities around the world. It is still growing I have heard. A physical game will improve the body, and playing a game will increase the kinds of skills appropriate to it. The goal of most games is to win. Modern dance creates new dances and new ways to dance. Dances are not considered winnable of losable. There are many games which require two people to play, and only a few games which require that both people win. Contact Improvisation explores the possibilities of physical cooperation. Each cooperation will have a unique result, a blend of classical physical principles and some of the variables of the human being; for instance, a person's spatial imagination, strength, flexibility, reflexes and their experience. Finding our regular salaried work, and adapting to it, we find our time constrained and variety of experience turns into routines. Therefore games. And therefore dance. Games and dance help us maintain our physical potential at higher (or should I say broader, or more various) levels than regular work tends to. Work available in urban areas is especially repetitive. A man with cerebral palsy, in a discussion about designing buildings for the disabled, said 'Everybody gets disabled someday'. And if not that, we get physically channelled. This is a kind of disuse. The dance game is useful for this situation. The body is another part of the mind.

Note: This was initially published on *Touch Sense Feel Move*, an accessible CDROM on the work of Touchdown Dance (Dymoke 2001). History. Section 2. Page 17.

Appendix 1

The Small Dance

The small dance is an attribute of Contact Improvisation (CI) that Paxton teaches as a somatic-oriented principle; it is a form of dialogue of mind and body that underlies the ability to track one's own movement using the kinaesthetic sense, even whilst dancing with a partner. The small dance is useful in CI to awaken the mind to the inner corporeal landscape; however, focus on this kinaesthetic experience recedes from consciousness with the added immediacy of sensory input from the external world, from touch and moving in space. In CI kinaesthetic feedback is received from physical contact with the partner as well as from moving in response to gravity and the floor (Banes 1987; Novack 1990). As with the small dance, the external environment provides the ensuing sense of connection and support for further movement exploration to take place (Pallant 2006; Koteen and Smith 2008; Dymoke 2014), see Chapter 5.

In the 'small dance' practice, Paxton had observed how the continual small adjustments of the body when standing 'still' gives a sense and awareness of the whole body. This perceptual process provided tools for Paxton; the body remembers and retains information laid down in the nervous system, and uses this inter-systemic intelligence together with mindful or consciously perceived 'use' of it (Hartley 1994; Juhan [1987] 2003; Bainbridge Cohen 2012).[94]

Like many others, I went on to experience the 'small dance' with Paxton in Touchdown Dance and other CI festivals, and in his intensive training 'Material for the Spine', in which the spine is the central core, the segmented axis that can move the whole body when it initiates even small movements. We stood for several minutes at a time, collectively tuning into the pull of gravity downwards through our bones and the anti-gravitational responses that prevent falling; we were reminded to notice the smallest adjustments of the body from within the torso, joints or tissues. This practice was specific, a means to perceive the inner spaces of the body, the deep internal adjustments of the bones and other connective tissues that maintain the body in an upright position, around its centre of gravity. It was a means to feel the innate reflexive responses to tangential forces passing through the body, and as such, it was an effective learning experience and readily applicable in the practice of CI, in which another moving body imposes force and receives my own.

With imbalances of weight or agility, the partnering experience adjusts to accommodate these physical differences. Accessibility occurs innately in the act of doing, as if there is some pre-understanding of fairness, of an intrinsic need for mutual accommodation for the experience to be purposefully effective and meaningfully affective for both.

Such changes were a shared experience for both the visually impaired and sighted participants in CI as they experienced new ways of sitting, standing and pathways between. Self-observation was nurtured and sensory feedback was discussed to provide information that was otherwise hidden from conscious awareness, and from view, to create time and space for change: 'In the act of releasing tension into a state of readiness and allowing the senses to expand first, we are making a space for the physical movement to follow' (Paxton and Kilcoyne 1993: 11).

The teaching process was without any traditional visual aids, using 'sensations and physical experience' as methods that the body understands. Far from the traditional learning environment, these methods were in direct response to the reality in the room; they addressed the lack of motility and the potential for mutability of the human body (Abrams 2007; Johnson 1995).

Appendix 2

The 'Hatching Chick' – And the 'birth' of the Membrane Concept

At *CI 25* in the United States, I sat with Steve after breakfast. I gave him an update on the progress of the work as we were giving workshops around the United Kingdom and had projects in Greater Manchester. My most poignant metaphor at the time was of the hatching chick; it held the legacy of the work I did in Denmark. I had held the image of a chick starting to feel ready for a change, pushing at the edges of the shell, sensing the light on the other side, the sounds, the movement the enticement of the world beyond. Then as the shell starts to crack there is no turning back, as the chick has to find its own way and its own support from another kind of 'shell' to contain it in the wider world. For me, the chick could symbolise potential, a participant or the work itself. Metaphorically, the kind of shell that goes with the 'chick' became more like a motile membrane than a brittle egg shell. The membrane creates a boundary to contain the 'core' self, the sense of self and of identity; it enables self-movement, to follow the flow or to flow back and forth from one experience to the next. The accrual of experiences creates a sense of who the self is becoming and what to let go of or leave behind. The self builds greater autonomy and connection to the world, with an ability to move through membranes that separate one place from another, from one state of awareness to another. The 'chick' may find the potential to grow into a more full-fledged adult and, to follow this analogy, after a workshop some are barely hatched, some are already full-fledged, and we all move forwards together, allowing space for both limitations and possibilities, for differences and availability for change.

This emergent metaphor gains further resonance as the membrane transfers from a symbol to a metaphysical and anatomical concept over time and forms an underpinning theoretical construct during the journey, to which I return over the course of this publication. The sphere of growth is a familiar concept, the caretaker of that sphere is another. There are times when the membrane is one of solidarity and co-responsibility when the individual person identifies and is supported by others. There are times when the membrane is hard to sustain, when others around us may take on a role to nurture, protect and support, to reassure us that all is well. These moments are significant moments of transition, of learning, of forming and reforming a sense of self – and here I refer to health and wellbeing, to the ability to receive, to act and feel at ease with others, to self-regulate during a process of change. As teachers and facilitators of an experiential approach to self-discovery, the membrane concept is a useful tool to define the space, the distance and the amount of stimulus that each person can usefully manage. The process of change is not for me to determine from outside looking in, it is for me to hold the space – so that as the body moves there is time for the mind to move and for the whole being to grow as one into itself and then out into the world beyond. Simply put the membrane is a place of containment, where we intuitively decide yes or no, or sense doubt and confusion, and then over time consider from experience to continue or regroup. The prevention of this process is to be denied self-realisation, to be oppressed, to be directed by conformity, by the social codes that Kelly (1984) defined as a restraint as '"straight" social codes and structures' (Kelly 1984: 9) and Hulton as 'social diktat'.

In the realm of the Touchdown Dance praxis and research, an ability or desire to engage with the space beyond the body, and so with others, is experienced as being relative to an ability or desire to move, and in turn, impacts on an ability or desire to change or grow. Mobility, autonomy and agency are all interlinked and witnessed in examples of the lived experience of teaching CI to those experiencing (self or externally imposed) restraint, exclusion or neglect. Personal stories also reveal how the benefits of this psycho-social experience may be relative to internal or subjective factors such as mood, or to the external, inter-subjective factors such as the intention and nature of the self–other relationship on offer.[95] Freedom of movement is experienced relative to the degree and nature of any impingement, which may be subjective, like innate or learned physical or perceptual limitations, or imposed, as in the form of rules or inappropriate environments. Either way, the receptive and intuitive bodily sense is affected in CI and regulated with the clarity, proximity and containment of the two-way, dialogical membrane concept. CI is a praxis that impacts positively upon mobility, autonomy and agency – on how 'I' present to the world at any moment. In the praxis of CI, it is evident that how we perceive the body determines how we use the body, and 'usage' may take years to learn. How we think about the body informs how we use it, and how we think and perceive our bodies in movement is contingent on access to both.

Timeline

1986 Founding of Skipping Visual Impairment.
1987 First seasonal workshops at Dartington College of Arts.
1988–89 Continuing of workshops. Blind Faith performance.
1991 The name Touchdown Dance is taken and the three-weekend workshop visit to Denmark takes place.
1992 Touchdown Dance setup as a Charity. Katy and Torben attend a workshop at DCA.
1992 Dance Umbrella. *Sixth Sense – Second Sight* and *Moving Prescription* performances.
1993 Publication of *On the Braille in the Body*, Journal of Dance Research.
1994 Katy Dymoke takes over as director of Touchdown Dance charity.
1998 Ray Chung co-creates the first performance in the North West.
1999 Resolution Festival performance of *I-radiate.*
2000 Capture commission of *SENSE-8* Ten-minute dance film. Creation of Touch Sense Feel Move, accessible CDROM. Publication of *What Dancers do that Other Health Workers Don't.* Dymoke, K. Jabadao publishing.
2002 Two-week workshop in CI with Karen Nelson and Lisa Nelson.
2002 Free Summer Dance on the South Bank Festival, Royal Festival Hall – in collaboration with Julyen Hamilton, and workshops for children. Work in process performance of *TACT.*
2006 *CLOSER* was created with Julyen Hamilton and Angus Balbernie. Tour of England then Scotland 2010 *APPARENTLY NORMAL* research and development followed by NW tour.
2017–2022 343 m/s devised with Angus Balbernie – performed in Glasgow and Kilmarnock 2017, Dartington; 2019; Voarte Festival Lisbon, Portugal; 2022 Ageless festival, Leeds.

Notes

1. I discuss this and the final project in Denmark in a chapter entitled *Touch – A natural language* (Dymoke 2000).
2. Body–Mind Centering® and BMC® are the service marks of Bonnie Bainbridge Cohen, used with permission.
3. I return to this in further detail through the book.
4. Such discussions manifested many years later in new policies by the Arts Council of England 2013. *Great art and Culture for Everyone*. https://www.artscouncil.org.uk/sites/default/files/download-file/Great_art_and_culture_for_everyone.pdf. Accessed 26 October 2018.
5. Quote from webpage https://www.independentliving.org/donet/148_national_disability_arts_forum.html. Accessed 12 February 2020.
6. The Foundation of Community Dance is now called People Dancing.
7. See her book *The Roots of Thinking*, Maxine Sheets-Johnstone (1990).
8. CI has been widely documented since 1975 in the *Contact Quarterly Journal – A Vehicle for Moving Ideas*, see https://contactquarterly.com/. Accessed 28 March 2023.
9. p. 41. Here I am quoting Peter Hulton, who I have since met in other research contexts, talking in a recorded interview about the ethos of training that was core to the DCA educational approach that he instigated in the 1980s, influenced by the Liberal Arts model of Black Mountain College in the United States. http://theatredanceperformancetraining.org/2018/10/peter-hulton-on-dartington-college-of-arts/. Accessed 19 March 2020. He published a series of papers by professionals as practitioners, called Theatre Papers.
10. p. 41. See Nelson (2014) for a historical overview of the history of Practice as Research (PaR) in academia. p. 42.
11. This discourse took place in philosophy and psychology, in the realm of phenomenology; such as Lacan (see Fink 1995), Judith Butler (2005), Francisco Varela and Jonathan Shear (1999), etc. p. 46.
12. Torben performed with Performance Training Workshops (PTW) and came on tour, he enjoyed the attention and agreed to be photographed and mentioned in articles. I heard that he passed away in 2019. p. 51.
13. This story is told in more detail in the methods in Chapter 2, p. 53.
14. The small dance is explored in more depth in Part 1.3 and Chapter 2 and Appendix 1.

15. See Chapter 3 for further discussion on touch and Chapters 4 and 5 on the optic sense of the body, p. 54.
16. This was confirmed by research undertaken at Saddlers Wells by the then director who was visually impaired, in 2006; https://touchdowndance.co.uk/past-workshop-events/2006-sadlers-wells/, p. 55. Accessed 28 March 2023.
17. Interoception – the perception of internal states from sensory receptors in viscera, that indicate hunger, thirst, discomfort, etc. and bring the body to mind, to attend to self-regulation and maintain balance. p. 58.
18. See preface – the medical model of disability prescribed the disability rather than the social barriers as the reason for exclusion and the label 'blind' objectifies the disabled person and denies them other identities. p. 60.
19. A time would come when visually impaired participants partnered with each other and learned more about non-visual movement methods – however freedom to orient and move into the space was more demanding of trust and required sighted 'spotters'.
20. This exercise and the following ones are given further explanation in Chapter 5. p. 61.
21. This is taken into more depth in Chapters 2 and 4.
22. I return to this in Chapters 2 and 4, and in Chapter 8, Holly Thomas gives her account. p. 63.
23. This concept is explored further in the inclusive methods Paxton and Kilcoyne devised in chapter. p. 65.
24. See Chapter 4 on spatial orientation and the different modalities. p. 66.
25. See modality-specific approaches in Chapter 4 (Figure 9). p. 70.
26. These are the names of visually impaired participants who have been named in previous publications so I keep to them. p. 74.
27. These aspects are identified by Petra Kuppers (a disabled artist and activist) as essential for inclusion to be effective (Kuppers 2011, 2014).
28. When visually impaired people came to move together the sense of empathy between them was heightened by the commonality of non-visual learning; however, this 50/50 principle still applied, see Chapter 5. p. 75.
29. This experiential approach to learning resonates with the non-dualist ontology of the phenomenologist Maurice Merleau-Ponty (Matthews 2006; Daly 2016) which is based on the premise that the body and mind are mutual aspects of the same unitary being and that movement is therefore experienced as the movement of both body *and* mind (Sheets-Johnstone 2000, 2010; Bainbridge Cohen 2012). Imbalance, however, occurs inwardly if the mind thinks of the body as 'other' as not 'me' and so dismissed of its innate intelligence, and emergent role in reflexive self-critique. p. 76.
30. Further discussion of the interrelationship of emotion and perception to consciousness and learning comes in the methods in Chapter 2.
31. These four foundational principles return in greater detail in Chapter 2, Part 2. p. 82.
32. This is explored more in Chapter 3 on Touch communication. p. 83.

33. See Chapter 2 for more detail on these movement ideas. p. 85.
34. See Chapter 2 for the impact of postmodern dance and CI. p. 87.
35. Independent dance artists are not affiliated to a school or company and work as teachers/artists/makers. p. 90.
36. Steven C. Rose (1979), *A Touch of Sensitivity,* Horizon Publications: BBC Education Unit (video). I return to touch in more detail in Chapter 4.
37. This is explored further in touch communication, Chapter 4. p. 96.
38. See Chapter 8 and Jamus Wood for his insights into this. p. 99.
39. See Chapter 3 for more detail of the workshop process. p. 104.
40. A sense of body image has been found to arise experientially from the combined senses of movement and touch in BMC® (Bainbridge Cohen 2012) and psychotherapy (Totton 2016). See also Schilder 1950, 1999 for a psychological perspective and Sheets-Johnstone (1990, 1999, 2000, 2010) for a phenomenological view of the tactile-kinesthetic foundation to consciousness and thought. See Chapter 8 for examples from Holly Thomas and Jamus Wood. p. 107.
41. The emancipatory aspect (Walmsley 2001; Totton 2018) was also significant, particularly with the involvement of learning disabled people lacking capacity. It shows the importance of their participation in research and impact on the quality of services they access, where the research involves other people involved in their daily lives and those with a duty of care (Gilbert 2004; Walmsley 2001, 2004; Redmond 2005; Williams et al. 2008, that work *with* rather than *on* learning disabled people (Redmond 2005; Nind 2008; McClimens and Allmark 2011).
42. A useful source of information on participatory research can be found on the National Institute for Health Research website: https://www.invo.org.uk.
43. See Chapter 8.
44. See Chapter 4 for more on touch.
45. For a definition of practice-led research, see Candy (2006: 1):

> There are two types of practice related research: practice-based and practice-led: 1. If a creative artefact is the basis of the contribution to knowledge, the research is practice-based. 2. If the research leads primarily to new understandings about practice, it is practice-led. p. 111.

46. See Nelson (1996: 65) CI is 'a touch revolution. It's a revolution against the tyranny of non-touch. It's a politic of movement from the inside out, organising to break the code of space and distance between people' she talks of the sexual inferences of touch being displaced, through the experience, by non-sexualised touch. p. 112.
47. See Fulkerson (1996: 41). CI practitioners who continued to develop CI contributed to the creation of an ideology 'with the aim of reconciling roles between men and women which had been for so long (and continue to be today) scandalously unequal in the dance world'. p. 115.

48. Evidently, society has marginalised and not given space for visually impaired people to take a stand against the lack of opportunity to participate in activities that a predominantly visual world deemed inaccessible to them. Heller and Gentaz state how research involving blind people has been a negative experience – as it focuses on what they cannot do rather on what they can do well. Scientific research uses systematic methods, demanding large 'sample' sizes for the research to be considered valid and applicable, and 'representative and normal samples' requiring participation across society; evidently visually impaired people can't constitute a sample group and so are excluded. Heller and Gentaz consider this situation to be a 'fabrication' and determine instead to 'adopt an object approach'; in their case 'to the study of haptics and touch perception in blind individuals' (Heller and Gentaz 2014: 220). CI provides a similar 'object approach', as through inclusive research salient and underlying issues arise and contribute to the types of discourses that advocate for change in the research environment. Research may then deal with issues that affect their quality of life; studies can be made available for them to participate in, which can then contribute to societal change (Walmsley 2002, 2004). Touchdown Dance operated in this considered and pragmatic way. p. 116.
49. The continuum concept, I reiterate, alludes to the space between polarities, between aspects of self, such as inner and outer, and is influenced by Jones (2009). p. 119.
50. See Cynthia Novack (1990) *Sharing the Dance; Contact Improvisation and American Culture,* in which she considers the anthropological and 'ethnohistorical' aspects of CI – to 'describe and analyse the history of a way of life and a way of dancing as part of culture' (Novack 1990: 16).
51. Other allied research-led practices are referenced in the accounts written about the work that also influenced the language used in teaching, and were generative of the praxis, including consciousness studies, reflexivity, phenomenology, somatic movement education and ontogenetics (which I return to in Chapters 4 and 5). p. 121.
52. See Batson and Wilson (2014) for further analysis of the process of attending to the somatic experience in dance. p. 125.
53. See Gilles Deleuze (1990), the Logic of Sense, for consideration of the dialogue between the surface events and the core of the body, and of touch as the sense to which the other senses refer back for empirical knowledge, which concurs with Bainbridge Cohen (2012) and the Body–Mind Centering® approach to the perceptual process. p. 127.
54. This is explored further in Chapter 4, where I consider the modality-specific methods and variable applications. p. 129.
55. At Dartington College of Arts. See Chapter 1. p. 26 onwards.
56. I discuss this principle in more detail in Dymoke (2014) and Chapter 4. p. 132.
57. The subtle training that we have undertaken to do this movement is in more detail in the modality chapter.

58. The full quote is

> Contact improvisation itself is based on a shared partnership, where the process of mutual co-operation is more important than any one person in the partnership Because of this it is easy to structure the workshop learning process in a way which reflects an equal and democratic partnership.
>
> (Paxton and Kilcoyne 1993: 15) p. 137.

59. See Chapter 8, the 'Backword' from Steve Paxton. p. 138.
60. See Walmsley (2001, 2004), Gilbert (2004) and Williams et al. (2008) for further information on new research methods involving marginalised and disenfranchised groups in which the researcher role involves participant input, and inclusive research practices such as involving learning disabled people in research to gain new understanding about their needs evolved. For mixed method research exploring multiple subjectivities, see Hansen et al. (2005), Cooper and McLeod (2011), and for the value of in-depth practice-led research in interdisciplinary contexts, see Fishman (1991) and McLeod (2015). p. 150.
61. This epiphany was not taken in detail in Chapter 1 as it reappeared to Paxton's memory in our correspondence in 2020, with salience to the touch topic.
62. See Appendix 2. p. 152.
63. The concept of a spherical space is familiar to sighted people who have trained in dance. The kinesphere refers to the sphere around the body that marks the periphery of the body in space. It stays with us as we move; it may decrease as we reduce our body size and position and expand as we open into space. The sphere is significant as the space in which we perceive the body as it moves, that supports Paxton and Kilcoyne to witness the participants and support them to explore their bodily sense. p. 153.
64. Many other research studies have emphasised the significance of touch, the tactile sense of the skin and of human-to-human contact, to the survival of the infant (Montagu 1986; Autton 1989; Bainbridge Cohen [1993] 2012), to healthy development of the self (Stern 1985, 2010; Schore 2003, 2004; Gallagher 2008; Totton 2015), to healthy continued growth and the ability to relate to others. The severity of abuse caused by touch deprivation has established the importance of touch to survival (Montagu 1986; Autton 1989; Juhan 2003), the inseparability of this primary sense from movement (Bainbridge Cohen 2018) and its essential role as a non-verbal language of communication (Piper and Stronach 2008; Dymoke 2014, 2018). p. 157.
65. *The Discursive Mind* (1994) the psychologists Rom Harre and Grant Gillet discuss the nature of research into human behaviour and the need to consider the worlds of each person involved and the conditioning of perception through societal, cultural and other influences that form a veil through which the world is perceived: 'There are only co-participants in the project of making sense of the world and our experience of it' (iHarre and Gillet 1994: 21).

66. I explore this topic in depth in an article 'Contact Improvisation: The non-erotic touch in an art sport' (Dymoke 2014). p. 162.
67. See Introduction and Chapter 1 for an overview of this historical injustice. See also the *#Just Ask Don't Grab* campaign, on Twitter founded by Dr Amy Kavanagh. p. 168.
68. See Figure 9 and Chapter 4. p. 169.
69. See Chapter 8 by Holly Thomas and Jamus Wood. p. 171.
70. See Sheets-Johnstone (2011: 41). This term is taken from 'Husserl's exposition of phenomenological methodology, thus a reduction of the natural attitude, the practice of free variation with respect to the topic of concern, and the discernment of the essential character and features of the topic of concern by way of the reduction and free variational practice'. p. 172.
71. There is a disparity and distance between opposites but between both is a field in which they are mutually informative and interdependent. p. 173.
72. See Chapter 1. p. 175.
73. See Depraz and Natalie (1999), 'The phenomenological reduction as praxis', *Journal of Consciousness Studies*, 6, pp. 95–110. Depraz adopts the 'epoché' – a Greek philosophical term for setting aside existing beliefs, assumptions and judgements, a form of psycho-phenomenology.
74. The document – *Embodying Cellular Consciousness* is from a study guide of the 'Senses and Perception' course in the BMC® Somatic Movement Education training program, Bainbridge Cohen (1994: 8). p. 182.
75. Deprivation is a form of neglect and is an underlying cause of self-deprivation and self-neglect (Montagu 1986; Field 2003; Dymoke 2018, 2019). p. 184.
76. See Appendix, in BMC®, the ability to yield into self provides the density with which to push away from other – e.g. the supporting surface, which establishes the interrelationship of self-other, subject/object and underlies the ability to then reach towards an object and bring (pull) it towards self. p. 193.
77. It is significant that Bannon and Holt's (2012) study took place in a higher education context, exclusive to visually impaired people, and in that context, touch is ethically debated from both sides of the argument, and they are very careful to present a balanced account – as I hope to do here. p. 196.
78. See Chapter 1. p. 202.
79. The membrane concept relates to relational fields and the limits of perceptual awareness, and they mark the boundary of consciousness around one or two individuals or a whole community. p. 205.
80. I return to motility in later chapter sections.
81. See Chapter 8, Holly Thomas shares her experience. p. 207.
82. Heller and Gentaz (2014) discuss the importance of intermodal sensory perception and so how difficult it is to draw down 'valid' results in practice-based research (Heller and Gentaz 2014: 207) as the variation between subjects is hard to ascertain. This supports our premise and ongoing exploration of the benefits gained for all in inclusive practice. p. 245.

83. Creative partnership was a government-backed programme that supported professional artists to work in schools to a specific scheme of work and time period. Touchdown Dance was a 'core partner' in Manchester and Salford schools as well as with Mid Pennine Arts. BBC Children in Need is a fundraising initiative. p. 251.
84. https://www.rnib.org.uk/rnibconnect/young-people/activities-you-can-do-with-your-child
85. In another study, *Sight Impairment at Age Eleven: Secondary Analysis of the Millennium Cohort Survey* (Harris et al. 2014), in the section on inclusion in social activities, such as sport, one-third of visually impaired children compared to one-fifth of sighted children did n't take part in physical activity on a weekly basis, and those with visual impairment and other impairments was higher at 35 per cent (Harris et al. 2014: 145). The other area of difference exceeding 10 per cent of the VI/sighted children surveyed was relative to spending unsupervised time with friends at weekends and emotional wellbeing – 67 per cent of visually impaired children with other disabilities were reported to 'have difficulty with emotions, concentration, behaviour or being able to get on with other' compared to 33 per cent of sighted children (Harris et al. 2014: 64). I pull out this statistic as the next group I consider are of this group and between the age of 8 and 14 years old. Also relevant is the fact that 47 per cent of the same group of children reported being 'clingy in new situations compared to 25 per cent of sighted children. Likewise, visually impaired children disliked school and are twice as likely to experience being 'bullied or pick on by other children' – 46 per cent of visually impaired and 39 per cent of visually impaired and another disability, compared to 24 per cent of sighted children (Harris et al. 2014: 104) and had low aspirations regarding secondary education and further/higher education (Harris et al. 2014: 103). Regarding tiredness, 30 per cent of children self-reported feeling tired all of the time, and 35 per cent of children with visual and 'other impairment' compared to 20 per cent of sighted children (Harris et al. 2014: 114); 44 per cent of children with sight impairment and another impairment do not work well independently compared with 10 per cent of children without sight impairment and 5 per cent of children with sight impairment only – which sets them up to feel dependent on others in learning situations'. p. 262.
86. In their article on the *Dynamics of Touch*, Carello and Turvey have a diagram on the interrelationship between proprioception and exteroception – (a) Perception is of environmental properties (exteroception), body properties (proprioception), the environment relative to the body (expropriоception) and the body relative to the environment (proexteroception). p. 264.
87. The 'mood scale' was created as a strategy for emotional expression for profoundly disabled young people back in 2013, using two balls, one on either side of the room, a yellow for happy sunshine and blue for sad, each child places themselves along the line either at the happy end or sad end, and then asked to speak their mood. The group then noticed each other's mood and many became able to communicate who had not been assessed as having such capacity. p. 268.

88. https://www.bild.org.uk/wp-content/uploads/2020/01/RP-tool-RRN-with-all-included.10.18.TT_.pdf. Accessed 18 May 2023. Safeguarding guidelines now identify such decisions as psychological restraint: depriving a person of choices, controlling them through not permitting them to do something, making them do something or setting limits on what they can do, without physically intervening. It includes the implicit use of threat and coercion. pp. 277–78.
89. Gallese, Sheets-Johnstone and Bainbridge Cohen are useful in this retrospective revisiting of the work, as a means to open up the meaning and learning that was experienced to a wider field of application and potential. p. 279.
90. Image Lab (US) is a group of independent artists who explore improvisation methods run by Lisa Nelson (colleague and performance partner of Steve Paxton). Lisa had developed work involving movement and touch and was influenced by BMC®, so a good match as a collaborator with Karen. p. 296.
91. The interview can be found at https://vocaleyes.co.uk/about/contact-us/. Accessed 1 February 2021. p. 315.
92. The first minute of the film can be found at https://touchdowndance.co.uk/sense-8/, (accessed 18 May 2023) the full on Vimeo or on DVD by contacting us. p. 366.
93. This was a separate visit to the one with Jamus Wood. We also returned another time as the three of us. p. 377.
94. Body–Mind Centering® and BMC® are the registered service marks of Bonnie Bainbridge Cohen. p. 389.
95. https://www.youtube.com/watch?v=_82Od5NM4LI. Paxton (2015) speech at the Walker institutes 75 anniversary. The host talks about Touchdown Dance as being ‘ahead of its time’ and quotes Paxton who had once stated ‘the fiction of cultured dance and the truth of improvisation’. Paxton talks of his experience in gymnastic tumbling and experiencing ‘the mind assessing body sense’. p. 392.
96. See Holly Thomas in Chapter 8 for her experiential accounts of this.

Bibliography

Abright, A. C. (1997), *Choreographing Difference*, Hanover: Wesleyan University Press.

Ackerman, D. (1990), *A Natural History of the Senses*, New York: Vintage Books.

Albright, A. Cooper (2010), 'Falling on screen', *International Journal of Screendance*, 1, pp. 21–26.

Albright, A. Cooper (2019), *How to Land: Finding Ground in an Increasingly Unstable World*, New York: Oxford University Press.

All Party Parliamentary Group on Arts, Health and Wellbeing Inquiry (2017), *Creative Health: The Arts For Health and Wellbeing*, http://www.artshealthandwellbeing.org.uk/appg-inquiry/Publications/Creative_Health_Inquiry_Report_2017.pdf. Accessed 26 October 2018.

Allegue, L. Kershaw, B., Jones, S. and Piccini, A. (eds) (2009), *Practice-as-Research in Performances and Screen*, Basingstoke: Palgrave Macmillan.

American Psychological Association definition, https://dictionary.apa.org.intersensory-perception. Accessed 1 March 2020 (in Chapter 1.3).

Aquino, A. T. and Lee, S. S. (2000), 'The use of nonerotic touch with children: Ethical and developmental considerations', *Journal of Psychotherapy in Independent Practice*, 1:3, pp. 17–30.

Arts Council of England (2013), 'Great art and culture for everyone', https://www.artscouncil.org.uk/sites/default/files/download-file/Great_art_and_culture_for_everyone.pdf. Accessed 26 October 2018.

Arts Council of England Corporate Plan (2015–18), https://www.artscouncil.org.uk/sites/default/files/download-file/Corporate-Plan_2015-18_Arts-Council-England.pdf. Accessed 26 October 2018.

Arts and Disability Ireland (2010), *Shift in Perspective: An Arts and Disability Resource Pack*, Dublin: Arts and Disability Ireland.

Asheri, S. (2009), 'To touch or not to touch; a relational body psychotherapy perspective', in L. Hartley (ed.), *Contemporary Body Psychotherapy: The Chiron Approach*, London and New York: Routledge, pp. 106–20.

Ashwill, B. (1992), 'Emery Blackwell: A dancer becoming', *Contact Quarterly Dance Journal*, 17:1, Winter, pp. 20–23.

Autton, N. (1989), *Touch, an Exploration*, London: Darton Longman and Todd.

Bainbridge Cohen, B. (2012), *Sensing Feeling and Action: The Experiential Anatomy of Body–Mind Centering®*, 3rd ed., Northampton: Contact Editions.

Banes, S. (1987), *Terpsichore in Sneakers: Post-Modern Dance*, Middletown: Wesleyan University Press.

Bannon, Fiona and Holt, Duncan (2012), 'Touch: Experience and knowledge', *Journal of Dance and Somatic Practices*, 3:1&2, pp. 215–27.

Barba, E. (2001), 'Theatre anthropology', in N. Savarese and E. Barba (eds), *A Dictonary of Theatre Anthropology: The Secret Art of the Performer*, London and New York: Routledge, pp. 8–20.

Barrett, E. and Bolt, B. (eds) (2007), *Practice as Research Approaches to Creative Arts Enquiry*, London: I.B. Tauris and Co Ltd.

Bartlett, K. (1996), 'Community dance coming of age', *MAiLOUT: Art work with people*. April/May, pp. 6–8.

Batson, G. and Wilson, M. (2014), *Body and Mind in Motion: Dance and Neuroscience in Conversation*, Bristol: Intellect.

Becker, C. (1994) (ed.), *The Subversive Imagination: Artists, Society and Social Responsibility*, London: Routledge.

Benjamin, A. (1995), 'Unfound movement', *Dance Theatre Journal*, 12:1, pp. 44–47.

Benjamin, A. (2002), *Making and Entrance: Theory and Practice for Disabled and Non-disabled Dancers*, Abingdon and New York: Routledge.

Boal, A. (1993), *Theatre of the Oppressed*, New York: Theatre Communications Group.

Brandstetter, G., Gerko. E. and Zubarik, S. (eds) (2013), *Touching and Being Touched: Kinesthesia and Empathy in Dance and Movement*, Berlin: De Gruyter.

Brazelton, K. and Barnard, B. T. (1990), *Touch the Foundation of Experience*, Madison: International University Press.

Brisenden, S. (1989), 'What is disability culture?', in E. Morrison (ed.) (1991), *Disability Arts in London: Anthology the First Five Years*, London: Masterfirm Ltd, pp. 9–10

Butler, J. (2005), *Giving and Account of Oneself*, New York: Fordham University Press.

Caldwell, C. (1997a), *Ethics and Techniques for touch in Somatic Psychology*, Wheaton: Quest Books.

Caldwell, C. (ed.) (1997b), *Getting in Touch: The Guide to New Body-Centered Therapies*, Wheaton: Quest Books.

Caldwell, P. (1996), *Getting in Touch*, Brighton: Pavilion Publishing.

Candy, L. (2006), 'Practice based research: A guide', https://www.creativityandcognition.com/wp-content/uploads/2011/04/PBR-Guide-1.1-2006.pdf. Accessed 1 November 2022.

Carello, C. and Turvey, M. (2015), 'Dynamic (effortful) touch', *Scholarpedia*, 10:4, p. 8242.

Classen, C. (ed.) (2005), *The Book of Touch*, Oxford: Berg.

Classen, C. (2012), *The Deepest Sense: A Cultural History of Touch*, Chicago: University of Illinois Press.

Colleran, M. (1988), 'Integration or segregation, do disabed people have the choice?', in E. Morrison (ed.) *Disability Arts in London: Anthology the First Five Years*, London: Masterfirm Ltd, p. 7.

Cooper, M. (2008), *Essential Research Findings in Counselling and Psychotherapy: The Facts are Friendly*, London: Sage.

Cooper, M. and McLeod, J. (2007), 'A pluralistic framework for counselling and psychotherapy: Implications for research', *Counselling & Psychotherapy Research*, 7, pp. 135–43.

Cooper, M. and McLeod, J. (2011a), *Pluralistic Counselling and Psychotherapy*, London: Sage.

Cooper, M. and McLeod, J. (2011b), 'Person-centered therapy: A pluralistic perspective', *Person-Centered and Experiential Psychotherapies*, 10, pp. 210–23.

Cooper, M. and Dryden, W. (2015), *The Handbook of Pluralistic Counselling and Psychotherapy*, London: Sage.

Cox, C. and Hayes, J. (1999), 'Experiences of administering and receiving therapeutic touch in intensive care', *Intensive and Critical Care Nursing*, 15:5, pp. 283–87.

Curtis, J. (2007), 'Contact improvisation and its influence on contemporary dance practice, panel discussion at Tanzfabrik, Berlin', *Contact Quarterly* webtexts, 40:2, June 29.

Daly, A. (2016), 'The ethical interworld, in Anya Daly', in *Merleau-Ponty and the Ethics of Intersubjectivity*, London: Palgrave Macmillan, pp. 281–303.

Damasio, A. (2000), *The Feeling of What Happens: Body, Emotion and the Making of Consciousness*, London: Vintage Books.

Deitch-Dey, M. (1998), 'Change through exchange: Contact improvisation', Unpublished M Phil thesis, Plymouth University.

Deleuze, G. (1990), *The Logic of Sense*, New York: Columbia University Press.

Deleuze, G. (1994), *Difference and Repetition* (trans. P. Patton), New York: Columbia University Press.

Deleuze, G. and Gauttari, F. (1984), *Anti-Oedipus*, London: The Athlone Press Ltd.

Deleuze, G. (1987), *A Thousand Plateaus*, Minneapolis: University of Minnesota Press.

Depraz, N. (1999a), 'The phenomenological reduction as praxis', *Journal of Consciousness Studies*, 6:2&3, pp. 95–110.

Depraz, N. (ed.) (1999b), 'Phenomenological reduction as praxis', in F. J. Varela and J. Shear (eds), *Investigating Phenomenal Consciousness; New Methodologies and Maps*, Thorveton: Imprint Academic, pp. 95–110.

Depraz, N., Varela, F. J. and Vermersch, P. (2003), *On Becoming Aware: A Pragmatics of Experiencing*, Amsterdam: John Benjamins BV.

Dowd, I. ([1981] 2007), *Taking Root to Fly*, 7th ed., New York: Dowd.

Dowler, L. (2019), 'Instinctive connections: Improvisation as a research methodology in health and care settings', in V. L. Midgelow (ed.), *The Oxford Handbook of Improvisation in Dance*, Oxford: Oxford University Press, pp. 445–62.

Dymoke, K. (1996a), 'A policy for all?' *MAiLOUT: Art Work with People*, April/May, pp. 14–15.

Dymoke, K. (1996b), *Touch, Sense, Feel, Move: Touchdown Dance*, CDROM, First edition, UK: Digital Marmelade.

Dymoke, K. (1998), 'Focus on the arts: Touchdown dance', *Deafblind Review*, 21, January, pp. 28–29.

Dymoke, K. (2000a), 'Dance your way to wonder', in *LOOK: National Federation of Families with Visually Impaired Children*, 2000, 32, p. 17.

Dymoke, K. (2000b), 'Touch, a natural language', in P. Greenland (ed.), *What Dancers Do That Other Health Workers Don't*, Leeds: Jabadao Publishing, pp. 83–97.

Dymoke, K. (2005a), 'Awakened by touch', *Animated,* Summer 2005, Foundation for Community Dance, https://www.communitydance.org.uk/DB/animated-library/awakened-by-touch?ed=14057. Accessed 1 November 2022.

Dymoke, K. (2005b), Journal notes.

Dymoke, K. (2014), 'Contact improvisation, the non-eroticized touch in an "art-sport"', *Journal of Dance and Somatic Practice,* 6:2, pp. 205–18.

Dymoke, K. (2017), 'The lost and found. Working with Basic Neurocellular Patterns (BMC)', in C. Trevarthan and S. Daniel (eds), *Rhythms of Relating in Children's Therapies, Connecting Creatively with Vulnerable Children*, London: Jessica Kingsley, pp. 172–87.

Dymoke, K. (2019), 'Touching the untouchables', *International Journal of Politics and Psychotherapy*, 17:3, pp. 1–12.

Early, F. (1999), 'Animated, summer. We who are still here', https://www.communitydance.org.uk/DB/animated-library/we-who-are-still-here?ps=UUpsbCkiyafcpo9nFTBJ1tC7G6Cr0v. Accessed 10 October 2019.

Edwards, D. J. A. (1979), 'The role of touch in interpersonal relations implications for psychotherapy', *South African Journal of Psycholog*, 11:1, pp. 29–37.

Eiden, B. (1998), 'The use of touch in psychotherapy', *Self and Society*, 25, pp. 6–13.

Etherington, K. (2007), 'Ethical research in reflexive relationship', *Qualitative Inquiry*, 13, pp. 599–616.

Field, T. (2003), *Touch*, Cambridge: MIT Press.

Fink, B. (1995), *The Lacanian Subject: Between Language and Jouissance*, Princetown: Princetown University Press.

Finkelstein, V. (1989), 'Disabled people and our culture development', in E. Morrison (ed.) *Disability Arts in London: Anthology the First Five Years*, London: Masterfirm Ltd, pp. 3–6.

Fishman, D. (1991), *The Case for a Pragmatic Psychology*, New York: New York University Press.

Foster, S. L. (1992), 'Dancing bodies', in J. Crary and S. Kwinter (eds), *Incorporations*, New York: Zone, pp. 480–95.

Foucault, M. (1971), *The Archaeology of Knowledge and the Discourse on Language*, New York: Pantheon Books.

Foucault, M. (1977), *Discipline and Punish*, New York: Pantheon.

Foucault, M. (1984), *Power/Knowledge*, Brighton: Harvester.

Foucault, M. ([1984] 1991), 'What is an author?', in P. Rabinov (ed.), *The Foucault Reader*, London: Penguin, pp. 101–20.

Foucault, M. ([1961] 1995), *Madness and Civilisation: A History of Insanity in the Age of Reason*, New York: Vintage Books.

Foucault, M. (1998), *The History of Sexuality. Vol 1: The Will to Knowledge*, London: Penguin.

Freire, P. (1976), *Education, the Practice of Freedom*, London: Writers and Readers Publishing Cooperative.

Freire, P. ([1970] 2007), *Pedagogy of the Oppressed*, New York: Continuum.

Freire, P. (2016), *Pedagogy in Process: The Letters to Guinea-Bissau*, London: Bloomsbury Academic.

Fulkerson, M. (1996), 'Taking the glove without the hand', *Contact Quarterly*, Winter/Spring, pp. 40–42.

Gallace, A. (2012), 'Living with touch', *The Psychologist*, 25, pp. 896–99.

Gallagher, S. (2005), *How the Body Shapes the Mind, 2005/2011*, New York: Oxford University Press.

Gallagher, S. (2008), *The Phenomenological Mind*, London: Routledge.

Gallagher, S. and Cole, J. (1995), 'Body schema and body image in a deafferented subject', *Journal of Mind and Behavior*, 16, pp. 369–90.

Gendlin, E. T. (1999), 'A new model', in F. Varela and J. Shear (eds), *The View from Within: First-Person Approaches to the Study of Consciousness*, Thorverton: Imprint Academic, pp. 232–37.

Gibson, J. J. (1966), *The Senses Considered as Perceptual Systems*, Boston: Houghton Mifflin.

Gleeson, M. and Timmins, F. (2004), 'Touch: A fundamental aspect of communication with older people experiencing dementia', *Nursing Older People*, 16:2, pp. 18–21.

Gregor, M. and Timmermann, J. (eds and trans.) (2011), *Immanuel Kant, Groundworks for the Metaphysics of Morals*, Cambridge: Cambridge University Press.

Grenn, C. and Nicoll, L. (2001), 'Therapeutic caring; a learning disability experience', *Complementary Therapies in Nursing and Midwifery*, 13, pp. 43–53.

Halliley, M. (1991), 'Touched by the experience. Mark Halliley reports on a dance form that puts the blind and the sighted on the same footing', *The Independent*, 22 October, n.pag.

Hanna, T. (1988), *Somatics, Reawakening the Mind's Control of Movement, Flexibility, and Health*, Cambridge: Da Capo Press.

Hanson, W. E., Creswell, J. W., Plano Clark, V. L., Petska, K. S. and Creswell, J. D. (2005), *Mixed Methods Research Designs in Counselling Psychology*, Lincoln: Faculty Publications, Department of Psychology.

Harre, R. and Gillett, G. (1994), *The Discursive Mind*, London: Sage.

Hart, P. (2010), 'Moving beyond the common touchpoint – Discovering language with congenitally deafblind people', Doctoral thesis, Dundee: University of Dundee.

Hartley, L. (1995), *Wisdom of the Body Moving*, Berkeley: North Atlantic Books.

Hartley, L. (2004), *Somatic Psychology: Body, Mind and Meaning*, London: Whurr Publishers.

Hartley, L. (2018), 'Dance in the 1970s - A crucible for *de-schooling*: Releasing into being danced', *Journal of Theatre, Dance and Performance Training*, 9:3, pp. 358–66, https://doi.org/10.1080/19443927.2018.1479295.

Harvey, D. (1989), *The Condition of Postmodernity*, Oxford: Blackwell Publishers.

Heller, M. A. and Gentaz, E. (2014), *The Psychology of Touch and Blindness*, New York and London: Psychology Press.

Heslin, R. and Alper, T. (1983), 'Touch: A bonding gesture', in J. Wiemann and R. Harrison (eds), *Nonverbal Interaction*, Beverly Hills and London: Sage, pp. 47–75.

Houston, S. (2009a), 'Dancing around exclusion: An examination of the issues of social inclusion within choreographic practice in the community', in J. Butterworth and L. Wildschut (eds), *Choreography in Contexts: A Critical Reader*, London: Routledge, pp. 205–19.

Houston, S. (2009b), 'The touch "taboo" and the art of contact: An exploration of Contact Improvisation for prisoners', *Research in Dance Education*, 10:2, pp. 97–113.

Hulton, P. (2018), 'Peter Hulton on Dartington College of Arts', *Theatre Dance and Performance Training*, 26 October, http://theatredanceperformancetraining.org/2018/10/peter-hulton-on-dartington-college-of-arts/. Accessed 16 June 2022.

Humphrey, D. (1959), *The Art of Making Dances* (ed. B. Pollack), New York: Grove Press.

Hunter, M. and Struve, J. (1997), *The Ethical Use of Touch in Psychotherapy*, London: Sage.

Hwachukwu, T. and Robinson, M. (2011), *The Role of Diversity in Building Adaptive Resilience*, England: Arts Council of England, http://culturehive.co.uk/wp-content/uploads/2013/07/Diversity_and_adaptive_resilience_public-1.pdf. Accessed 26 October 2018.

Johnson, D. H. (2000), 'Intricate tactile sensitivity; A key variable in western integrative body-work', *Progress in Brain Research*, 122, pp. 479–90.

Johnson, D. H. (2006), 'The primacy of experiential practices in body psychotherapy', in G. Marlock and H. Weiss (eds), *Handbook of Body Psychotherapy*, Stuttgart: SchattauerGMbH, Verlag fur Medizin und Naturwissenschaffen, pp. 117–25.

Jones, S. (2009), *Practice-as-Research in Performances and Screen* (eds L. Allegue, B. Kershaw S. Jones and A. Piccini), Basingstoke: Palgrave Macmillan.

Jordan, S. (1992), *Striding Out*, London: Dance Books.

Kaltenbrunner, T. (2004), *Contact Improvisation: Moving – Dancing – Interaction*, 2nd ed., Oxford: Meyer & Meyer Ltd.

Kampe, T. (2015), 'The art of making choices: The Feldenkrais Method as a soma-critique', in S. Whatley, N. Garrett Brown and K. Alexander (eds), Attending to Movement: Somatic Perspectives on Living in this World, pp. 77– 90.

Kelly, O. (1984), *Community, Art and the State: Storming the Citadels*, London: Comedia and Owen Kelly.

Kemp, R. (2012), *Embodied Acting: Neuroscience Tells Performance*, Oxford: Routledge.

Keogh, M. (2002), *The Art of Waiting: Essays on Contact Improvisation*, North Easton: Martin Keogh.
Keogh, M. (2018), *The Practice of Contact Improvisation Dancing Deeper Still: Version X*, USA: Intimately Rooted Books.
Keriac and Pritchard, M. (1981), 'Politics and contact: A discussion', *Contact Quarterly*, 7:1, Fall, pp. 34–38.
Kilcoyne, K. (1989), 'Skipping visual impairment', *The New Beacon: The Journal of Blind Welfare*, 73:860, pp. 18–21.
Kilcoyne, K. (1990a), 'Physical conversations: Research into movement for visually-impaired people', *New Beacon, Art to Share*, 74:874, pp. 99–101.
Kilcoyne, K. (1990b), 'Skipping visual impairment', *The New Beacon Journal of Blind Welfare*, LXXV:875, pp. 56–67.
Kilcoyne, K. (1991), 'Common sense: Making use of the sense of touch', *The British Journal of Visual Impairment*, 9:2, pp. 47–49.
Kilcoyne, K. (1993a), 'Sixth sense – Second sight', *The New Beacon: The Journal of Blind Welfare*, 77:907, pp. 117–20.
Kilcoyne, K. (1993b), 'Sixth sense – Second sight: Report of a dance event', *New Beacon, Journal of Blind Welfare*, 77:907, pp. 117–20.
Koteen, D. and Smith, N. (2008), *Caught Falling: The Confluence of Contact Improvisation, Nancy Stark-Smith, and Other Moving Ideas*, Holyoke: Marcus Printing.
Kristeva, J. (n.d.), 'The future of a revolt', http://www.kristeva.fr/the-future-of-a-revolt.html. Accessed 7 March 2020.
Kristeva, J. (2003), *Intimate Revolt: The Powers and Limits of Psychoanalysis*, New York: Columbia University Press.
Kuppers, P. (2011), *Disability Culture and Community Performance: Find a Strange and Twisted Shape*, Houndmills, New York: Palgrave.
Kuppers, P. (2014), *Studying Disability Arts and Culture: An Introduction*, Basingstoke: Palgrave MacMillan.
Kuypers, P. (2018), *Contact Improvisation*, Brussels: Contredanse.
Lansley, J. and Early, F. (eds) (2011), *The Wise Body: Conversations with Experienced Dancers*, Bristol: Intellect.
Lawson, H. (1985), *Reflexivity: The Post-modern Predicament*, London: Hutchinson.
Lee, R. (2001), 'Dancing nation: Four people. Four stories. Four communities', *Animated Autumn*, Leicester: Foundation for Community Dance, pp. 23–26.
Levine, P. (2008), *Healing Trauma*, Boulder: Sounds True Inc.
Levine, P. (2010), *In an Unspoken Voice*, Berkely: North Atlantic Books.
Lomas, Christine (1998), 'Art and the community: Breaking the aesthetic of disempowerment', in S. Shapiro (ed.), *Dance, Power, and Difference: Critical and Feminist Perspectives on Dance Education*, Champaign: Human Kinetics, pp. 149–70.
MaCabe, K. (1997), 'The use of contact improvisation in integrated dance', Unpublished dissertation (Professional Diploma for Community Dance), Laban Centre.

Macdonald, A. (2018), 'Touch consolation and loss in this is for you', *Journal of Dance and Somatic Studies*, 9:2, pp. 255–68.

Manetti, D. and Caiani, S. Z. (eds), 'Agency: From embodied cognition to free will', *Humana Mente: Quarterly Journal of Philosophy*, p. 15.

Manning, Erin (2007), *Politics of Touch: Sense, Movement, Sovereignty*, Minneapolis: University of Minnesota Press.

Martz, E. and Lindy, J. (2010), 'Exploring the trauma membrane concept', in Springer (ed.), *Trauma Rehabilitation after War and Conflict*, Munich: Springer Science + Business Media, http://media.axon.es/pdf/85137_2.pdf. Accessed 22 May 2023.

Massay-Stokes, M. (2001), 'Communication, expressing feelings, and creative problem solving', in J. Robert-McComb (ed.), *Eating Disorders in Women and Children, Prevention, Stress Management, and Treatment*, Boca Raton, London, New York and Washington, D.C.: CRC Press, pp. 291–315.

Matthews, E. (2006), *Merleau-Ponty: A Guide for the Perplexed*, London, New York: Continuum Publishers.

Maturana, H. and Varela, F. (1980), *Autopoiesis and Cognition: The Realization of the Living*, Boston: Reidel.

McDonagh, D. ([1970] 1990), *The Rise and Fall of Modern Dance*, London: Dance Books.

McEwan, B., and Jonsson, S. (2008), 'Relational violence, the darkest side of haptic communication', in L. K Guerrero and M. Hecht (eds), *The Nonverbal Communication Reader*, 3rd ed., Prospect Heights: Waveland Press, pp. 232–41.

McLeod, J. (2013), 'Developing pluralistic practice in counselling and psychotherapy: Using what the client knows', *The European Journal of Counselling Psychology*, 2:1, pp. 51–64.

McNiff, J. and Whitehead, J. (2009), *Doing and Writing Action Research*, Los Angeles and London: Sage.

Meekums, B. (1993), 'Research as an act of creation', in H. Payne (ed.), *Handbook of Inquiry in the Arts Therapies*, London: Jessica Kingsley Publishers, pp. 130–37.

Meekums, B. (2002), *Dance Movement Therapy: A Creative Psychotherapeutic Approach*, London: Sage.

Meekums, B. (2006), 'Embodiment in dance movement therapy training and practice', in H. Payne (ed.), *Dance Movement Therapy: Theory, Research and Practice*, 2nd ed., London: Routledge, pp. 167–83.

Meekums, B. (2010), 'Moving towards evidence for dance movement therapy: Robin Hood in dialogue with the King', *The Arts in Psychotherapy,* 37:1, pp. 35–41.

Meekums, B. (2014), 'Becoming visible as a profession in a climate of competitiveness: The role of research', *Body Movement and Dance in Psychotherapy: An International Journal for Theory, Research and Practice*, 9:3, pp. 123–37.

MENCAP (n.d.), https://www.mencap.org.uk/about-us/mencaps-history. Accessed 15 April 2019.

Merleau-Ponty, M. (1962), *The Phenomenology of Perception*, London: Routledge and Kegan Paul.

Merleau-Ponty, M. (1968), *The Visible and the Invisible*, Evanston: Northwestern University Press.

Millett-Gallant, A. (2017), *Re-membering: Putting Mind and Body Back Together Following Traumatic Brain Injury*, Chapel Hill: Wisdom House Books.

Millett-Gallant, A. and Howie, E. (eds) (2016), *Disability and Art History*, New York: Routledge.

Mitra, R. (2018), 'Talking politics of contact improvisation with Steve Paxton', *Dance Research Journal*, 50:3, pp. 5–18.

Montagu, A. (1986), *Touching; The Human Significance of the Skin*, 3rd ed., New York and London: Harper and Row.

Morrison, E. (ed.) (1991), 'Introduction, the story continues', in *Disability Arts in London, Anthology the First Five Years*, London: Masterfirm Ltd, p. 2.

Nelson, K. (1996), 'Touch revolution. *Giving Dance', Contact Quarterly a Vehicle for Moving Ideas*, 21:1, Winter/Spring, p. 65.

Nelson, R. (2013), *Practice as Research in the Arts: Principles, Protocols, Pedagogies, Resistances*, Basingstoke: Palgrave Macmillan.

Noe, A. (2002), *Is the Visual World a Grand Illusion*, Thorverton: Imprint Academic.

Noe, A (2009), *Out of Our Heads; Why You Are Not Your Brain and Other Lessons from the Biology of Consciousness*, New York: Hill and Wang.

Novack, C. (1990), *Sharing the Dance: Contact Improvisation and American Culture*, Wisconsin: The University of Wisconsin Press.

Novack, C. (2007), 'Contact improvisation, an introduction', in P. Kuppers and G. Robertson (eds), *The Community Performance Reader*, London and New York: Routledge, pp. 224–26.

Novack, C. J. (1990), *Sharing the Dance, Contact Improvisation and American Culture*, Madison, London: University of Wisconsin Press.

Nwachukwu, T. and Robinson, M. (2011), *The Role of Diversity in Building Adaptive Resilience*, England: Arts Council of England, http://culturehive.co.uk/wp-content/uploads/2013/07/Diversity_and_adaptive_resilience_public-1.pdf. Accessed 10 June 2021.

Oliver, J. (2013), 'The social model of disability: Thirty years on', *Disability & Society*, 28:7, pp. 1024–26.

Paley, J. (2002), 'Virtues of autonomy, T. Kantian ethics of care', *Nursing Philosophy*, 3:2, pp. 133–43.

Patterson, M. (2007), *The Senses of Touch: Haptics, Affelcts and Technologies*, Oxford: Berg.

Paxton, S. (2008), 'Backwords', in D. Koteen and N. Smith (eds), *Caught Falling; The Confluence of Contact Improvisation, Nancy Stark-Smith, and Other Moving Ideas*, Holyoke: Marcus Printing, pp. 86–89.

Paxton, S. (2010a), personal e-mail correspondence, 5 January 2010, n.pag.

Paxton, S. (2010b), e-mail, 4 January.

Paxton, S. (2010c), e-mail, 4 January.

Paxton, S. (2010d), e-mail, 5 January.

Paxton, S. (2013), e-mail, 8 September.

Paxton, S. (2015), Talking Dance. https://www.youtube.com/watch?v=_82Od5NM4LI. Accessed 30 October 2019.

Paxton, S. (2018), *Gravity*, Brussels: Contredanse Editions.

Paxton, S. (2019a), personal e-mail correspondence, n.pag.

Paxton, S. (2019b), recorded skype interview transcribed by K. Dymoke, n.pag.

Paxton, S. (2020a), personal e-mail correspondence, n.pag.

Paxton, S. (2020b), Talking Dance (transcript of presentation to dance students on body – Cloning in dance), *Contact Quarterly*, YouTube, https://walkerart.org/calendar/2014/talking-dance-steve-paxton; https://www.youtube.com/watch?v=_82Od5NM4LI. Accessed 22 May 2023.

Paxton, S. and Dymoke, K. (2001), 'Touchdown dance correspondance', in K. Nelson (ed.), *Vu du Corps. Mouvement et Perception. Nouvelles de Danse nr 48–49*, Brussels: Contredanse, pp. 144–84.

Paxton, S. and Kilcoyne, K. (1990a), 'Physical conversations; Research into movement for visually-impaired people', *The New Beacon: The Journal of Blind Welfare*, 74:874, pp. 99–101.

Paxton, S. and Kilcoyne, K. (1990b), 'Skipping visual impairment', *The New Beacon: The Journal of Blind Welfare*, 74:876, pp. 56–57.

Paxton, S. and Kilcoyne, K. (1992), 'A common sense', *Contact Quarterly a Vehicle for Moving Ideas*, 17:1, pp. 48–50.

Paxton, S. and Kilcoyne, K. (1993), 'On the Braille in the body: An account of the Touchdown Dance integrated workshops with the visually impaired and the sighted', *Dance Research: The Journal of the Society for Dance Research*, 11:1, Spring, pp. 3–51.

Pethybridge, R. (2014), 'Relative proximity: Reaching towards an ethics of touch in cross-generational dance practice', *Journal of Dance & Somatic Practices*, 6:2, pp. 175–87.

Phelan, P. (2005), 'Reconsidering identity politics, essentialism, and dismodernism: An afterword', in C. Sandahl and P. Auslander (eds), *Bodies and Commotion, Disability and Performance*, Michigan: University of Michigan Press, p. 352.

Piper, H. and Stronach, I. (2008), *Don't Touch! The Educational Story of a Panic*, Abingdon: Routledge.

Prestidge, M. (1997), 'Beyond the Fringe', *Animated – the National Community Dance Magazine*, Autumn, Suffolk: Micropress Printers, Foundation for Community Dance, pp. 14–15.

Ptashek, A. (1992), 'The moving body, and integrated movement course', *Contact Quarterly Dance Journal*, 17:1, pp. 40–43.

Rainer, Y. ([1965] 2006), 'No manifesto', *Feelings Are Facts: A Life,* Cambridge, London: The MIT Press, pp. 168–78.

Robinson, M. (2011), 'The business case for diversity', *Arts Professional*, https://www.artsprofessional.co.uk/magazine/246/feature/business-case-diversity-building-adaptive-resilience. Accessed 26 October 2018.

Robinson, M. (2013), 'New ACE strategy; keep calm and carry on', *Arts Professional*, https://www.artsprofessional.co.uk/magazine/blog/new-ace-strategy-keep-calm-and-carry. Accessed 25 October 2018.

Rowan, J. (2010), *Personification: Using the Dialogical Self in Psychotherapy and Counselling*, Hove: Routledge.

Sarco-Thomas, M. and Zammit, R. (2020), 'Contact improvisation and the health sciences: Mapping the terrain', in M. Sarco-Thomas (ed.), *Thinking Touch in Partnering and Contact Improvisation: Philosophy, pedagogy, Practice*, Cambridge: Cambridge Scholars Press, pp. 91–115.

Schilder, P. (1935), *The Image and Appearance of the Human Body*, New York: International Universities Press.

Schore, A. (1994), *Affect Regulation and the Origin of the Self: The Neurobiology of Emotional Development*, Hove: Lawrence Erlbaum Associates.

Schore, A. (2003a), *Affect Dysregulation and Disorders of the Self*, New York and London: W. W. Norton and Co.

Schore, A. (2003b), *Affect Regulation and the Repair of the Self*, New York and London: W. W. Norton and Co.

Shapiro, L. (ed.) (2014), *The Routledge Handbook of Embodied Cognition*, London: Routledge.

Sheets-Johnstone, M. (1990), *The Roots of Thinking*, Philadelphia: Temple University Press.

Sheets-Johnstone, M. (1999), *The Primacy of Movement*, Philadelphia: John Benjamin Publishing Company.

Sheets-Johnstone, M. (2000), 'Kinetic tactile-kinesthetic bodies: Ontological foundations of apprenticeship learning', *Human Studies*, 23:4, pp. 343–70.

Sheets-Johnstone, M. (2010), 'Kinesthetic experience; understanding movement inside and out', *Body Movement and Dance in Psychotherapy*, 5:2, pp. 111–27.

Sheets-Johnstone, M. (2008a), *The Roots of Morality*, University Park: The Pennsylvania State University Press.

Sheets-Johnstone, M. (2008b), 'On the hazards of being a stranger to oneself', *Psychotherapy and Politics International*, 6:1, pp. 17–29.

Sheets-Johnstone, M. (2009), *The Corporeal Turn: An Interdisciplinary Reader*, Exeter: Imprint Academic.

Sheets-Johnstone, M. (2011), 'From movement to dance', in S. Gallagher and D. Zahavi (eds), *Phenomenology and the Cognitive Sciences*, vol. 11, no. 1, New York: Springer Science+Business Media B.V., pp. 39–57.

Sheets-Johnstone, M. ([1966] 2015), *The Phenomenology of Dance*, Philadelphia: Temple University Press.

Sherrington, Charles (1906), *The Integrative Action of the Nervous System*, New Haven: Yale University Press.

Siddal, J. and Rist, R. (2002), 'Hands on, hands off. Teaching through touch is a delicate matter', *Animated the Community Dance Magazine*, People Dancing, Spring, https://www.communitydance.org.uk/DB/animated-library/hands-on-hands-off?ed=14068. Accessed 23 March 2020.

Siegel, E. V. (1984), *Dance-Movement Therapy, Mirror of Our Selves, The Psychoanalytic Approach*, New York: Human Sciences Press.

Smith, E. W. L., Clance, P. R. and Imes, S. (eds) (1997), *Touch in Psychotherapy, Theory, Research, and Practice*, New York: Guildford Publications.

Smith, H. and Dean, T. (eds) (2009), *Practice-Led Research and Research-Led Practice in the Creative Arts*, Edinburgh: Edinburgh University Press.

Stern, D. N. (1985), *The Interpersonal World of the Infant: A View from Psychoanalysis and Developmental Psychology*, New York: Basic Books.

Stern, D. N. (2004), *The Present Moment in Psychotherapy and Everyday Life*, New York and London: W. W. Norton and Co.

Stern, D. N. (2010), *Forms of Vitality: Exploring Dynamic Experience in Psychology, the Arts, Psychotherapy, and Development*, Oxford: Oxford University Press.

Stinson, S. W. (2004), 'My body/myself; lessons from dance education', in L. Bresler (ed.), *Knowing Bodies, Moving Minds – Towards Embodied Teaching and Learning*, Dordrecht: Kluwer Academic Publishers, pp. 153–67.

Stronach, I. (2010), *Globalising Education, Educating the Local: How Methods Made Us Mad*, Abingdon: Routledge.

Stronach, I., Frankham, J., Bibi-Nawaz, S., Cahill, G., Cui, V., Dymoke, K. and Mohd-Khir, M. (2014), 'The X-factor: Performing reflexivity as indeterminate praxis', *Qualitative Inquiry*, 19:4, pp. 288–97.

Tanner, L. (2019), 'Promoting consent to touch in personal care', in I. A. James and L. Gibbons (eds), *Communication Skills for Effective Dementia Care: A Practical Guide to Communication and Interaction Training*, London and Philadelphia: Jessica Kingsley Publishers.

Tew, E. (1996), 'It feels like flying', *The New Beacon: Journal of Blind Welfare*, June, pp. 39–43.

Thomson, C. (1994), '"Dance and the concept of community", focus on community dance', *Dance and the Child International Journal*, 3, pp. 20–30.

Thomson, C. (2006), 'We are a dancing nation', *Animated*, Spring, http://www.communitydance.org.uk/DB/animated-library/we-are-a-dancing-nation.html?ed=14046. Accessed 23 August 2013.

Totton, N. (ed.) (2000), *Psychotherapy and Politics: Psycho-Politics: Entering the Twenty-First Century*, London: Sage.

Totton, N. (2003), *Body Psychotherapy an Introduction*, Maidenhead: Open University Press.

Totton, N. (ed.) (2005a), *New Dimensions in Body Psychotherapy*, Maidenhead: Open University Press.

Totton, N. (2005b), 'Do bodies tell the truth?' Speech at the Association of Chiron Psychotherapists, Meeting in the Flesh, 23 April, Cambridge.

Totton, N. (2015), *Embodied Relating: The Ground of Psychotherapy*, London: Karnac Books.

Truman, R. (1999), 'Contact improvisation and the middle manager; what Unipart can learn from a democratic dance form', *Contact Connection,* Issue 16, Autumn, pp. 2–4.

Tuan, Yi-Fu (2005), 'The pleasures of touch', in C. Classen (ed.), *The Book of Touch*, Oxford: Berg, pp. 74–79.

Tufnell, M. (2010), *Dance, Health and Wellbeing: Pathway to Practice for Dance Leaders Working in Health and Care Settings*, Leicester: Foundation for Community Dance.

Turner, V. (1974), *Dramas, Fields and Metaphors: Symbolic Action in Human Society*, New York: Cornel University Press.

Turner, V. (1982), *From Ritual to Theatre: The Human Seriousness of Play*, New York: PAJ Publications.

Unkovich, G., Butté, C. and Butler, J (eds) (2017), *Dance Movement Psychotherapy with People with Learning Disabilities: Out of the Shadows, into the Lights*, London: Routledge.

van der Kolk, B. A. (1994), 'The body keeps the score', *The Harvard Review of Psychiatry*, 1, pp. 253–65.

van der Kolk, B. A. (2014), *The Body Keeps the Score: Mind, Brain and Body in the Transformation of Trauma*, New York: Viking.

Varela, F. and Shear, J. (eds) (1999), *The View from Within. First-Person Approaches to the Study of Consciousness*, Thorverton: Imprint Academic.

Varela, F. J. and Shear, J. (1999), 'First person methodologies: What, why, how?', *Journal of Consciousness Studies*, 6:2&3, pp. 1–14.

Varela, F. J., Thompson, E. and Rosch, E. (1991), *The Embodied Mind: Cognitive Science and Human Experience*, Cambridge: MIT Press.

Vasey, S. ([1989] 1991), 'Joking apart', in E. Morrison (ed.) (1991), *Disability Arts in London: Anthology the First Five Years*, London: Masterfirm Ltd, p. 15

Walmsley, J. (2001), 'Normalisation, emancipatory research and inclusive research in learning disability', *Disability and Society*, 16:2, pp. 187–205.

Walmsley, J. (2004), 'Inclusive learning disability research; the (non-disabled) researchers' role', *The British Journal of Learning Disabilities*, 32, pp. 65–71.

Warshaw, R. ([1997] 1981), 'What are we teaching?', in N. S. Smith, H. Jastremsky and M. Gustavsen (eds), *Contact Quarterly's Contact Improvisation Sourcebook: Collected Writings and Graphics from Contact Quarterly Dance Journal 1975–1992*, Northampton, Mass: Contact Editions, pp. 88–90.

Weber, A. and Varela, F. (2002), 'Life after Kant: Natural purposes and the autopoietic foundations of biological individuality', *Phenomenology and the Cognitive Sciences*, 1:2, pp. 97–125.

Westland, G. (2011), 'Physical touch in psychotherapy: Why are we not touching more?', *Body, Movement and Dance in Psychotherapy*, 6:1, pp. 17–29.

Williams, V., Marriott, A. and Townsley, R. (2008), *Shaping Our Future*, Bristol: Nora Fry Research Centre, University of Bristol.

Printed in the USA
CPSIA information can be obtained
at www.ICGtesting.com
JSHW060456010823
45645JS00001B/1

9 781789 388367